Planar Slow-Wave Structures: Applications in Traveling-Wave Tubes

Online at: https://doi.org/10.1088/978-0-7503-5764-7

IOP Series in Electromagnetics and Metamaterials

Series Editor
Akhlesh Lakhtakia, *Pennsylvania State University*

Series in Electromagnetics and Metamaterials
The Series on Electromagnetics and Metamaterials, published by IOP, is an innovative and authoritative source of information on a fundamental science that has been enabling a multitude of transformative technologies for two centuries and more. The electromagnetic spectrum extends from millihertz waves to microwaves to terahertz radiation to ultraviolet light and even soft x-rays. In each spectral regime, different classes of materials have different kinds of electromagnetic response characteristics. Much has been discovered and much has been technologically exploited, but even more remains to be discovered and even more remains to be put to use for diverse applications.

Electromagnetics is an evermore vibrant arena of techno-scientific research. This is amply exemplified by the huge current interest in metamaterials. By virtue of carefully designed and engineered morphology, metamaterials exhibit response characteristics that are either completely absent or muted in their constituent materials.

Each book in the series offers an extended essay on a foundational topic: an emerging topic; a currently hot topic; and/or a tool for metrology, design, and application. Ranging from 60 to 120 pages, books are written by internationally renowned experts who have been charged with making the content not only authoritative, but also easy to understand, thereby offering more synthesis and depth than a typical review article in a journal.

Illustrated in full color for both ebook and printed copies, these short books are easily searchable in the ebook format. The series is thus more modular and dynamic than traditional handbooks and more coherent than contributed volumes.

This series is edited by Akhlesh Lakhtakia, the Charles Godfrey Binder (Endowed) Professor of Engineering Science and Mechanics at the Pennsylvania State University. Initial topics targeted in the series include symmetries of Maxwell equations, homogenization of bianisotropic materials, metamaterials and metasurfaces, transformation optics, nanophotonics for medicine and biology, single photons, radiation sources, optical bolometry, magnetic resonance imaging, and detection and imaging of buried objects. Additional topic suggestions are welcomed and will be promptly considered and decided upon. As part of the IOP digital library, students and professors at purchasing institutions will have unlimited access to the ebooks for classroom and research usage.

A full list of the titles published in this series can be found at: https://iopscience.iop.org/bookListInfo/iop-series-on-electromagnetics-and-metamaterials#series.

Planar Slow-Wave Structures: Applications in Traveling-Wave Tubes

Chen Zhao
School of Electronics and Information Engineering, Nanjing University of Information Science and Technology, Nanjing, China

Sheel Aditya
Department of Electrical Engineering, Indian Institute of Technology Delhi, Delhi, India

and

School of Electrical & Electronic Engineering, Nanyang Technological University, Singapore

IOP Publishing, Bristol, UK

ISBN 978-0-7503-5764-7 (ebook)
ISBN 978-0-7503-5762-3 (print)
ISBN 978-0-7503-5765-4 (myPrint)
ISBN 978-0-7503-5763-0 (mobi)

DOI 10.1088/978-0-7503-5764-7

Version: 20241101

IOP ebooks

British Library Cataloguing-in-Publication Data: A catalogue record for this book is available from the British Library.

Published by IOP Publishing, wholly owned by The Institute of Physics, London

IOP Publishing, No.2 The Distillery, Glassfields, Avon Street, Bristol, BS2 0GR, UK

US Office: IOP Publishing, Inc., 190 North Independence Mall West, Suite 601, Philadelphia, PA 19106, USA

By Sheel Aditya: Dedicated to my family, teachers, and students.

Contents

Preface

Millimeter-wave (30–300 GHz) and terahertz (300–3000 GHz) frequencies have applications in communications, electronic warfare, the detection of concealed weapons and hazardous chemicals, medical imaging, etc. It is very important to develop sources that efficiently generate high power at these frequencies. Although tremendous progress has been made in solid-state devices in the last few decades, these have an inherent disadvantage in terms of efficiency, since, in these devices, the charge carriers move in a solid medium, and the efficiency drops as the temperature rises. On the other hand, in vacuum electron devices (VEDs), an electron stream moves in vacuum, and the operation is more robust against temperature rise. For this reason, most communications satellites and airborne radar systems use traveling-wave tube amplifiers (TWTAs), which are a very popular type of VED. The current commercial value of the VED market is estimated to be more than 1 billion USD. Among the different types of VEDs, traveling-wave tubes (TWTs) stand out due to their large bandwidth, high efficiency, and moderate power levels.

An important component of a TWT is a slow-wave structure (SWS), which traditionally has been a circular helix. The circular helix offers wideband operation and high amplification per unit length. However, the circular helix is difficult to mass-produce; it needs to be precision wound one piece at a time. Even this becomes challenging at millimeter-wave frequencies due to the submillimeter diameter of the helix.

In order to tackle the issue of the difficult fabrication of circular helices, a number of SWSs have been proposed and developed. Among these SWSs, planar SWSs stand out, since they can be mass-produced with precision using printed circuit or microfabrication techniques. Planar SWSs often operate at relatively low voltages, which can further reduce device complexity. Moreover, most planar SWSs can accommodate a sheet-shaped electron beam, thus leading to relatively high amplification and/or compact size. Planar SWSs can thus enable low-cost mass production of VEDs at millimeter-wave frequencies.

Planar SWSs, such as the microstrip meander-line SWS, the planar helix, the interdigital SWS, the coplanar SWS, and their variations, have been proposed and actively studied in recent years. The purpose of this book is to describe the various aspects of planar SWSs, including their basic theory, design methods, structural variations, fabrication techniques, and their applications in TWTs.

This book begins with an introduction to TWTs and SWSs. The main chapters of this book focus on microstrip meander line and planar helix SWSs. These chapters describe the configurations of these SWSs, methods of analysis used to derive propagation characteristics (dispersion characteristics), techniques for broad banding (dispersion shaping), the application of these structures in TWTs, and the simulation of the performance of such TWTs at different frequencies. These chapters include a novel and simpler analysis method for planar helices, namely the 'effective dielectric constant' method. Also included is an accurate method for determining the interaction impedance of SWSs; interaction impedance is a measure of the strength of the interaction between an electron beam and electromagnetic wave in a TWT.

To demonstrate the feasibility of fabricating planar SWSs, this book describes the fabrication and measured results of several microstrip meander-line and planar helix SWSs at different frequencies ranging from the C/X, Ka, and V bands to the W band. One of the highlights of this book is a description of the 'on-wafer' measurement of the cold-test S-parameters of some SWSs using coplanar waveguide (CPW) probe stations. This avoids the need to dice individual SWSs from a wafer or substrate; it also enables fast evaluation of a large number of microfabricated SWSs.

In addition to their application in TWT amplifiers, planar SWS-based TWTs can also be used as oscillators by designing them as backward-wave oscillator (BWOs). BWOs are attractive due to the spectral purity of their output and the electronic tunability of their output frequency. Some planar SWSs for BWO applications are introduced in this book. The combination of oscillator and amplifier operations within a single SWS is also described.

Acknowledgments

We gratefully acknowledge the encouragement, motivation, and help received from various people in the preparation and completion of this book. It was the encouragement of Professor Baidyanath Basu that initiated this project as a title under the IOP Series in Electromagnetics and Metamaterials. As the series editor, Professor Basu provided invaluable support at various stages of the project. The IOP Publishing team patiently responded to our queries and promptly provided the requisite information and support from the beginning up to the publication of this book. We are thankful to all the researchers whose work has been cited in this book. In particular, we are indebted to the members of our research group at Nanyang Technological University, Singapore, whose work forms the backbone of this book. Since no book of this nature can capture all the material relevant to the topics covered in this book, we apologize to those researchers whose work has not been mentioned in this book.

We could not have completed this book without the support of our respective spouses and families. Heartfelt thanks to them.

Chen Zhao
Sheel Aditya

Author biographies

Chen Zhao

Chen Zhao received his BEng degree in 2011 from the University of Electronic Science and Technology of China (UESTC), Chengdu, China and his PhD degree from Nanyang Technological University, Singapore in 2016. From August 2015 to December 2017, he worked as a research scientist at Temasek Lab., National University of Singapore. In December 2017, he joined Nanjing University of Information Science and Technology as a distinguished professor. His current research interests include millimeter-wave TWTs, the modeling and characterization of SWSs, and the advanced fabrication of VEDs. He has published more than 50 journal and conference papers and holds 15 patents in the field of VEDs. He was a recipient of the 2018 Jiangsu Innovative and Entrepreneurial Doctor award and the 2020 Jiangsu Six Talent Peaks award.

Sheel Aditya

Sheel Aditya received his BTech and PhD degrees in electrical engineering from the Indian Institute of Technology (IIT) Delhi, India, in 1974 and 1979, respectively. From 1980 to 2001, he held academic positions ranging from lecturer to professor in the Department of Electrical Engineering, IIT Delhi. From 2001 to 2019, he was an associate professor at the School of Electrical and Electronic Engineering, Nanyang Technological University, Singapore. He has published more than 270 journal and conference papers on topics such as planar microwave waveguides and antennas, integrated optics, microwave photonics, optical fiber communication, and microfabricated RF VEDs. He has been involved in several research and development projects and is the coholder of five patents in India, Singapore, and USA. He was an editor (Electromagnetics) for the *IETE Journal of Research* (India) from 2008 to 2021. He continues to frequently review submissions to IEEE journals. He has been a fellow of the IETE (Institution of Electronic and Telecommunication Engineers, India) since 1994 and a life senior member of the IEEE (Institute of Electrical and Electronic Engineers, USA) since 2018. He can be reached at sheel.aditya@ieee.org.

IOP Publishing

Planar Slow-Wave Structures: Applications in Traveling-Wave Tubes

Chen Zhao and Sheel Aditya

Chapter 1

Introduction

1.1 Introduction to vacuum electron devices

For many decades, vacuum electron devices (VEDs) have played an important role in society. They have been used as high-power microwave (HPM) amplifiers or oscillators in numerous applications such as radio/TV broadcasting, satellite broadcasting and communications, radars, military systems, medical/biomedical applications, high-energy particle acceleration, microwave heating, etc [1]. Their working frequency can range from below 100 MHz up to hundreds of gigahertz or even terahertz. The output power of VEDs can range from hundreds of milliwatts to several megawatts [2]. For some special applications, for example, HPM weapons, the peak power may exceed gigawatts [1].

Many types of VEDs are in use currently. Magnetrons, one of the earliest microwave tubes, commonly provide kilowatts of power and are widely used in microwave ovens. Klystrons with high output power are used in radars as well as in particle accelerators. Traveling-wave tubes (TWTs) are used in satellites due to their wide bandwidth, high gain and efficiency, and long lifetime. Gyrotrons can easily operate at hundreds of gigahertz without necessitating tube size reduction.

Solid-state power amplifiers (SSPAs) are in strong competition with VEDs. While the basic principle of both types of devices is to convert the kinetic energy of electrons into radio-frequency (RF) power, the two types of devices have relative merits and demerits. In SSPAs based on bipolar junction transistors (BJTs) or field-effect transistors (FETs), the electron stream flows in a solid semiconductor material. SSPAs, and semiconductor-based devices in general, have the advantages of easy mass production and lower noise than most VEDs. They are replacing VEDs in some applications, such as civilian communications and relatively low-frequency systems. However, in some other applications, for example, satellite communications, VEDs still show many advantages compared to SSPAs. First, SSPAs can only

doi:10.1088/978-0-7503-5764-7ch1

work with relatively low driving voltages and low power levels. In SSPAs, a significant amount of the kinetic energy of the charge carriers is converted to waste heat, causing a temperature rise in the device. Besides, SSPAs are not able to operate under high-temperature (over 200 °C) conditions due to the chemical degradation of the semiconducting channel as well as carrier mobility reduction [3]; as a result, SSPAs require bulky heat sinks. In contrast, in VEDs, as the electron stream flows in vacuum, the output power can be much higher than in SSPAs, and the typical efficiency is also better. Moreover, VEDs can operate at higher temperatures and are less susceptible to radiation damage in space. For these reasons, VEDs continue to be widely used and are irreplaceable in most high-power applications. Of course, it is possible to combine the advantages of SSPAs and VEDs in a single module, as has been done successfully in microwave power modules (MPMs) [1].

1.2 Traveling-wave tubes

TWTs are one of the most widely used types of VEDs. TWTs work as power amplifiers at frequencies ranging from below 1 GHz to above 100 GHz. The continuous-wave (CW) output power can range from a few watts to thousands of watts. The peak pulse power can reach over a megawatt. The efficiency of TWTs used in space usually exceeds 60%. Although magnetrons or klystrons can achieve much higher output power, TWTs display better linearity and, more importantly, broadband capability. TWTs can achieve octave or even decade bandwidths in practice [2].

TWTs find application in many kinds of communication systems. In satellite communication systems (downlink from satellites, uplink from ground-based, airborne, or shipboard transmitters to satellites), TWTs are becoming more and more dominant due to their higher bandwidth and linearity. Besides, TWTs turn out to have a much longer operating life (>15 years) and higher reliability and efficiency (>60%) than solid-state devices. TWTs are also the most widely used VEDs in radar and electronic countermeasure (ECM) systems, thanks to their wide bandwidth, high gain, and low noise [1].

1.2.1 Brief history of TWTs

The idea of the TWT was first patented by a Russian, Andrei V Haeff, in 1933; he was then a doctoral student at the Kellogg Radiation Laboratory at Caltech. The device consisted of a helical slow-wave structure (SWS) with two electron beams flowing outside the helix. However, at that time, as the electron beam was poorly focused, this device had very poor efficiency.

In 1941, Rudolph Kompfner, a researcher at the Physics Department, University of Birmingham, worked on improving the beam–wave interaction in klystrons in order to increase the range of radars. To achieve this goal, he proposed using the continuous interaction of electrons with a traveling wave instead of the discrete interaction of electrons only within short gaps. Around the same time, the invention of the precision electron gun provided well-focused electron beams. In 1943, Kompfner built his first TWT using a helical SWS with a pencil electron beam traveling through the center of the helix, which is quite close to the TWT we know today. Later, Dr John R Pierce from Bell Labs discovered the potential of TWTs in communication applications. He worked out the theory of TWTs, explaining mathematically how the electromagnetic wave interacts with the electrons and becomes a growing wave along the SWS. It should be noted that the backward-wave oscillator (BWO), also called the carcinotron, was also invented by Kompfner. With these pioneering works, TWTs quickly became popular around the world, and theoretical as well as experimental work on TWTs has been expanding ever since.

1.2.2 Basic operating principle of TWTs

The operating principle of TWTs is based on Cherenkov radiation. When high-speed charge particles pass through a medium at a speed greater than the phase velocity of light in the medium, electromagnetic radiation is produced. This is basically how an electron beam transfers energy to an electromagnetic wave in a TWT. However, it requires making the electrons' speed higher than that of the electromagnetic wave. The most commonly used method is to use an SWS in which the electromagnetic wave propagates more slowly than in air.

A schematic diagram of a TWT is shown in figure 1.1. As shown in this figure, a TWT usually consists of six major parts: an electron gun, an SWS, focusing magnets, a collector, RF input/output, and a vacuum envelope. The electron gun produces an electron beam and accelerates the beam to the required velocity. A low-power RF signal enters the tube from the input port. When the electron beam and electromagnetic wave have similar velocities (velocity synchronism), beam–wave interaction occurs. When the condition for Cherenkov radiation is satisfied, the electromagnetic wave absorbs energy from the electron beam while traveling along the tube. Finally, an amplified electromagnetic wave is obtained at the output port. The focusing magnets are used to form an axial magnetic field in order to counter the effect of repulsive radial space charge forces between the electrons. Periodic permanent magnet (PPM) focusing is commonly used due to its advantages of high

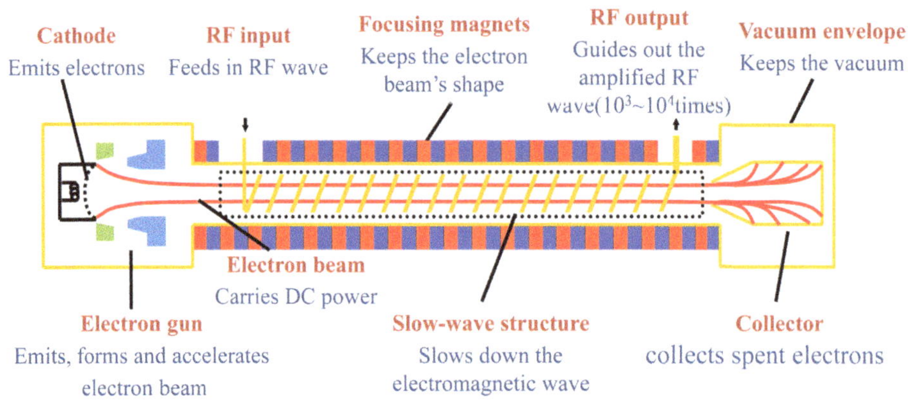

Figure 1.1. Simplified schematic diagram of a TWT.

field strength, low weight, and compact size [4]. After passing through the SWS, the electrons still possess some kinetic energy, since the beam power cannot be fully converted into electromagnetic power. A collector recovers energy from the spent electron beam. A multistage depressed collector is usually used to increase the total efficiency of the TWT [5]. The whole structure, including the electron gun, the SWS, and the collector, is enclosed in an evacuated envelope.

1.3 Slow-wave structures

In a TWT, it is important that the electron beam has a velocity close to the phase velocity of the electromagnetic wave. However, the electromagnetic wave usually travels much faster than the electron beam. In order to achieve velocity synchronism, an SWS is needed. An SWS is thus a key part of a TWT. An SWS is a kind of waveguide that is able to support an electromagnetic wave with a phase velocity much lower than the speed of light.

Periodic structures are usually able to support waves with low phase velocity. Periodic structures commonly used as SWSs in TWTs include helices [6–8], coupled cavities [9–11], periodically loaded waveguides [12], folded waveguides [13–16], ladder circuits [17–19], etc. Among these, the most commonly used type of SWS is the circular helix because of its wide bandwidth, ease of dispersion control, and high interaction impedance. The interaction impedance represents the strength of the beam–wave interaction in an SWS. Figure 1.2 shows one period of the circular helix, which is placed at the center of a cylindrical metal shield. In order to fix the circular helix SWS firmly inside the metal shield, supporting dielectric rods are often needed.

1.3.1 Cold-test parameters of SWSs

1.3.1.1 Dispersion diagram

The cold-test parameters refer to the electromagnetic properties of the SWS without the electron beam. These parameters include the propagation constant, phase velocity, dispersion, interaction impedance, reflection and transmission coefficients, etc. The dispersion characteristics are the most important property of SWSs. In

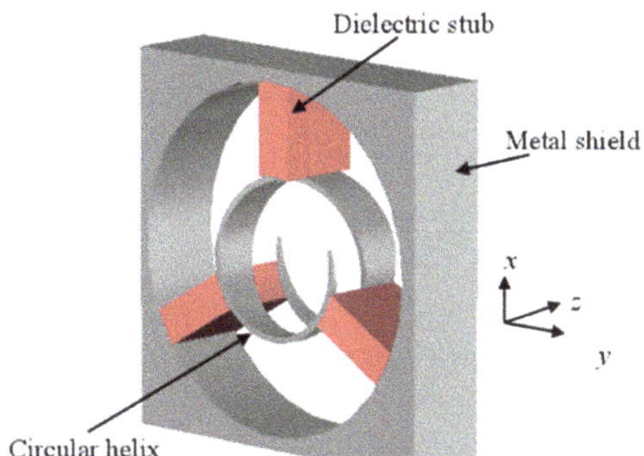

Figure 1.2. One period of a circular SWS [18].

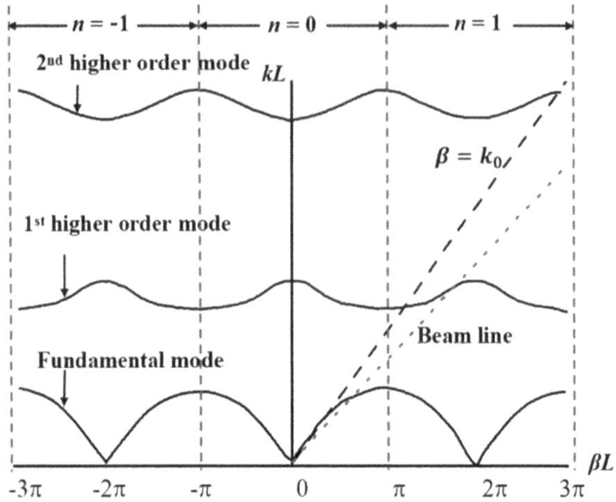

Figure 1.3. Dispersion diagram of a typical periodic structure.

order to make the TWT operate with high gain over a broad frequency range and low unwanted oscillations, it is important to study how the propagation constant, phase velocity, and group velocity of different modes change with frequency.

A dispersion diagram or k–β diagram is a useful tool for presenting the dispersion characteristics of an SWS. Figure 1.3 presents the dispersion characteristics of a typical periodic structure (normalized wave number, kL, versus normalized phase constant, βL); k is the wave number, $k = \omega\sqrt{\varepsilon\mu}$, $\omega = 2\pi f$, $\varepsilon = \varepsilon_0\varepsilon_r$, $\mu = \mu_0\mu_r$, L is the period, and β is the phase constant. Some useful parameters, including the phase velocity ($v_p = \omega/\beta$) and the group velocity ($v_g = d\omega/d\beta$), can easily be obtained from this diagram.

We can see that, as in most waveguides, there is more than one mode supported by a periodic structure and there are passbands and stopbands formed by different modes. Usually, the fundamental mode is of the most interest, and higher-order modes should be avoided. The $\beta = k_0$ line is the free-space phase constant line; any mode below this line has the slow-wave property. As can be seen in this figure, the phase velocity of the fundamental mode is slower than the speed of light in free space.

For a given k, multiple values of β exist; these arise from the space harmonic characteristic of periodic structures [2]:

$$\beta_n = \beta_0 + \frac{2\pi n}{L} \tag{1.1}$$

where β_0 is the propagation constant of the fundamental mode and n is an integer. Besides, both forward waves and backward waves are supported by a periodic structure. For the forward wave, the phase velocity has the same sign as that of the group velocity. On the contrary, for the backward wave, the phase velocity has a different sign from that of the group velocity. A dotted line is shown in the figure to represent the phase constant for the electron beam. A TWT operates at a point where the beam line intersects with one of the waves. Oscillations may occur when the beam line intersects with the backward wave; this should normally be avoided in a TWT. However, the interaction with the backward wave can also be used to realize a microwave source known as a BWO.

1.3.2.2 Interaction impedance

Another important parameter of an SWS is the interaction impedance, which represents the strength of the beam–wave interaction. The symbol for interaction impedance is K_c, and its unit is ohms. The interaction impedance is defined by Pierce as [20]:

$$K_c = \frac{|E_{z,\,n}|^2}{2P\beta_n^2} \tag{1.2}$$

where $E_{z,n}$ is the axial component of the electric field of the nth space harmonic at the axis of the SWS and P is the average power flow of the electromagnetic wave flowing in the z-direction (shown in figure 1.2). $E_{z,n}$ can be calculated as follows:

$$E_{z,\,n} = \frac{1}{L}\int_0^L E_z e^{j\beta_n z} dz \tag{1.3}$$

where L is the length of one period of the SWS.

P can be calculated using the Poynting vector [2]:

$$P = \frac{1}{2} Re \int \left(\vec{E} \times \vec{H}^*\right)_z dS. \tag{1.4}$$

The power propagating through the structure can also be related to the group velocity v_g through the following:

$$P = Wv_g \tag{1.5}$$

where W is the energy stored per unit axial length of the structure.

1.3.2 Hot-test parameters of SWSs

The hot-test parameters describe the performance of the TWT in the presence of an electron beam; they include the output power, gain, efficiency, etc. With a bunched beam, the output power of the electromagnetic wave increases exponentially along the tube. As the length of the tube increases, the output power reaches a saturation point, after which the output power drops. The RF efficiency η of a TWT without a collector can be calculated as follows [21]:

$$\eta = \frac{P_{\text{out}}}{V_{\text{beam}} \times I_{\text{beam}}} \tag{1.6}$$

where P_{out} denotes the output RF power; V_{beam} and I_{beam} denote the voltage and the current of the electron beam, respectively.

In some applications, such as ECMs and radar, the TWT is generally operated at the saturation point, where it reaches maximum efficiency [1]. On the other hand, in communications applications, the TWT is often operated in the linear region.

1.4 Brief review of existing SWSs

1.4.1 Circular helix

Even though there are large numbers of SWSs with different characteristics, the circular helix is still the most widely used SWS in TWTs, especially for communications and ECM applications. The circular helix has the widest bandwidth among all SWSs. Its interaction impedance is also unmatched by most other SWSs. Figure 1.4 shows the configuration of a circular helix loaded with dielectric support rods. The helix is formed by winding a tape conductor with a radius a and period L. Dielectric rods are used to support the helix inside the metal shield. The supporting rods are usually made of dielectric materials with high thermal conductivity, such as beryllium oxide (BeO) and anisotropic pyrolytic boron nitride (APBN) in order to transfer heat away from the helix efficiently. The pitch angle of the circular helix φ is controlled by both the period and radius and is defined as [6]:

$$\varphi = \tan^{-1}\frac{L}{2\pi a} \tag{1.7}$$

1.4.2 Coupled cavity

The coupled cavity SWS is also an important SWS used for TWT applications. As shown in figure 1.5, the coupled cavity consists of cavities stacked together with coupling apertures. The electromagnetic wave propagates along the structure through the coupling apertures. The beam travels through the beam tunnel at the center of the structure. When the electron beam has a similar velocity to that of the electromagnetic wave, the beam is modulated at the gaps, and the electromagnetic wave is amplified.

The dimensions of the cavities and the types of coupling apertures have a great influence on the properties of the SWS. The coupling aperture is able to couple the electromagnetic wave from one cavity to the next with a phase shift of ψ. The phase shift of one period of the cavity can be considered to be a combination of the phase shift caused by the coupling aperture and the propagation path from one cavity to

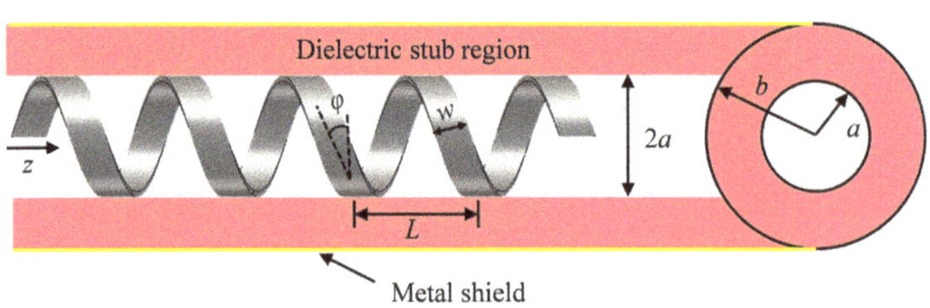

Figure 1.4. Schematic of a circular helix with dielectric support rods and a metal shield.

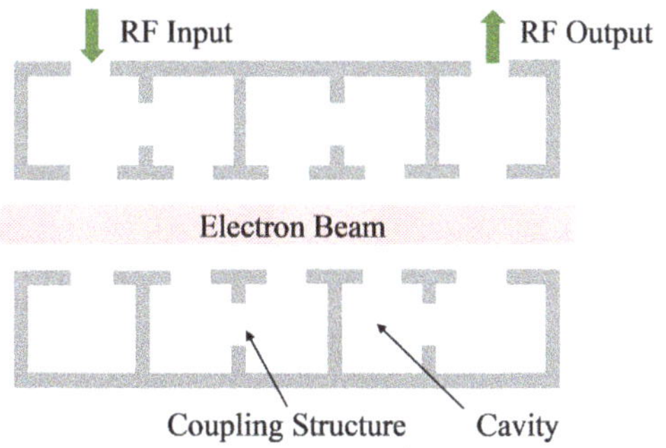

Figure 1.5. Coupled cavity SWS.

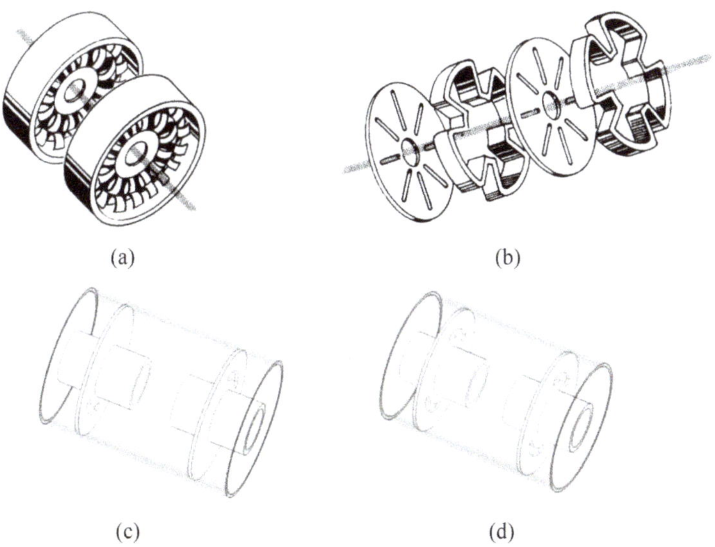

Figure 1.6. (a) Centipede coupling structure; (b) cloverleaf structure; (c) staggered slot structure; and (d) aligned slot structure. Parts (a) and (b) © [1957] IEEE. Reprinted, with permission, from [23]. Part (d) reprinted from [22].

the other. There are basically two types of SWSs with different coupling structures. When ψ is in the range from 0° to 180°, the structure is called a fundamental forward-wave SWS. Typical fundamental forward-wave SWSs include the centipede structure and the cloverleaf structure (figures 1.6(a) and (b)). With proper design of the SWS, the beam velocity can be made high to obtain the maximum output power and gain from the TWT. On the other hand, when ψ is in the range from 180° to 360°, the structure can be considered a fundamental backward-wave SWS or space harmonic SWS. Typical fundamental backward coupled cavity SWSs include staggered slot and aligned slot backward-wave SWSs (figures 1.6(c) and (d)) [22]. For backward-wave applications, the beam velocity must be low to use the fundamental mode of the SWS. This limits the maximum output power obtained.

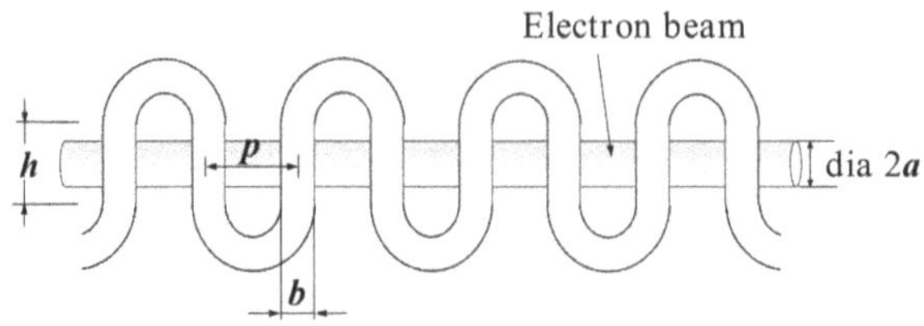

Figure 1.7. Folded waveguide SWS.

The electron beam and the electromagnetic wave in a coupled cavity SWS interact discontinuously at the gaps; in this respect, the coupled cavity SWS is unlike the helical SWS. Also, the bandwidth is much smaller than that of the helical SWS. On the other hand, with an all-metal structure, the power handling and thermal dissipation capability of the coupled cavity SWS are much higher than that of the helical SWS.

1.4.3 Folded waveguide

The folded waveguide can be treated as a serpentine rectangular waveguide with an electron-beam tunnel at the center of the waveguide (figure 1.7). The size of the beam tunnel is small, with a cutoff frequency higher than the operating frequency of the waveguide, so that the electromagnetic wave only propagates along the waveguide. The folding of the waveguide induces an additional phase shift of π per period. As a result, the phase seen by the electrons is:

$$\beta_n p = \beta_g\left(h + \frac{\pi p}{2}\right) + (2n + 1)\pi \tag{1.8}$$

where β_n is the propagation constant of the nth spatial harmonic and β_g is the propagation constant for the TE_{10} mode of the rectangular waveguide. As can be seen, when $n = 0$, the fundamental mode is a backward wave. Thus, for forward-wave applications, the beam velocity should be low in order to synchronize with the phase velocity of the forward wave. The operating principle here is similar to that of the fundamental backward-wave coupled cavity. The folded waveguide has the advantages of wide bandwidth, high output power, and easy fabrication with computer numerical control (CNC) machining techniques. Thus, it is very suitable for millimeter-wave or terahertz applications.

1.4.4 Staggered vane SWS

The staggered vane SWS has its origin in the traditional double-vane SWS. The configuration of the staggered vane SWS is presented in figure 1.8(a). The staggered vanes are located on the top and bottom walls of the rectangular waveguide. The double-vane SWS has aligned vanes, and the electric field in the beam tunnel is perpendicular to the direction of beam travel, leading to a very low interaction impedance. The electric field of the staggered vane SWS, on the other hand, has both transverse and longitudinal components, ensuring strong interaction with the

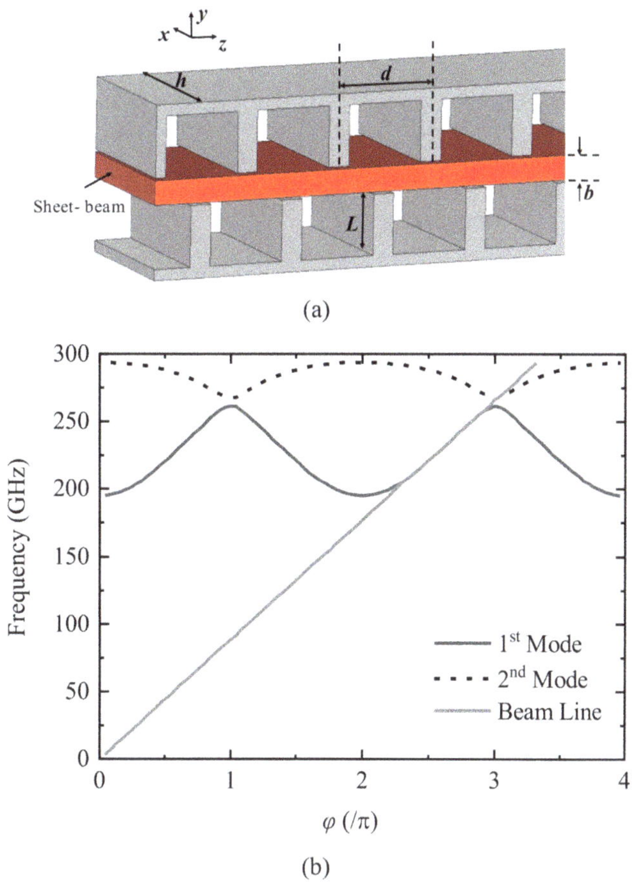

Figure 1.8. (a) Staggered vane SWS. (b) Dispersion properties of the staggered vane SWS. Part (a) reprinted from [24].

electromagnetic wave [24]. This structure also has the advantage of high bandwidth, as the SWS is less dispersive than the traditional double-vane SWS. A typical dispersion diagram for a staggered vane SWS is shown in figure 1.8(b). The fundamental mode of the SWS is a fast wave, while the first harmonic of the SWS is a slow wave. For forward-wave applications of a TWT based on this SWS, the beam velocity should synchronize with the phase velocity of the first harmonic of the SWS.

1.4.5 Sine waveguide

The sine-waveguide SWS was first proposed by the University of Electronic Science and Technology of China (UESTC) in 2012. As shown in figure 1.9, the structure evolves from a rectangular waveguide, with the top and bottom walls undulating up and down along the propagation direction of the waveguide. The operating principle and performance of the sine waveguide are quite close to those of the staggered vane

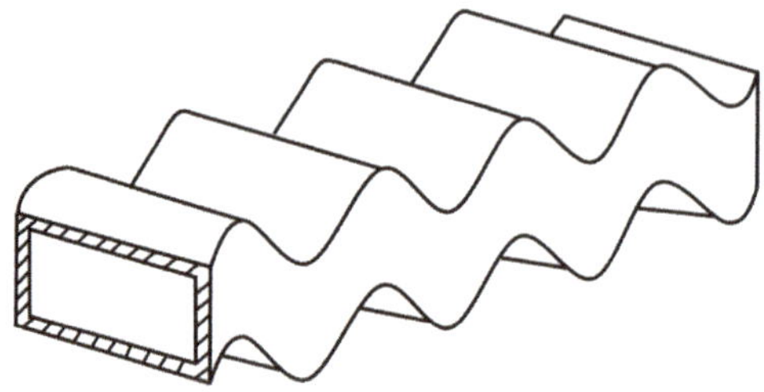

Figure 1.9. Sine-waveguide SWS.

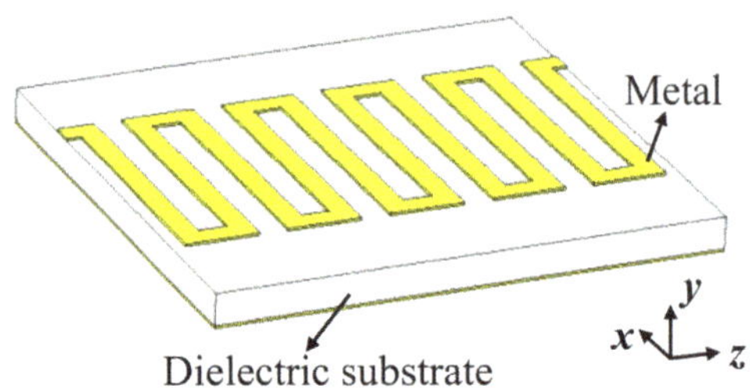

Figure 1.10. Meander-line SWS.

SWS. However, unlike the staggered vane SWS, where vanes stick out into the rectangular waveguide as discontinuities, the sine waveguide has a much smoother shape, which leads to low-loss transmission. The structure has a simple configuration which can easily be fabricated with microfabrication techniques, making it suitable for millimeter-wave or terahertz-frequency applications.

1.5 Planar SWSs

1.5.1 Microstrip meander-line slow-wave structures

The microstrip meander-line (MML) SWS is formed from a microstrip line laid out in a serpentine shape (figure 1.10). The development of the MML TWT was carried out by the companies Varian and RCA in the early 1950s and 1960s. In the 1950s, P N Butcher from Stanford University and F Paschke of Princeton University analyzed the electromagnetic transmission characteristics of MML structures using complex function analysis and the equivalent circuit method, respectively [25, 26]. Since the dispersion of the MML is stronger than that of the helical SWS and its interaction impedance is relatively low, it did not receive much attention for a long time. The related research only remained at the laboratory stage. In recent years, due to the demand for high-power millimeter-wave and submillimeter-wave microwave sources, MMLs have become a popular topic of research because of their advantages of easy fabrication and good thermal performance. Various aspects of the MML SWS are described in detail in chapters 2, 5, 6, and 9.

1.5.2 Planar helical SWSs

As the operating frequency increases, the dimensions of various parts of a TWT, including the SWS, reduce. Table 1.1 shows the radius of helical SWSs for different frequency ranges [27]. It can be seen that the radius of the helix is only tens of micrometers at 300 GHz. At such frequencies, conventional fabrication techniques are no longer adequate. This adversely affects attempts to miniaturize helix TWTs, lower their cost, and push up their frequency of operation. Besides, with such small dimensions, aligning the different parts of a TWT becomes very challenging. Even a small error in alignment may cause a significant deterioration in performance. Furthermore, the electron beam may hit the SWS and the dielectric supports when there is misalignment or the focusing magnetic field is not strong enough. This may cause dielectric charging in the TWT, which, in turn, may greatly affect the beam flow or even cause dielectric breakdown.

Although the circular helix is very widely used, it and its variations cannot easily be fabricated using printed circuit or microfabrication techniques. These methods could reduce the cost of fabrication and enable the precise fabrication of the small structures required for TWTs operating at millimeter-wave frequencies (30–300 GHz) and higher [28].

To address the issue of difficulty in fabrication, a planar version of the helix was proposed to ease the microfabrication of devices intended for operation at millimeter-wave and higher frequencies. In 2008, a rectangular helical structure was proposed by UESTC, and its dispersion characteristics were analyzed using the field analysis method [29]. Although this structure has a planar configuration, it is not suitable for printed circuit or microfabrication techniques because all four metal parts of the helical turn have a certain inclination angle. In 2009, a planar helix with straight-edge connections (PH-SEC) was proposed by the research group at Nanyang Technological University. The configuration of this SWS is presented in figure 1.11. As can be seen, the top and bottom strips are inclined at an angle of $\pm\Psi_1$ with respect to the y-axis. The straight-edge connections subtend a zero angle, $\Psi_2 = 0$, with respect to the x-axis. In practical applications, this structure can be supported by dielectric substrates. The PH-SEC retains the wideband properties of

Table 1.1. Radii of circular helices at different frequencies [27].

Frequency band/power level	Helical radius (mm)
C band (3.4–4.2 GHz/200 W)	2.30
K_u band (10.7–13.0 GHz/250 W)	0.8
K/K_a band (18–32 GHz/100 W)	0.4
Q band (37–42 GHz/50 W)	0.24
V band (60–65 GHz/30 W)	0.16
W band (95 GHz/10 W)	0.1
Submillimeter band (300 GHz/1 W)	0.032

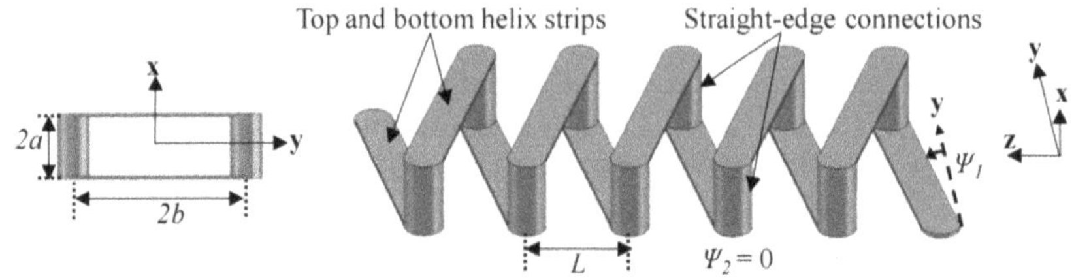

Figure 1.11. Planar helix with straight-edge connections (PH-SEC) without a supporting substrate.

the circular helix and also offers the advantage of easy fabrication due to the straight-edge connections between the top and bottom arrays of inclined conductors. Apart from the advantage of easy fabrication, a sheet beam can be used in a PH-SEC with a high aspect ratio, which can offer the advantages of increased gain and output power of the TWT or BWO. In addition, this structure can provide good heat dissipation, since the metal strips have a larger contact area with the support substrate. Various aspects of the PH-SEC are described in detail in chapters 4–8.

1.6 Microfabrication of SWSs and challenges

1.6.1 Microfabrication methods for SWSs

The relatively recent development of microfabrication techniques has enabled the fabrication of SWSs with extremely small dimensions. Microfabrication techniques allow one to fabricate metal and dielectric structures with good precision and surface finish. These techniques can also provide the advantages of mass production by making it possible to fabricate repeatable devices in large numbers, greatly reducing the cost of fabrication. Moreover, with microfabrication, it is possible, in principle, to place the electron gun, the SWS, and the collector on the same substrate. In this way, the alignment of different components can be improved significantly.

Some microfabrication techniques have successfully been applied to VEDs, including nano-CNC milling; deep reactive ion etching (DRIE); laser ablation; Lithographie, Galvanik, und Abformung (LIGA); wet/dry etching; etc. Most of these processing technologies are based on growth, deposition, and etching techniques and are therefore more suitable for fabricating SWSs with planar configurations. Microfabrication techniques were mostly developed for the semiconductor industry. The fabrication process of vacuum electronic devices with very small feature sizes is still under development, requiring further research and exploration.

1.6.1.1 All-metal SWSs

For all-metal SWSs operated at millimeter or terahertz frequencies, the most commonly used fabrication methods are nano-CNC, DRIE, X-LIGA, UV-LIGA, and additive manufacturing technologies. Some typical all-metal SWSs and the corresponding processing methods are shown in figure 1.12.

LIGA is a lithographic process that is used to fabricate high-aspect-ratio structures. Conventional LIGA technology uses deep X-ray lithography and is

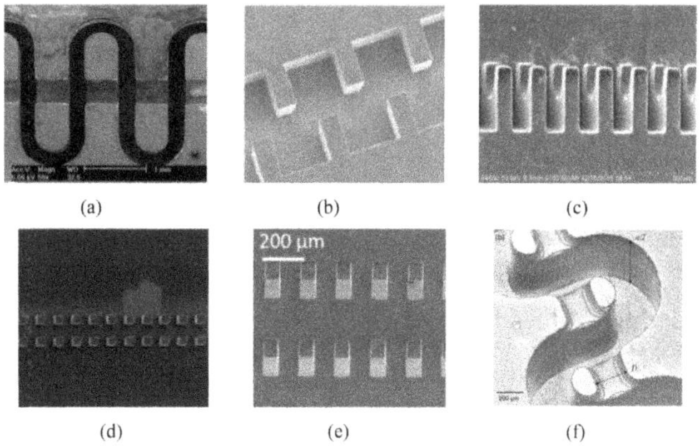

Figure 1.12. Microfabrication of all-metal SWSs: (a) 95 GHz folded waveguide (X-LIGA); (b) 220 GHz staggered vane SWS (UV-LIGA); (c) 346 GHz grooved single-grating SWS (UV-LIGA); (d) 300 GHz double corrugated waveguide (UV-LIGA); (e) 346 GHz double corrugated waveguide (nano-CNC); and (f) W-band folded waveguide (3D printing). Part (a) © [2006] IEEE. Reprinted, with permission, from [30]. Part (b) reprinted from [31]. Parts (c) and (e) © [2018] IEEE. Reprinted, with permission, from [32]. Part (d) Reproduced from [33]. © IOP Publishing Ltd. All rights reserved. Part (f) © [2019] IEEE. Reprinted, with permission, from [34].

capable of processing structures with a depth-to-width ratio greater than 500 and a thickness higher than 1500 μm. In 2003, Seoul National University used a two-step X-LIGA processing method to produce 100 GHz folded waveguide and coupled cavity structures with machining tolerances of less than 2 μm and a surface roughness of 20–70 nm [30]. Since X-LIGA requires expensive X-ray sources and complicated masks, UV-LIGA, which is based on UV sources and SU-8 photoresists, has been widely adopted. To date, UV-LIGA technology has been successfully used by several universities and institutes, such as UC Davis, Seoul National University, Lancaster University, and the Beijing Vacuum Electronics Research Institute, to realize folded waveguides, staggered vane SWSs, and double corrugated waveguides for use at frequencies from 100 to 350 GHz [32, 33, 35, 36].

CNC milling is a subtractive manufacturing method that uses computers to automate fabrication. Nano-CNC technology has been developed in recent years for the fabrication of millimeter-wave and terahertz SWSs. Neville C Luhmann's team at UC Davis first proposed the use of nano-CNC technology to fabricate terahertz SWSs and successfully fabricated a variety of SWSs using the Mori Seiki NN1000 nano-CNC milling machine, including staggered vane SWSs (220, 260, and 346 GHz), double corrugated waveguides (346 GHz), and folded waveguides (270 GHz), achieving dimensional tolerances of 1–5 μm and a surface roughness of 40–100 nm [37]. In 2018, the National Institute of Science and Technology in Korea fabricated G-band (110 GHz–300 THz) and Y-band (325–500 GHz) folded waveguides using a ROBONANO α-OiB nano-CNC milling machine and achieved a minimum surface roughness of 11.17 nm in the Y-band by improving the surface roughness with lubricants [38]. In 2022, UESTC made the first successful attempt to fabricate 1 THz improved sinusoidal waveguides using nano-CNC technology [39]. Nano-CNC technology can achieve high machining accuracy and surface roughness, but it does not support mass production, as it requires frequent tool changes to maintain high-precision fabrication.

With the development of high-resolution 3D printing techniques, there have been some attempts to fabricate SWSs using 3D printing techniques. In the study described in [34], folded waveguides for use in the W-band and the D-band were fabricated using a 'P4 mini' 3D printer from EnvisionTEC, Inc. This machine projects 2D images in sequence into a tray of liquid photopolymer resin, after which the structure is electroplated with copper. The fabricated SWS shows excellent RF performance in the upper millimeter W frequencies well above 100 GHz. However, the poor vacuum compatibility and the low melting temperature of the photo-polymer limit the application of this fabrication technique. In [40], a 35 GHz folded waveguide was fabricated using the direct metal laser sintering (DMLS) technique with both aluminum and chromium cobalt. However, the loss was relatively high due to the high roughness of the structure's surface. Most recently, Elve Inc. from UC Davis has reported an additive manufacturing technique for fabricating SWSs for millimeter-wave TWTs [41]. Using this approach, circuits at different frequencies can be fabricated quickly and consistently at scale. An E-band TWT has been developed based on a circuit fabricated in this manner. This device can provide a saturated output power of 100 W within the frequency range of 81 to 86 GHz.

1.6.1.2 Dielectric-substrate-based SWSs

In 2009, the University of Wisconsin fabricated a W-band MML using a micro-electromechanical systems (MEMS)-related fabrication technique. A meandering silicon structure with a high aspect ratio (3:1) was first fabricated using the DRIE technique (figure 1.13(a)). Selective metallization of 2 μm was then carried out only

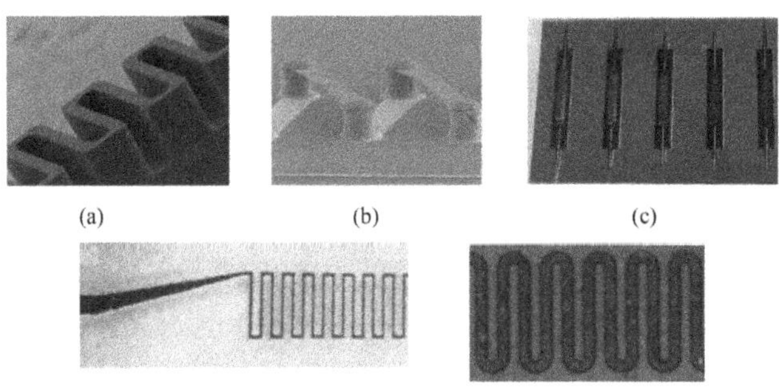

Figure 1.13. Microfabrication of the dielectric-substrate-based SWS: (a) W-band MML SWS (DRIE + UV-LIGA); (b) W band PH-SEC (UV-LIGA + lift-off); (c) K_a-band MML (photolithography); (d) V-band MML (laser ablation); and (e) W-band MML SWS (photolithography). Part (a) © [2019]. IEEE. Reprinted, with permission, from [42]. Part (b) © [2011] IEEE. Reprinted, with permission, from [43]. Part (c) © [2009] IEEE. Reprinted, with permission, from [44]. Part (d) © [2018] IEEE. Reprinted, with permission, from [45]. Part (e) © [2022] IEEE. Reprinted, with permission, from [46].

on top of the high-aspect-ratio structure [42]. Because the substrate material is only present below the metallic microstrip structure, the dielectric loading effect of the structure is lower, and the interaction impedance is increased. However, due to the high loss of the silicon substrate, the measured circuit loss was as high as 30 dB for 22 periods of the SWS.

In 2011, the authors studied and fabricated a W-band PH-SEC SWS on a 750 μm high-resistivity silicon wafer using UV-LIGA and lift-off techniques (figure 1.13(b)) [43, 47]. In 2018, we fabricated a V-shaped MML SWS on a quartz wafer using photolithography (figure 1.13(c)) [44]. Due to the low conductivity of the seed layer, its circuit loss was also relatively high.

Also in 2018, N M Ryskin's group from the Russian Academy of Sciences proposed a simple and effective method for the fabrication of an MML SWS by sputtering copper onto the surface of a quartz wafer and then shaping the meander line with laser ablation (figure 1.13(d)) [45, 48]. The fabricated SWSs showed good surface roughness with low loss. However, the laser ablation burned the edges of the metal strips, affecting the smoothness of the structure and making it difficult to obtain high accuracy in the frequency bands above 100 GHz. In 2022, L Yue from UESTC used photolithography and electroplating to fabricate a U-shaped meander-line SWS with a conformal dielectric substrate layer for use in the W-band. The circuit loss turned out to be less than that of traditional MML SWSs (figure 1.13(e)) [46].

1.6.2 Challenges for microfabricated SWSs

There are many challenges and issues that must be considered in the microfabrication process. Some of these issues are shared by all types of SWSs. Other issues are critical for dielectric-loaded SWSs, such as circular helix, PH-SEC, and meander-line SWSs. Some of the important challenges are listed below.

1.6.2.1 Dimensional and alignment accuracy

For an SWS working at millimeter-wave or terahertz frequencies, the dimensional accuracy of the structure becomes very important, since a small error may lead to a significant change, such as a frequency shift or additional reflection. In the fabrication process, each step may suffer from variations and errors. If each fabrication step is not well-controlled, the performance may be seriously affected. In any case, the effects of such variations and errors need to be considered at the design stage.

Accurate alignment of different parts of the SWS assembly is also a very important issue [49]. Helical SWSs have three parts: the metal helix, the supporting dielectric rods, and the metal shield. The three parts are fabricated separately and assembled together. Misalignment of these parts may seriously degrade the performance of the circuit. In a similar manner, for microfabricated SWSs involving multilayer construction, accurate alignment of the different layers is very important.

1.6.2.2 Attenuation

At high frequencies, the skin depth of conductors is very small. For example, at 100 GHz, the skin depth of copper is only around 200 nm. Limited by the metallization process, the roughness of the metal surface may be close to or larger than the skin depth of the conductor. In this case, a conductive structure fabricated for high-frequency applications will have higher resistivity and higher loss than for low-frequency applications [47]. The root-mean-square (RMS) roughness of the conductor surface can vary significantly with different fabrication processes. For instance, the two-step X-LIGA used to produce the folded waveguide SWS in [50] is able to achieve a surface roughness of 20–70 nm. In a W-band PH-SEC [47], the top and bottom metal strips were fabricated using a lift-off process and electroplating, respectively; the two layers had RMS roughnesses of 13.65 and 143.40 nm, respectively.

Loss in an SWS may also arise from dielectric materials. As the PH-SEC structure involves supporting dielectric substrates, it is important that the dielectric material has a low loss tangent at millimeter-wave frequencies.

1.6.2.3 Heat dissipation

With decreased area available for heat transfer and increased loss from reduced skin depth, heat transfer is a significant issue for microfabricated SWSs. Moreover, the small dimensions required for high-frequency operation cause difficulties in electron-beam alignment and focusing. If the electron beam is not focused well, more heat is generated by collisions between electrons and the structure. The heat thus generated limits the maximum output power of the TWT. For structures with dielectric loadings, the contact area and the thermal conductivity of the dielectric play an important role in heat dissipation [51].

1.6.2.4 Dielectric loading

Dielectric support structures are used in helical TWTs in order to support the SWS inside the metal shield and conduct heat away from the SWS. Vacuum-compatible materials with high thermal conductivity are often used to increase thermal dissipation. However, the dielectric support often has a loading effect, reducing the phase velocity as well as the interaction impedance. To reduce the dielectric loading effect, it is necessary to reduce the amount of dielectric or use materials with a low dielectric constant.

In traditional circular helix TWTs, thin rectangular support rods made of beryllium oxide (BeO, $\varepsilon_r = 6.5$) or APBN ($\varepsilon_r = 5.1$) are most widely used, as shown in figure 1.4 [6]. For some recently microfabricated SWSs, silicon ($\varepsilon_r = 11.9$) [42, 47], diamond ($\varepsilon_r = 5.7$) [50, 51], or quartz ($\varepsilon_r = 3.75$) [44, 52] have been used as supporting dielectric materials.

1.6.2.5 Dielectric charging

Normally, the magnetic focusing in TWTs prevents the axially-flowing electrons from spreading in the radial direction, but some issues such as an inadequate magnetic focusing field and/or misalignment of the electron gun (off-axis or inclined with respect to the axis) may cause electrons to hit the surrounding SWS. The charge of the electrons that land on the metallic SWS is conducted away. However, the charge of the electrons that land on the dielectric support structures may accumulate there and create a voltage difference between the SWS and the dielectric. This voltage may become so high as to cause dielectric breakdown. Even if this is not the case, a high voltage on the dielectric affects the electron motion, leading to the defocusing of the electron beam and in turn causing more electrons to hit the structure. This problem becomes more severe at millimeter-wave or terahertz frequencies, where precise alignment of the various parts and good control of the magnetic field are more difficult to achieve.

1.7 Methods of analysis and simulation for SWSs

1.7.1 Methods of analysis

1.7.1.1 Dispersion characteristics

A number of analytical methods have been developed to obtain the dispersion characteristics of SWSs such as the circular helix SWS, folded waveguide, and meander-line SWS, etc. These methods are very useful, as they can provide an insight into the operation of these SWSs. However, some of these analytical models rely on certain approximations and assumptions, so the accuracy is not very high. Furthermore, for some SWSs with complex configurations, it is difficult to study the dispersion property analytically.

1.7.1.2 Hot-test parameters

The linear gain of a TWT can be estimated using the Pierce theory of TWTs [20, 53] from the following parameters of the SWS and the electron beam:

$$C = \sqrt[3]{\frac{K_c I_{\text{beam}}}{4 V_{\text{beam}}}} \tag{1.9}$$

where C is the small signal gain parameter, K_c is the coupling or interaction impedance, and V_{beam} and I_{beam} are the voltage and current of the electron beam, respectively.

$$N = \frac{\beta_e l}{2\pi} \tag{1.10}$$

N is the number of guide wavelengths and β_e is the beam propagation constant.

$$b = \frac{\beta_0 - \beta_e}{\beta_e C} \tag{1.11}$$

b is a measure of the degree of synchronism between the beam and the electromagnetic wave, which has a phase constant of β_0.

$$d = \frac{\alpha}{\beta_e C} \tag{1.12}$$

d denotes circuit attenuation and is proportional to the attenuation constant α.

$$QC = \frac{\omega_q^2}{4C^2\omega^2} \tag{1.13}$$

QC is the space charge parameter and ω_q is the effective plasma frequency.

δ_1, δ_2, and δ_3 are the roots of the cubic equation given next. These roots are functions of b (the degree of sychronism), d (circuit attenuation), and QC (space charge):

$$\delta^2 = \frac{1}{(-b + jd + j\delta)} - 4QC. \tag{1.14}$$

These three roots correspond to three different modes of wave propagation that exist in a TWT. δ_1 corresponds to a growing wave, δ_2 corresponds to a decaying wave, and δ_3 corresponds to a normal propagating wave. The gain can be calculated as follows:

$$\text{gain} = A + BCN \tag{1.15}$$

where

$$A = 20\log_{10}\left|\left(1 + \frac{4QC}{\delta_1^2}\right)\left(\frac{1}{(1 - \delta_2/\delta_1)(1 - \delta_3/\delta_1)}\right)\right| \tag{1.16}$$

and

$$B = 54.6x_1 \tag{1.17}$$

in which x_1 is the real part of δ_1.

For a simple case, ignoring space-charge effects and circuit loss, and using a synchronous beam ($QC = b = d = 0$), the linear gain of the tube can be simply calculated as:

$$G = -9.54 + 47.3CN. \tag{1.18}$$

(Note: We use the symbols b and d here for the sake of consistency with the literature; elsewhere in this book, these symbols represent other quantities).

1.7.2 Methods of simulation

Simulation tools are vital for the design of TWTs, since the theoretical analyses always require certain simplifying assumptions to reduce the complexity of analysis. It is important for a simulation code to predict the performance with high accuracy, especially when developing an actual working device. The performance parameters include the cold-test parameters (dispersion characteristics and interaction impedance) and the hot-test parameters (gain, power, and efficiency). In recent years, the development of simulation codes has been driven heavily by the industry to accelerate the design process and reduce the overall development cost [54].

1.7.2.1 Dispersion characteristics and attenuation constant

There are basically two ways to obtain the characteristic properties of periodic structures using full-wave simulators. One method is to carry out a numerical analysis by imposing Floquet periodic boundaries on the sides of one period of the SWS and turning it into a periodic eigenmode problem [53, 55]. In this method, usually varying phase difference values are considered between the periodic boundaries, and the corresponding operating (eigen) frequencies are calculated. For lossless structures, the obtained eigenfrequency values are real. On the other hand, when the metal and dielectric losses are considered, the corresponding eigenfrequency values are complex:

$$\omega = \omega_r + j\omega_i \tag{1.19}$$

where ω_r and ω_i are the real and imaginary parts of the eigenfrequency, respectively. The circuit loss α (Np m^{-1}) at ω_r can be calculated as:

$$\alpha(\omega_r) = \frac{\omega_i}{\nu_g} \tag{1.20}$$

where v_g is the group velocity, which can be calculated from $d\omega_r/d\beta$. For low-loss and moderate-loss problems, the imaginary part of the frequency ω_i is much smaller than the real part of the frequency ω_r. In addition, using the relationship between the quality factor Q and the complex eigenfrequencies [56]:

$$Q = \frac{|\omega|}{2\omega_i} \approx \frac{\omega_r}{2\omega_i} \tag{1.21}$$

the circuit loss can also be calculated as:

$$\alpha(\omega_r) = \frac{\omega_r}{2\nu_g Q}. \tag{1.22}$$

This method shows good accuracy and has been successfully implemented in commercial software such as CST Microwave Studio (MWS) and Ansys High-Frequency Structure Simulator (HFSS). It is also widely used by researchers studying the dispersion properties of SWSs. However, most eigenmode simulation methods usually require a shielding boundary, and a structure with an open environment is not supported.

An alternative method of calculating the dispersion properties of a periodic SWS is to apply excitations to the input and output of the periodic structure of the SWS and calculate the transmission and reflection coefficients with full-wave simulations. Then, all the properties of the periodic SWS, including both the complex propagation constant ($\gamma = \alpha + j\beta$) and the complex characteristic impedance ($Z_0 = Z_R + jZ_I$) can be calculated with the ABCD matrix of one period of the structure using:

$$\cosh(\gamma p) = \frac{A + D}{2} \tag{1.23}$$

$$Z_0 = \sqrt{\frac{B}{C}}. \tag{1.24}$$

This method has been successfully used to extract the propagation properties of periodic structures [57–60]. However, the coupling between unit cells is not taken into account, as the structure is truncated by input/output couplers in the simulations. For SWSs with low coupling between unit cells, one period of the SWS is enough to extract the characteristic properties accurately. On the other hand, for SWSs with high coupling between unit cells, the property extracted from one period without the coupling between adjacent units clearly differs from that for the infinite periodic SWS. In order to capture the properties of infinite periodic structures more faithfully, an increased number of periods of the structures often needs to be considered.

The schematic view of a general periodic SWS of N periods is presented in figure 1.14. Let the transmission matrices of the whole structure of N periods and one unit cell be $[M_c]$ and $[M_p]$, respectively. The transmission matrix $[M_c]$ can easily be obtained from numerical simulations. For an SWS circuit consisting of N periods, $[M_c]$ can be expressed in a diagonalized form:

$$[M_c] = [M_p]^N = C\begin{pmatrix} \lambda_{1,\,N} & 0 \\ 0 & \lambda_{2,\,N} \end{pmatrix}C^{-1} = C\begin{pmatrix} \lambda_1^N & 0 \\ 0 & \lambda_2^N \end{pmatrix}C^{-1} \tag{1.25}$$

$$[M_p] = C\begin{pmatrix} \lambda_1 & 0 \\ 0 & \lambda_2 \end{pmatrix}C^{-1} \tag{1.26}$$

where λ_1 $(0 < |\lambda_1| < 1)$ and λ_2 $(|\lambda_2| > 1)$ are the eigenvalues of $[M_p]$ which represent the waves propagating in the positive and negative directions, respectively. The eigenvalues are closely related to the propagation constant and the complex propagation constant according to the relationship [2]:

$$\lambda_i = e^{-\gamma_i p} \quad (i = 1,\, 2). \tag{1.27}$$

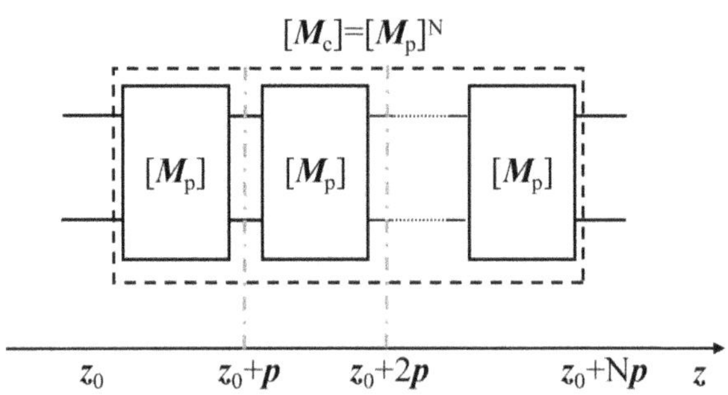

Figure 1.14. Schematic view of a general periodic SWS of N periods.

The eigenvalues of the waves propagating in both directions can be calculated as follows:

$$\gamma_i = \alpha_i + j\beta_i$$
$$= -\frac{\ln|\lambda_{i,\,N}|}{Np} - j\left[\frac{\mathrm{Arg}(\lambda_{i,\,N})}{Np} + \frac{2\pi m}{Np}\right] = -\frac{\ln|\lambda_{i,\,N}|}{Np} - j\left[\frac{\mathrm{unwrap}\{\mathrm{Arg}(\lambda_{i,\,N})\}}{Np}\right] \quad (1.28)$$

where the function Arg (·) is the principal argument of a complex number with values from $-\pi$ to $+\pi$. There are N different solutions, where m is an arbitrary integer varying from zero to $N-1$. The correct value of γ can be recovered by choosing the correct value of m. For most SWSs, such as the circular helix, the folded waveguide, and the MML, the dispersion properties of the SWSs at frequencies close to the starting frequency of operation are already known. Based on the general properties, we can easily obtain the value of m by unwrapping the phase. With the eigenvalues, the transmission matrix of one period of the SWS $[M_p]$ can be calculated using equation (1.26). Finally, the propagation constant and characteristic impedance can both be evaluated using equations (1.27) and (1.28).

1.7.2.2 Hot-test parameters

The hot-test parameters can be simulated with particle-in-cell (PIC) simulations. The charged particles can be included in the PIC simulations to obtain the hot-test parameters. The equation of motion and the Lorentz force are included in the calculation to consider the mutual coupling between the charged particles and the electromagnetic fields. The PIC simulation normally requires substantial computational resources due to the large number of charged particles present in the calculation domain [54]. Several PIC simulation codes have been developed for plasma physics, including ARGUS [54], MAGIC [61], Christine [62], VORPAL [63], CST MAFIA, and CST PIC [64].

1.7.3 Methods of measurement

1.7.3.1 Measurement of dispersion characteristics

For SWSs, it is necessary to measure the dispersion characteristics to evaluate their suitability for TWT applications. A graph of frequency versus the propagation constant or phase velocity is often used to represent the dispersion characteristics. Usually, only the fundamental space harmonic of the SWS is considered for measurement. When more space harmonics are involved, the graph is often known as a Brillouin diagram.

The traveling-wave method and the resonance method are often used to measure the dispersion characteristics. In the traveling-wave method, a probe is used to measure the phase at different locations by moving the probe along the direction of propagation of the wave. The phase constant β can be calculated as follows:

$$\beta = \frac{\Delta\varphi}{\Delta L} \quad (1.29)$$

where $\Delta\varphi$ and ΔL are the phase difference and the distance between two different locations, respectively. This method is more readily applicable to open structures that are easily accessible. For enclosed SWSs, an alternative approach is to measure the phase difference between the transmission coefficients (S_{21}) of SWSs with different numbers of periods n_1 and n_2 ($n_1 > n_2$). The phase constant can then be calculated using (1.29) with ΔL given by:

$$\Delta L = (n_1 - n_2)p \tag{1.30}$$

where p is the period.

The resonance method is another method that is usually used for the measurement of dispersion characteristics; it is applicable to all SWSs that possess transverse symmetry (most coupled cavity SWSs, ladder lines, ring–bar topologies, etc). In this method, the resonant frequencies of a short-circuited SWS consisting of N periods are measured. At the resonant frequency, the phase constant satisfies:

$$\beta = \frac{\theta}{p} = \frac{\pi}{Np}n \tag{1.31}$$

where θ is the phase change of one period and n is an integer representing the number of resonances [65].

1.7.3.2 Measurement of interaction impedance

The measurement of interaction impedance is often carried out using the perturbation method, where a small perturbation dielectric rod is placed along the axis of the SWS. The interaction impedance K_0 can be calculated using [66]:

$$K_0 = \frac{2}{\omega\beta(\varepsilon_r - 1)\varepsilon_0\pi r^2}\frac{\Delta\beta}{\beta} \tag{1.32}$$

where ω is the angular frequency, ε_r is the relative permittivity of the dielectric, r is the radius of the dielectric rod, and $\Delta\beta$ is the change of phase constant when the rod is inserted.

This approach can also be applied to the resonant perturbation method as follows [66]:

$$K_0 = \frac{2}{\omega\beta(\varepsilon_r - 1)\varepsilon_0\pi r^2}\frac{v_p}{v_g}\frac{\Delta f}{f} \tag{1.33}$$

where v_p and v_g are the phase velocity and group velocity, respectively.

1.8 Outline of the book

This book focuses on planar SWSs and their variations, particularly for applications in TWTs. The main chapters of the book describe the evolution of two of the planar SWSs, namely the MML and the planar helix. In contrast to other SWSs, these two types of SWSs are typically fabricated on dielectric substrates using photolithography, justifying their classification as planar SWSs. The recognition and demonstration of their potential for application are described in detail.

Chapter 1 introduces VEDs, TWTs, and SWSs. The characteristics of the SWS play a very important role in determining the performance of a TWT. Some existing SWSs and their properties are described briefly. Methods of analysis, simulation, and measurement are also described concisely.

In chapter 2, the MML SWS is introduced. The operating principle and theoretical analysis used to obtain the dispersion properties are discussed. The hot-test parameters and practical performance of TWTs based on meander-line SWSs are also described.

Chapter 3 presents initial studies of the planar helix, modeling it as a pair of unidirectionally conducting (UC) screens which are infinitely wide in the transverse direction. The guiding properties of a pair of UC screens with and without periodicity are analyzed. Also described are the dispersion characteristics of the planar helix in the presence of an electron beam; these characteristics indicate amplification for one of the forward waves.

In chapter 4, a practical planar helix structure with finite cross-sectional dimensions is introduced. This is achieved by incorporating transverse confinement in the form of straight-edge connections which connect the two screens at the sides, leading to what we call PH-SEC. This modification allows practical realization of PH-SECs using printed circuit techniques for deployment at relatively low microwave frequencies (1–30 GHz) or using microfabrication techniques for deployment at millimeter-wave frequencies (30–300 GHz).

In chapter 5, dispersion control techniques for planar SWSs are described. The effects of some of the main parameters of the MML SWS on the dispersion properties are presented, followed by the dispersion control techniques achieved with both metal vanes and coplanar ground planes in PH-SEC SWSs. Finally, we describe a practical design for a PH-SEC SWS intended to operate in the Ka band with low dispersion.

In chapter 6, some variations of both the meander-line SWS and the PH-SEC SWS are described to solve some of the issues of conventional planar SWSs. As a variation of the MML SWS, the logarithmic spiral and the angular log-periodic meander-line SWS are presented; these are intended for low operating beam voltages. Some other variations of the MML SWS, such as the V-shaped meander-line SWS and the 3D U-shaped meander-line SWS, are proposed to increase the interaction impedance of the SWS. Some variations of the PH-SEC SWS are then presented. A planar ring–bar SWS and an unconnected pair of PH-SECs are used to reduce the risk of backward-wave oscillation in TWTs. A modified PH-SEC structure that has a low risk of dielectric charging is also introduced.

In chapter 7, the applications of planar SWSs for BWOs are described. First of all, fundamental backward-wave circuits, including the interdigital SWS and the coplanar SWS, are introduced for millimeter-wave BWO applications. Next, a W-band BWO based on a PH-SEC is described. Finally, an efficiency-enhancement technique for BWOs is introduced and demonstrated for a PH-SEC-based BWO.

In chapters 8 and 9, some examples of design and fabrication techniques for PH-SEC and meander-line SWSs are presented. The fabrication processes for SWSs intended for use in different frequency bands are described in detail. The properties of some of these structures are demonstrated through both simulation and measurement.

References and further reading

[1] Barker R J, Booske J H, Luhmann N C and Nusinovich G S 2005 *Modern Microwave and Millimeter-Wave Power Electronics* (Piscataway, NJ: IEEE)

[2] Collin R E 1992 *Foundations for Microwave Engineering* (New York: McGraw-Hill)

[3] Kopp Bruce A, Billups. A J and Luesse M H 2001 Thermal analysis and considerations for gallium nitride microwave power amplifier packaging *Microw. J.* **44** 72–82

[4] True R 1984 A theory for coupling gridded gun design with PPM focussing *IEEE Trans. Electron Devices* **31** 353–62

[5] Kosmahl H G 1982 Modern multistage depressed collectors—a review *Proc. IEEE* **70** 1325–34

[6] Gilmour A S 1994 *Principles of Traveling Wave Tubes* (Boston, MA: Artech House)

[7] Kory C L 1996 Three-dimensional simulation of helix traveling-wave tube cold-test characteristics using MAFIA *IEEE Trans. Electron Devices* **43** 1317–9

[8] Chemin D *et al* 2001 A three-dimensional multifrequency large signal model for helix traveling wave tubes *IEEE Trans. Electron Devices* **48** 3–11

[9] Christie V L, Kumar L and Balakrishnan N 2002 Improved equivalent circuit model of practical coupled-cavity slow-wave structures for TWTs *Microw. Opt. Technol. Lett.* **35** 322–6

[10] Wilson J D 2001 Design of high-efficiency wide-bandwidth coupled-cavity traveling-wave tube phase velocity tapers with simulated annealing algorithms *IEEE Trans. Electron Devices* **48** 95–100

[11] Freund H P, Antonsen T M, Zaidman E G, Levush B and Legarra J 2002 Nonlinear time-domain analysis of coupled-cavity traveling-wave tubes *IEEE Trans. Plasma Sci.* **30** 1024–40

[12] Du C H and Liu P K 2009 A lossy dielectric-ring loaded waveguide with suppressed periodicity for gyro-TWTs applications *IEEE Trans. Electron Devices* **56** 2335–42

[13] Han S T *et al* 2004 Low-voltage operation of Ka-band folded waveguide travelling-wave tube *IEEE Trans. Plasma Sci.* **32** 60–6

[14] Booske J H *et al* 2005 Accurate parametric modeling of folded waveguide circuits for millimeter-wave traveling wave tubes *IEEE Trans. Electron Devices* **52** 685–94

[15] So J K *et al* 2006 Experimental investigation of micro-fabricated folded waveguide backward wave oscillator for submillimeter application *31st Int. Conf. on Infrared and Millimeter Waves and 14th Int. Conf. on Terahertz Electronics* 315

[16] Kory C, David J, Tran H T, Ives L and Chernin D 2005 Folded waveguide circuit optimizations using Christine 1D *IEEE Conf. Record-Abstracts. 2005 IEEE Int. Conf. on Plasma Science* (IEEE: Piscataway, NJ)

[17] Kory C L and Wilson J D 1995 Novel high-gain, improved-bandwidth, finned-ladder V-band traveling-wave tube slow-wave circuit design *IEEE Trans. Electron Devices* **42** 1686–92

[18] James B G and Kolda P 1986 A ladder circuit coupled-cavity TWT at 80–100 GHz *1986 Int. Electron Devices Meeting* 32 (IEEE: Piscataway, NJ)

[19] Wintucky E G *et al* 2002 A high efficiency, miniaturized Ka band traveling wave tube based on a novel finned ladder RF circuit design *3rd IEEE Int. Vacuum Electronics Conf.* (IEEE: Piscataway, NJ)
[20] Pierce J R 1950 *Traveling-Wave Tubes* (New York: Van Nostrand)
[21] Gilmour A S 1986 *Microwave Tubes* (Boston, MA: Artech House)
[22] Gilmour A S 2011 *Principles of Klystrons, Traveling Wave Tubes, Magnetrons, Cross-Field Ampliers, and Gyrotrons* (Norwood, MA: Artech House)
[23] Chodorow M and Craig R 1957 Some new circuits for high-power traveling-wave tubes *Proc. IRE* **45** 1106–18
[24] Shin Y M, Barnett L R and Luhmann N C 2009 Phase-shifted traveling-wave-tube circuit for ultrawideband high-power submillimeter-wave generation *IEEE Trans. Electron Devices* **56** 706–12
[25] Butcher P N 1957 The coupling impedance of tape structures *Proc. IEEE B: Radio Electron. Eng* **104** 177–87
[26] Paschke F 1958 A note on the dispersion of interdigital delay lines *RCA Rev.* **19** 418–22
[27] Chong C K and Menninger W L 2010 Latest advancements in high-power millimeter-wave helix TWTs *IEEE Trans. Plasma Sci.* **38** 1227–38
[28] Ives R L 2004 Microfabrication of high-frequency vacuum electron devices *IEEE Trans. Plasma Sci.* **32** 1277–91
[29] Fu C, Wei Y, Wang W and Gong Y 2008 Dispersion characteristics of a rectangular helix slow-wave structure *IEEE Trans. Electron Devices* **55** 3582–9
[30] Shin Y M *et al* 2006 Experimental investigation of 95 GHz folded waveguide backward wave oscillator fabricated by Two-Step LIGA *IEEE Int. Vacuum Electronics Conf.* 419–20
[31] Baig A *et al* 2011 Design, fabrication and RF testing of near-THz sheet beam TWTA *Terahertz Sci. Technol.* **4** 181–207
[32] Feng J *et al* 2018 Fabrication of a 0.346-THz BWO for plasma diagnostics *IEEE Trans. Electron Devices* **65** 2156–63
[33] Malekabadi A and Paoloni C 2016 UV-LIGA microfabrication process for sub-terahertz waveguides utilizing multiple layered SU-8 photoresist *J. Micromech. Microeng.* **26** 095010
[34] Cook A M, Joye C D and Calame J P 2019 W-band and D-band traveling-wave tube circuits fabricated by 3D printing *IEEE Access* **7** 72561–6
[35] Shin Y-M, Barnett L R, Gamzina D, Luhmann N C, Field M and Borwick R 2009 Terahertz vacuum electronic circuits fabricated by UV lithographic molding and deep reactive ion etching *Appl. Phys. Lett.* **95** 181505
[36] Paoloni C *et al* 2016 THz backward-wave oscillators for plasma diagnostic in nuclear fusion *IEEE Trans. Plasma Sci.* **44** 369–76
[37] Gamzina D *et al* 2016 Nano-CNC machining of sub-THz vacuum electron devices *IEEE Trans. Electron Devices* **63** 4067–73
[38] Lee I, Choi W, Shin J and Choi E 2018 Microscopic analyses of electrical conductivity of micromachined-folded waveguides based on surface roughness measurement for terahertz vacuum electron devices *IEEE Trans. Terahertz Sci. Technol.* **8** 710–8
[39] Yang R *et al* 2022 Design and experiment of 1 THz slow wave structure fabricated by nano-CNC technology *IEEE Trans. Electron Devices* **69** 2656–61
[40] Anderson J P, Ouedraogo R and Gordon D 2014 Fabrication of a 35 GHz folded waveguide TWT circuit using rapid prototype techniques *2014 39th Int. Conf. on Infrared, Millimeter, and Terahertz waves (IRMMW-THz)* 1–2

[41] Gamzina D and Kowalczyk R 2023 Novel design and manufacturing techniques revitalize mm-wave TWTs *Microw. J.* **66** 20–32 https://www.microwavejournal.com/articles/39929-novel-design-and-manufacturing-techniques-revitalize-mmwave-twts?page=1
[42] Sengele S, Jiang H, Booske J H, Kory C L, van der Weide D W and Ives R L 2009 Microfabrication and characterization of a selectively metallized W-band meander-line TWT circuit *IEEE Trans. Electron Devices* **56** 730–7
[43] Chua C, Tsai J, Tang M, Aditya S and Shen Z 2011 Microfabrication of a planar helix with straight-edge connections slow-wave structure *Adv. Mater. Res.* **254** 17–20
[44] Wang S, Aditya S, Xia X, Ali Z and Miao J 2018 On-wafer microstrip meander-line slow-wave structure at Ka-band *IEEE Trans. Electron Devices* **65** 2142–8
[45] Ryskin N M *et al* 2018 Planar microstrip slow-wave structure for low-voltage V-band traveling-wave tube with a sheet electron beam *IEEE Electron Device Lett.* **39** 757–60
[46] Yue L *et al* 2022 A high interaction impedance microstrip meander-line with conformal dielectric substrate layer for a W-band traveling-wave tube *IEEE Trans. Electron Devices* **69** 5826–31
[47] Chua C *et al* 2011 Microfabrication and characterization of W-band planar helix slow-wave structure with straight-edge connections *IEEE Trans. Electron Devices* **58** 4098–105
[48] Ryskin N M *et al* 2021 Development of microfabricated planar slow-wave structures on dielectric substrates for miniaturized millimeter-band traveling-wave tubes *J. Vac. Sci. Technol.* B **39** 013204
[49] Wei W, Wei Y, Wang Y, Wang W and Gong Y 2015 A study of the effects of helix misalignment on the cold parameters of a sheath helix slow-wave structure *IEEE Trans. Electron Devices* **62** 1334–41
[50] Shin Y M *et al* 2006 Microfabrication of millimeter wave vacuum electron devices by two-step deep-etch X-ray lithography *Appl. Phys. Lett.* **88** 091916
[51] Yan S, Yao L and Yang Z 2008 Effect of thermal strain in helical slow-wave circuit on TWT cold-test characteristics *IEEE Trans. Electron Devices* **55** 2278–81
[52] Guo G, Zhang T, Zeng J, Yang Z, Yue L and Wei Y 2022 Investigation and fabrication of the printed microstrip meander-line slow-wave structures for D-band traveling wave tubes *IEEE Trans. Electron Devices* **69** 5229–34
[53] Basu B N 1996 *Electromagnetic Theory and Applications in Beam-Wave Electronics* (Singapore: World Scientific)
[54] Antonsen T M, Mondelli A A, Levush B, Verboncoeur J P and Birdsall C K 1999 Advances in modeling and simulation of vacuum electronic devices *Proc. IEEE* **87** 804–39
[55] Collin R E 1990 *Field Theory of Guided Waves* 2nd edn (New York: Wiley) https://www.wiley.com/en-us/Field+Theory+of+Guided+Waves%2C+2nd+Edition-p-9780879422370
[56] Tsuji M, Shigesawa H and Takiyama K 1983 On the complex resonant frequency of open dielectric resonators *IEEE Trans. Microw. Theory Tech.* **31** 392–6
[57] Zhu L 2003 Guided-wave characteristics of periodic coplanar waveguides with inductive loading∼unit-length transmission parameters *IEEE Trans. Microw. Theory Tech.* **51** 2133–8
[58] Valerio G, Paulotto S, Baccarelli P, Burghignoli P and Galli A 2011 Accurate bloch analysis of 1-D periodic lines through the simulation of truncated structures *IEEE Trans. Antennas Propag.* **59** 2188–95
[59] Eberspacher M A and Eibert T F 2013 Dispersion analysis of complex periodic structures by full-wave solution of even–odd-mode excitation problems for single unit cells *IEEE Trans. Antennas Propag.* **61** 6075–83

[60] Mesa F, Rodríguez-Berral R and Medina F 2018 On the computation of the dispersion diagram of symmetric one-dimensionally periodic structures *Symmetry* **10** 307

[61] Chen X, Toh W K and Lindsay P A 2004 Physics of the interaction process in a typical coaxial virtual cathode oscillator based on computer modeling using MAGIC *IEEE Trans. Plasma Sci.* **32** 1191–9

[62] Abe D K, Levush B, Antonsen T M, Whaley D R and Danly B G 2002 Design of a linear C-band helix TWT for digital communications experiments using the CHRISTINE suite of large-signal codes *IEEE Trans. Plasma Sci.* **30** 1053–62

[63] Nieter C and Cary J R 2004 VORPAL: a versatile plasma simulation code *J. Comput. Phys.* **196** 448–73

[64] CST STUDIO SUITE www.3ds.com/products-services/simulia/products/cst-studio-suite

[65] Horsley A W and Pearson A 1966 Measurement of dispersion and interaction impedance characteristics of slow-wave structures by resonance methods *IEEE Trans. Electron Devices* **ED-13** 962–9

[66] Kory C L and Dayton J A 1998 Computational investigation of experimental interaction impedance obtained by perturbation for helical traveling-wave tube structures *IEEE Trans. Electron Devices* **45** 2063–71

[67] Dayton J A, Kory C L, Mearini G T, Malta D, Lueck M and Gilchrist K 2009 Applying microfabrication to helical vacuum electron devices for THz applications *2009 IEEE Int. Vacuum Electronics Conf., IVEC 2009* 41–4

[68] Kory C L *et al* 2009 95 GHz helical TWT design *2009 IEEE Int. Vacuum Electronics Conf., IVEC 2009* (Piscataway, NJ: IEEE) 125–6

[69] Weiland T 1985 On the unique numerical solution of maxwellian eigenvalue problems in three dimensions *Part. Accel.* **17** 227–42

Chapter 2

Microstrip meander-line slow-wave structure

2.1 Introduction

The circular helix is widely recognized as the most popular slow-wave structure (SWS) for wideband traveling-wave tube (TWT) applications. However, as the operating frequency increases to the millimeter-wave or terahertz range, the circular helix becomes increasingly difficult to fabricate due to its reduced size. The microstrip meander-line (MML) SWS is a planar SWS that can be fabricated easily and can effectively operate at millimeter-wave and higher frequencies.

Figure 2.1 depicts the configuration of the MML SWS. The structure consists of a meandering metal strip with width w and thickness t, positioned on the top surface of a dielectric substrate. The thickness of the substrate is h. A conductive ground plane is located on the bottom surface of the substrate. The substrate material has a dielectric constant of ε_r and normally possesses a low loss tangent. Vacuum-compatible dielectric materials such as boron nitride, beryllium oxide, quartz, and diamond are often used for the substrate.

The microstrip line is a widely used microwave transmission line that supports a quasi transverse electromagnetic (TEM) wave. Due to the meandering configuration of the metal strip, the electromagnetic wave follows a longer propagation path, resulting in an increased phase shift in the direction of transmission. This property leads to the slow-wave characteristics of the MML SWS. A detailed analysis of the electromagnetic properties of the MML SWS is discussed in this chapter.

The MML SWS exhibits several advantages over the helical SWS. First of all, it has a planar configuration that is compatible with printed circuit or microfabrication techniques. These techniques allow for the fabrication of highly precise structures with small feature sizes, which is particularly advantageous for applications at millimeter-wave or terahertz frequencies. In addition, microfabrication techniques enable mass production and integration, leading to a reduction in the cost

doi:10.1088/978-0-7503-5764-7ch2

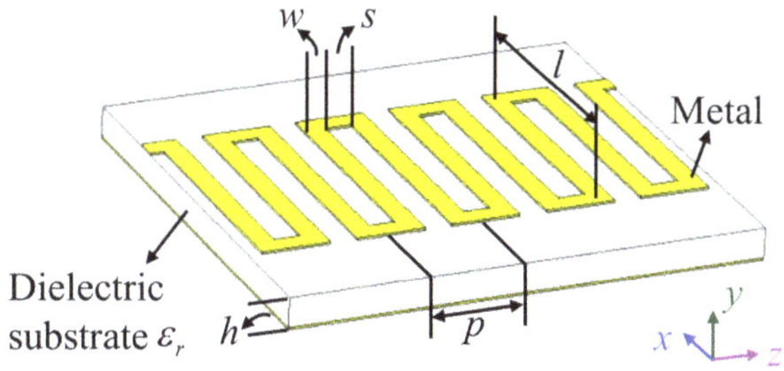

Figure 2.1. Configuration of an MML SWS.

of the fabrication process. Additionally, this structure can readily operate with a sheet electron beam. Such compatibility results in increased beam current that enhances the output power of the TWT. Last but not least, the MML SWS has better thermal dissipation capability than the helical SWS because the dielectric substrate has better contact with the metal strip and the metal enclosure.

Despite the advantages of the MML SWS, it is important to note that the dielectric substrate also introduces certain limitations. First of all, the electric field is more concentrated in the dielectric material than in the vacuum above the SWS. This generally leads to reduced interaction impedance, resulting in low gain and output power when the SWS is used in a TWT. In addition, the dielectric material causes dispersion in the SWS, i.e. the phase velocity can change significantly with frequency. This decreases the operating bandwidth of the TWT. Furthermore, there is a risk of electrons landing on the dielectric substrate and causing charge accumulation if the electron beam is not well focused. If the voltage due to charge accumulation exceeds the threshold breakdown voltage, the substrate material may break down.

In view of these problems, many variations of the MML SWS have been proposed to improve its performance. Some of these variations are described in detail in chapter 6 of this book. This chapter mainly focuses on the traditional MML SWS. The chapter is organized as follows. Section 2.2 addresses the guiding properties of the meander-line SWS. It starts with a general description of the model of the microstrip line. This is followed by methods of analysis of the MML SWS based on field theory as well as the equivalent circuit approach. Section 2.3 presents an example of an MML SWS applied in a TWT; it includes a detailed design procedure, the electromagnetic guiding properties of the SWS, the design of the input/output couplers, and particle-in-cell (PIC) simulation results.

2.2 Propagation properties of the meander-line SWS

2.2.1 Microstrip line

The MML SWS is based on the microstrip transmission line. The configuration of the microstrip line is presented in figure 2.2. It consists of a metal strip with a width w and a thickness t on the top surface of a substrate and a ground plane on the

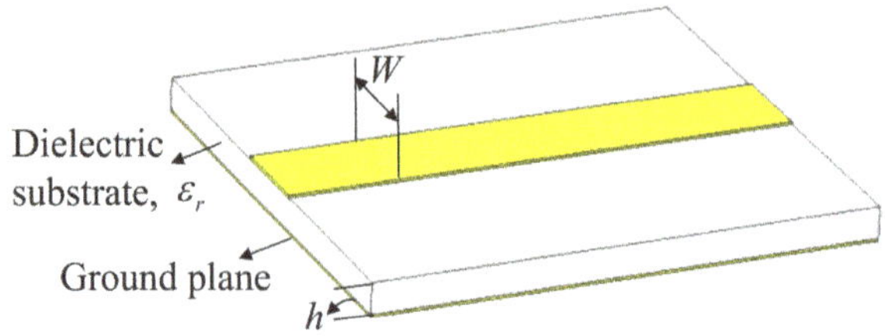

Figure 2.2. Configuration of a microstrip line.

bottom surface of the substrate. The thickness of the dielectric substrate is *h*. Similar to the coaxial line and waveguide, the microstrip line is commonly used as a transmission line in RF circuits. Compared with other transmission lines, the microstrip line has the advantages of compact size, low weight, wide bandwidth, and ease of integration with solid-state devices. As a result, it has gained extensive usage in microwave components such as filters, couplers, antennas, amplifiers, etc. In this subsection, the properties and design formulas of the microstrip line are summarized briefly.

2.2.1.1 Wave propagation in microstrip lines

When an electromagnetic wave is traveling along a microstrip line, a portion of the electromagnetic field is within the substrate, while the rest of the electromagnetic field is above the substrate. In other words, the electromagnetic wave is propagating in an inhomogeneous medium. Under this condition, strictly speaking, the TEM mode of propagation is not supported, as the longitudinal components of both the electric field and the magnetic field are also present. However, as the longitudinal components of the field in the microstrip line are much smaller than the transverse components, the fundamental mode of the microstrip line can be approximately treated as the TEM mode. This propagation mode is called the quasi-TEM mode [1, 2]. Under this approximation, the properties of the microstrip line over the operating frequency range are the same as for the static (DC) case. Good approximations can be obtained by using the quasi-static model of the microstrip line for operating frequencies from zero to a few gigahertz.

As the electromagnetic wave travels at different speeds in the dielectric and in vacuum, the phase velocity of the electromagnetic wave is slower than that in vacuum and faster than that in the dielectric. It is often convenient to use the approximation of a single homogeneous material with an effective dielectric constant ε_{eff} to replace the two-layer inhomogeneous medium in the microstrip line; the ε_{eff} value lies between the dielectric constant of the dielectric and that of vacuum.

2.2.1.2 Properties of microstrip lines

Some design formulas are available in the research literature for the effective dielectric constant and the characteristic impedance of a microstrip line. These parameters can be approximately calculated as follows:

$$\varepsilon_{\text{eff}} = \begin{cases} \dfrac{\varepsilon_r + 1}{2} + \dfrac{\varepsilon_r - 1}{2}\left\{\dfrac{1}{\sqrt{1 + 12h/W}} + 0.04\left(1 - \dfrac{W}{h}\right)^2\right\} & \dfrac{W}{h} \geqslant 1 \\ \dfrac{\varepsilon_r + 1}{2} + \dfrac{\varepsilon_r - 1}{2}\dfrac{1}{\sqrt{1 + 12h/W}} & \dfrac{W}{h} \leqslant 1 \end{cases} \tag{2.1}$$

$$Z_0 = \begin{cases} \dfrac{\eta}{2\pi\sqrt{\varepsilon_{\text{eff}}}} \ln\left(\dfrac{8h}{W} + 0.25\dfrac{W}{h}\right) & \dfrac{W}{h} \geqslant 1 \\ \dfrac{\eta}{\sqrt{\varepsilon_{\text{eff}}}}\left\{\dfrac{W}{h} + 1.393 + 0.667 \ln\left(\dfrac{W}{h} + 1.444\right)\right\}^{-1} & \dfrac{W}{h} \leqslant 1 \end{cases} \tag{2.2}$$

where η is the wave impedance of free space. These formulas provide results with an accuracy better than 1% [3].

The dielectric loss and conductor loss can be calculated as follows:

$$\alpha_d = \frac{8.686 k_0 \varepsilon_r(\varepsilon_{\text{eff}} - 1)\tan\delta}{2\sqrt{\varepsilon_{\text{eff}}}(\varepsilon_r - 1)} \text{ dBm}^{-1} \tag{2.3}$$

$$\alpha_c = \frac{8.686 R_s}{Z_0 W} \text{ dBm}^{-1} \tag{2.4}$$

where R_s is the surface resistivity of the conductor and $\tan\delta$ is the loss tangent of the dielectric material.

2.2.1.3 Dispersion in microstrip lines

The dielectric in the microstrip line causes dispersion. This means that the phase velocity of an electromagnetic wave propagating in the microstrip line changes with frequency. Accordingly, the effective dielectric constant changes with frequency as well. Equation (2.1) is obtained under the quasi-static approximation, which is only applicable for low microwave frequencies. For high-frequency applications, dispersion has to be considered. An empirical formula for the effective dielectric constant, considering the effect of dispersion $\varepsilon_{\text{eff}}(f)$, has been developed by Kobayashi based on data from full-wave numerical simulations [4] and can be expressed as:

$$\varepsilon_{\text{eff}}(f) = \varepsilon_r - \frac{\varepsilon_r - \varepsilon_{\text{eff}}}{1 + (f/f_{50})^m} \tag{2.5}$$

where

$$f_{50} = \frac{f_{TM_0}}{0.75 + (0.75 - 0.332\varepsilon_r^{-1.73})W/h} \tag{2.6}$$

$$f_{TM_0} = \frac{c}{2\pi h\sqrt{\varepsilon_r - \varepsilon_{\text{eff}}}} \tan^{-1}\left(\varepsilon_r\sqrt{\frac{\varepsilon_{\text{eff}} - 1}{\varepsilon_r - \varepsilon_{\text{eff}}}}\right) \tag{2.7}$$

$$m_0 = 1 + \frac{1}{1 + \sqrt{W/h}} + 0.32(1 + \sqrt{W/h})^{-3} \tag{2.8}$$

$$m_c = \begin{cases} 1 + \dfrac{1.4}{1 + W/h}\left\{0.15 - 0.235e^{\frac{-0.45f}{f_{50}}}\right\} & (\dfrac{W}{h} \leqslant 0.7) \\ 1 & (\dfrac{W}{h} \geqslant 0.7) \end{cases} \tag{2.9}$$

$$m = m_0 m_c \leqslant 2.32. \tag{2.10}$$

In these equations, ε_{eff} is the quasi-static value obtained from (2.1). When the product $m_0 m_c$ is greater than 2.32, the parameter m is set to 2.32. Equation (2.5) can provide an accuracy better than 0.6% for $0.1 < W/h \leqslant 10$, $1 < \varepsilon_r \leqslant 128$, and any value of h/λ.

2.2.2 Field theory of the MML SWS

Before applying a meander-line SWS in a TWT, it is important to obtain the electromagnetic wave propagation properties of the SWS, such as the phase propagation constant, phase velocity, and interaction impedance. Field theory serves as an effective way of obtaining these properties. There have been a number of studies on the field analysis of MML SWSs [5–10]. This section provides a detailed introduction to the field theory under the quasi-static condition [6]. The effect of the metal shield is considered in this method. The metal strips are assumed to be perfectly conducting and to have infinitesimal thickness. A method that considers the effect of the metal thickness is described in [10].

2.2.2.1 Generic structure of parallel metal strips

The field analysis of an MML SWS starts from a generic structure of an infinite array of parallel metal strips. The configuration of the generic structure is shown in figure 2.3. On the top surface of the substrate is an infinite array of metal strips. The

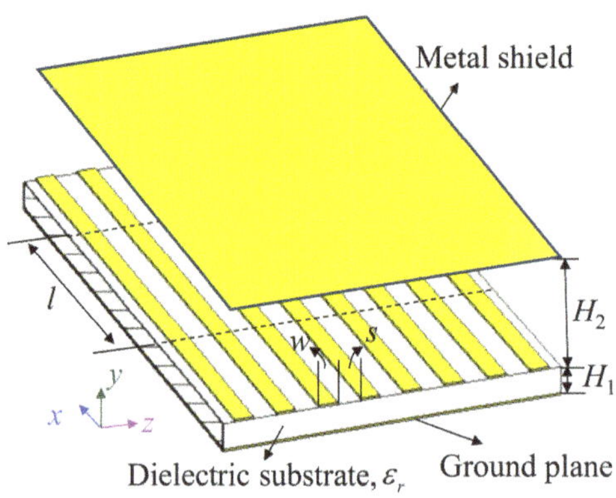

Figure 2.3. Configuration of an infinite array of parallel metal strips.

width of each metal strip is w, and the gap between adjacent strips is s. The substrate has a thickness of H_1 and is grounded by a bottom ground plane. Additionally, a metal shield is positioned above the top surface of the substrate at a height of H_2. The MML SWS in figure 2.1 can be considered to be formed by truncating the generic structure with width l and alternately connecting the ends of adjacent strips. The field analysis starts by obtaining the solution for the wave propagating on the generic structure.

It is assumed that a normal mode is supported in the generic structure. As the structure is uniform with respect to x, the wave propagates with constant amplitude and a phase constant of k in the x-direction. The propagation in the structure can be characterized by an effective dielectric constant ε_{eff}. Then, the wave function $\Phi(x, y, z)$ satisfies the Helmholtz equation:

$$\nabla^2\Phi + (\omega^2/c^2)\varepsilon_{\text{eff}}\Phi = 0 \tag{2.11}$$

Furthermore, it is commonly assumed that the normal mode propagating in the x-direction can be treated as a TEM mode [11]. As the structure is uniform in the x-direction, the wave function can be written in the form:

$$\Phi(x, y, z) = U(y, z)e^{\pm j\beta x} \tag{2.12}$$

where U represents the electric potential. Substituting the expression for Φ into (2.11) yields:

$$\frac{\partial^2 U}{\partial y^2} + \frac{\partial^2 U}{\partial z^2} + (k_0^2\varepsilon_{\text{eff}} - \beta^2)U = 0. \tag{2.13}$$

Further, for a TEM mode,

$$\beta^2 - k_0^2\varepsilon_{\text{eff}} = 0 \tag{2.14}$$

so that (2.13) can be simplified to

$$\frac{\partial^2 U}{\partial y^2} + \frac{\partial^2 U}{\partial z^2} = 0. \tag{2.15}$$

As the structure is periodic in the z-direction, according to Floquet theory, it has a constant phase change of φ per period in the z-direction and U can be expressed as:

$$U(y, z) = f(y, z)e^{j\varphi z/p} \tag{2.16}$$

where $f(y, z)$ is a periodic function with p which can be expanded using a Fourier series:

$$f(y, z) = \sum_{m=-\infty}^{+\infty} F_m(y)\exp\left(\frac{j2\pi mz}{p}\right). \tag{2.17}$$

Using (2.16) and (2.15), we can get:

$$\frac{\partial^2 f}{\partial y^2} + \frac{\partial^2 f}{\partial z^2} + 2j\frac{\varphi}{p}\frac{\partial f}{\partial z} - \frac{\varphi^2}{p^2}f = 0. \tag{2.18}$$

The Fourier coefficients $F_{\mathrm{m}}(y)$ satisfy:

$$\frac{d^2F_m}{dy^2} - \beta_m^2 F_m = 0 \tag{2.19}$$

where $\beta_m = \frac{\varphi + 2\pi m}{p}$. Solving this differential equation, it is easy to find that F_{m} is proportional to $e^{j\beta_m y}$. Using (2.16), (2.17), and (2.19), we can obtain U as follows:

$$U(y, z) = \begin{cases} \sum\limits_m C_m e^{j\beta_m z} \sinh|\beta_m|\, y, & 0 \leqslant y \leqslant H_1 \\ \sum\limits_m D_m e^{j\beta_m z} \sinh|\beta_m|(H_2 - y), & H_1 \leqslant y \leqslant H_2 \end{cases} \tag{2.20}$$

where C_m and D_m are amplitude coefficients. These coefficients need to be obtained in order to derive the solution for the potential U. Boundary conditions enable one to obtain these coefficients. First, as U is continuous across the surface of the substrate, C_m and D_m satisfy:

$$\frac{D_m}{C_m} = \frac{\sinh|\beta_m|\, H_1}{\sinh|\beta_m|(H_2 - H_1)}. \tag{2.21}$$

Then, as each period contains two metal strips, even and odd modes exist due to the different combinations of potential on the two strips. These can be represented by two symmetries:

$$f_o(y, -z) = -f_o^*(y, z)$$

$$f_e(y, -z) = +f_e^*(y, z) \tag{2.22}$$

where o and e stand for odd and even, respectively, and * denotes the complex conjugate.

Using the expressions presented in (2.20) and considering the conditions in (2.21) and (2.22), further calculations of these coefficients could be performed using numerical techniques, such as the finite-difference method [12] or Green's functions [6]. Here, we briefly introduce the calculation process of a Green's function. In [11], a Green's function was used to represent the potential of a very narrow, uniformly charged substrip. In this calculation, the strip in the generic structure is divided into N pieces of narrow substrips, and the Green's function is used to obtain the potential and charge of each subsection. The total charge Q for each of the modes, odd and even, can be calculated as:

$$Q = \sum_{n=1}^{N} \lambda_n \tag{2.23}$$

where λ_n is the charge on each substrip. The capacitance per strip can be obtained from:

$$C = \frac{Q}{V}. \tag{2.24}$$

The calculation of capacitance is carried out twice. The first calculation obtains the capacitance C_k in the presence of the dielectric material. The second calculation obtains the capacitance C_0 without the dielectric. The effective dielectric constant can be calculated as follows:

$$\varepsilon_{\text{eff}} = \frac{C_k}{C_0}. \tag{2.25}$$

The phase velocity and characteristic impedance Z_0 can be obtained from:

$$v_p = \frac{c}{\sqrt{\varepsilon_{\text{eff}}}} \tag{2.26}$$

$$Z_0 = \frac{1}{cC_0\sqrt{\varepsilon_{\text{eff}}}}. \tag{2.27}$$

From the previous three equations, we can derive the electromagnetic properties of the generic structure, including the effective dielectric constant, the propagation constant, the phase velocity, and the characteristic impedance for both the even and odd modes. These quantities are represented by the subscripts e and o, respectively, for the even and odd modes.

2.2.2.2 Dispersion characteristics of meander-line SWSs

Knowing the parameters of wave propagation of the generic structure of parallel metal strips, the next step is to obtain the dispersion of the meander-line SWS. As illustrated in figures 2.1 and 2.3, the meander-line SWS can be realized by truncating a generic structure with width l and connecting the ends of adjacent strips. In the current analysis, the electrical length of the connections is disregarded. More complex models of the connections could be included in the theory.

In each period of the meander-line SWS, there are two metal strips. Denoting the left strip by 1 and the right strip by 2, we can write expressions for the voltages and currents on these two strips:

$$\begin{cases} V_1(x) = [a_+e^{j\beta_e x} + a_-e^{-j\beta_e x} - b_+e^{j\beta_o x} + b_-e^{-j\beta_o x}]e^{-\frac{j\varphi}{4}} \\ V_2(x) = [a_+e^{j\beta_e x} + a_-e^{-j\beta_e x} + b_+e^{j\beta_o x} - b_-e^{-j\beta_o x}]e^{+\frac{j\varphi}{4}} \\ I_1(x) = \dfrac{1}{Z_e}[a_+e^{j\beta_e x} - a_-e^{-j\beta_e x}]e^{-\frac{j\varphi}{4}} + \dfrac{1}{Z_o}[-b_+e^{j\beta_o x} - b_-e^{-j\beta_o x}]e^{-\frac{j\varphi}{4}} \\ I_2(x) = \dfrac{1}{Z_e}[a_+e^{j\beta_e x} - a_-e^{-j\beta_e x}]e^{-\frac{j\varphi}{4}} + \dfrac{1}{Z_o}[b_+e^{j\beta_o x} + b_-e^{-j\beta_o x}]e^{+\frac{j\varphi}{4}} \end{cases} \tag{2.28}$$

where a_+, a_-, b_+ and b_- are the amplitudes of the even and odd modes propagating in the $+x$ direction and the $-x$ direction, respectively. Simple boundary conditions can be imposed on the voltages and currents on the connecting links at the sides of the meander line, as follows:

$$\begin{cases} V_1(0.5l) = V_2(0.5l) & I_1(0.5l) = -I_2(0.5l) \\ V_2(-0.5l) = V_3(-0.5l) & I_2(-0.5l) = -I_3(-0.5l). \\ V_3(x) = V_1(x)e^{j\varphi} & I_3(x) = I_1(x)e^{j\varphi} \end{cases} \tag{2.29}$$

The application of these boundary conditions in the expressions of voltage and current leads to the characteristic (or dispersion) equation:

$$\tan^2\frac{\varphi}{4} = \frac{Z_o}{Z_e}\begin{cases} \tan\beta_e(l/2)\tan(\beta_o l/2) \\ \cot(\beta_e l/2)\cot(\beta_o l/2) \end{cases} \tag{2.30}$$

where the upper and lower equations represent the forward wave and the backward wave, respectively.

The dispersion equation derived from the analysis is calculated for the following parameters: $\varepsilon_r = 6.5$, $H_1 = 1.5875$ mm, $l = 12.4968$ mm, $w = 1.143$ mm, $s = 1.4478$ mm, and $H_2 = 4.1275$ mm.

The dispersion diagram is presented in figure 2.4. The solid line represents the analytical result. The first stopband can be observed in the frequency range of 4.1 to 5.1 GHz. The fabrication of this structure on a beryllium oxide substrate with the aforementioned dimensions was also carried out at the Lincoln Laboratory using photolithographic techniques. The measured dispersion curve is also represented by small circles in figure 2.4. The measured result is in very good agreement with the analytical results; this experiment also reveals the location of the first stopband.

2.2.2.3 Characteristic impedance

To achieve good wave transmission in an SWS, it is important to obtain the value of the characteristic impedance of the SWS to ensure that the characteristic impedance

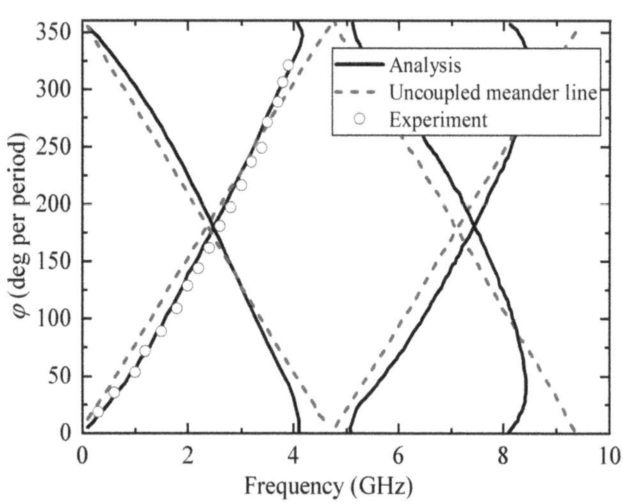

Figure 2.4. Dispersion diagram for the meander line. © [1974] IEEE. Reprinted, with permission, from [6].

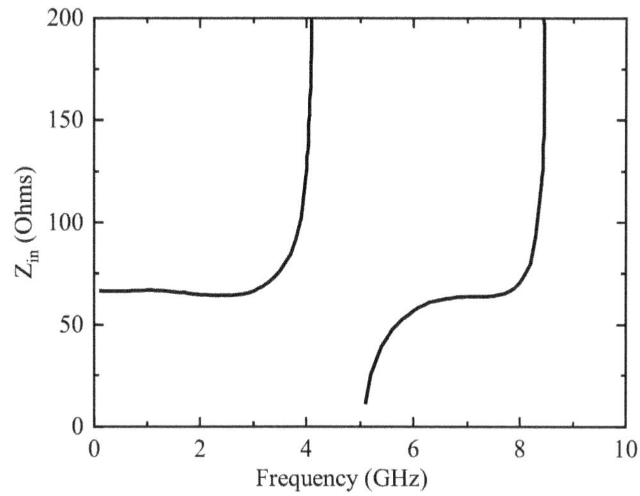

Figure 2.5. Input impedance function Z_{in} for the MML SWS. © [1974] IEEE. Reprinted, with permission, from [6].

of the couplers matches that of the SWS. The solution of the eigenmode equation indicates that $a_+ = a_- = a$, $b_+ = b_- = b$, where a and b satisfy:

$$\frac{b}{a} = \mp\sqrt{\frac{Z_o \sin \beta_e l}{Z_e \sin \beta_o l}}. \tag{2.31}$$

In (2.31), the minus sign stands for the forward wave and the plus sign stands for the backward wave. The characteristic impedance can be calculated as follows:

$$Z_{in} = \frac{V_1(0)}{I_1(0)} = \begin{cases} \sqrt{Z_e Z_o \dfrac{\tan(\beta_o l/2)}{\tan(\beta_e l/2)}} \\ -\sqrt{Z_e Z_o \dfrac{\cot(\beta_o l/2)}{\cot(\beta_e l/2)}} \end{cases} \tag{2.32}$$

The calculated characteristic impedance of the meander-line SWS is presented in figure 2.5. The characteristic impedance is relatively stable at lower frequencies. It increases rapidly as the frequency increases. A stopband from about 4–5 GHz can be observed; the impedance goes to infinity within the stopbands.

2.2.2.4 Interaction impedance

The interaction impedance of the SWS is defined by Pierce as:

$$K_n = \frac{|E_{z,n}|^2}{2\beta_n^2 P} \tag{2.33}$$

where K_n is the interaction impedance of the nth space harmonic, $E_{z,n}$ is the on-axis longitudinal electric field of the nth space harmonic, β_n is the propagation constant of the nth space harmonic and P is the power flow. The power flow can be evaluated by multiplying the group velocity by the stored energy per period. The group velocity can be obtained from the dispersion diagram using $v_g = d\omega/d\beta$. The stored energy can be got from the capacitance. The power flow is given by:

$$P = \frac{2l}{p} v_g \left[\frac{|a|^2}{v_e Z_e}\left(1 + \frac{\sin \beta_e l}{\beta_e l}\right) + \frac{|b|^2}{v_o Z_o}\left(1 - \frac{\sin \beta_o l}{\beta_o l}\right) \right] \tag{2.34}$$

The electric field within the meander-line SWS can be directly calculated by differentiating the potential Φ obtained from the earlier calculations. The interaction impedance can then be obtained from equation (2.33). Some results were given in [6]. However, these results are probably incorrect [12]. Some results obtained using the same equations are given in the appendix worksheet of [12], in which the electric potentials were obtained using the finite-difference method. The calculated results in [12] are a better match for the measured interaction impedance in [13].

2.2.3 Equivalent circuit theory of meander-line SWSs

The equivalent circuit theory provides another way of obtaining the dispersion characteristics of the meander-line SWS. The accuracy of the equivalent circuit analysis may not be as good as that of the field analysis. On the other hand, it is a straightforward and convenient way to derive the electromagnetic properties of the MML SWS without complex numerical evaluations. In this section, a simple equivalent circuit analysis for the MML SWS is introduced. Since no field expressions are involved in this method, only the dispersion characteristics and the characteristic impedance can be calculated; the interaction impedance and the attenuation constant cannot be obtained.

2.2.3.1 Dispersion diagram

In an infinitely long periodic SWS, the voltage–current relationship between the input and output of the nth segment can be expressed using the following equation:

$$\begin{bmatrix} V_n \\ I_n \end{bmatrix} = [T] \begin{bmatrix} V_{n+1} \\ I_{n+1} \end{bmatrix} = e^{\gamma p'} \begin{bmatrix} V_{n+1} \\ I_{n+1} \end{bmatrix} \tag{2.35}$$

where V_n, I_n, V_{n+1}, and I_{n+1} are the voltage and current at the input and output terminals of the equivalent transmission line, respectively. $\gamma = j\beta + \alpha$ is the complex propagation constant for the periodic structure, p' is one period of the SWS, and $[T]$ is the transmission matrix. As depicted in figure 2.6, the MML SWS can be regarded as a cascaded connection of nine separate sections, including five sections of uniform microstrip transmission lines (A_1–A_5) and four sections of discontinuities caused by right-angle bends (B_1–B_4). As shown in figure 2.7, the transmission matrix $[T]$ of one period of the SWS can be obtained by cascading the transmission matrices of all nine sections:

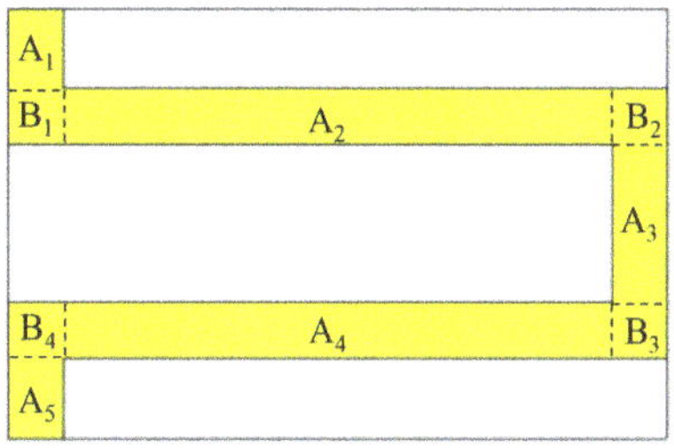

Figure 2.6. Segmented representation of the MML SWS.

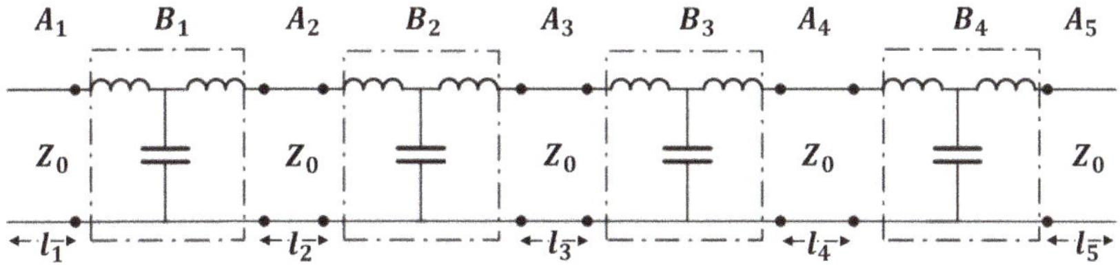

Figure 2.7. Equivalent circuit model of one period of the MML SWS.

$$[T] = \begin{bmatrix} A & B \\ C & D \end{bmatrix} = [A_1][B_1][A_2][B_2][A_3][B_3][A_4][B_4][A_5] \tag{2.36}$$

Segments A_1 to A_5 can be considered to be uniform transmission lines. Based on transmission line theory, the transmission matrix for A_1–A_5 can be expressed as [2]:

$$[A_i] = \begin{bmatrix} \cos \beta_0 l_i & jZ_0 \sin \beta_0 l_i \\ j \sin \beta_0 l_i / Z_0 & \cos \beta_0 l_i \end{bmatrix} (i = 1, 2 \ldots 5) \tag{2.37}$$

where β_0 is the propagation constant, Z_0 is the characteristic impedance of the microstrip line, and l_i is the path length of each segment of the uniform microstrip line. The expression for the propagation constant β_0 at different frequencies can be calculated using:

$$\beta_0 = \frac{2\pi f}{c} \sqrt{\varepsilon_{re}(f)} \tag{2.38}$$

where c is the speed of light in free space, f represents the frequency, and $\varepsilon_{re}(f)$ is the dispersive effective dielectric constant of the dielectric material, which can be calculated using equation (2.5).

The discontinuities B_1–B_4 caused by the 90° bending of the microstrip line can be represented by an equivalent circuit that consists of capacitances and inductances. This equivalent circuit is illustrated in figure 2.8. The closed-form expressions for the equivalent circuit elements are as follows [14]:

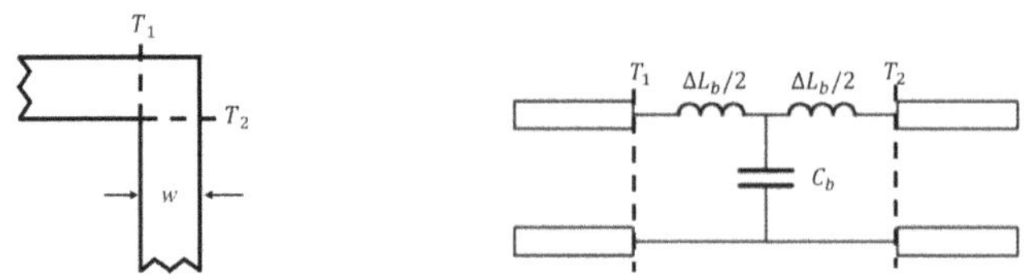

Figure 2.8. Equivalent circuit for right-angle bends.

$$\frac{C_b}{w}(pF/m) = \begin{cases} \dfrac{(14\varepsilon_r + 12.5)w/h - (1.83\varepsilon_r - 2.25)}{\sqrt{w/h}} + \dfrac{0.02\varepsilon_r}{w/h} & \dfrac{w}{h} < 1 \\ (9.5\varepsilon_r + 1.25)w/h + 5.2\varepsilon_r + 7.0 & \dfrac{w}{h} \geqslant 1 \end{cases} \quad (2.39)$$

$$\Delta L_b/h(nH/m) = 100(4\sqrt{w/h} - 4.21) \quad (2.40)$$

where ΔL_b and C_b are the equivalent inductance and capacitance in the equivalent circuit. The transmission matrix of the 90° right-angle bends can be calculated using:

$$[B_j] = \begin{bmatrix} 1 & X_{\Delta L_b/2} \\ 0 & 1 \end{bmatrix} \begin{bmatrix} 1 & 0 \\ X_{C_b}^{-1} & 1 \end{bmatrix} \begin{bmatrix} 1 & X_{\Delta L_b/2} \\ 0 & 1 \end{bmatrix} \quad (j = 1, 2, \ldots, 4) \quad (2.41)$$

where $X_{\Delta L_b/2}$ and X_{C_b} are the reactances of the equivalent circuit inductance and the capacitance, respectively. By substituting (2.37) and (2.41) into (2.35) and calculating the cascade matrix, the frequency-dependent characteristics of the SWS can be obtained.

In order to compare and validate the dispersion characteristics of the SWS obtained in this manner, simulations were performed for the MML SWS using the CST MWS Eigenmode Solver for a structure with the same dimensions as those given in section 2.2.2.2. Figure 2.9 displays the dispersion diagram of the SWS, presenting both the simulation and analysis results. The analysis results are a good match for the simulations of the fundamental mode (mode 1) of the SWS. At the same time, it is important to note that there are some discrepancies between the analysis and simulation curves for mode 2. These discrepancies arise because the coupling between the metal strips is not considered in the analysis. The equivalent circuit model assumes an idealized scenario in which each metal strip is independent and does not interact with its neighboring strips.

2.2.3.2 Characteristic impedance

The effective characteristic impedance, namely the Bloch impedance, can be represented by:

$$Z_0 = \sqrt{B/C} \quad (2.42)$$

where B and C are components of the transmission matrix for one period of the SWS. Given the characteristic impedance, designers can choose suitable types of

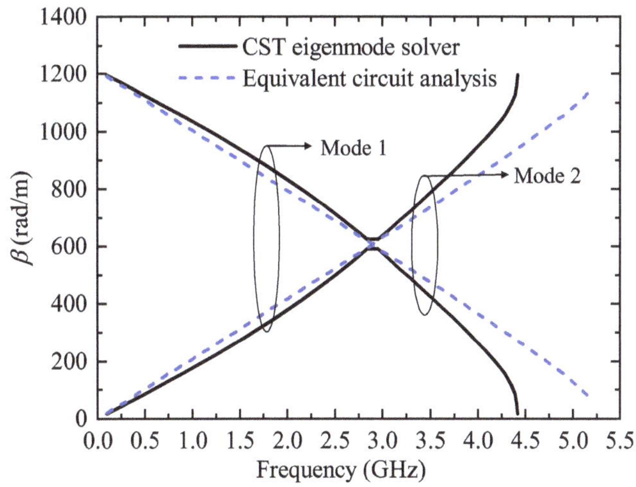

Figure 2.9. Dispersion diagram for the MML SWS obtained from simulation and analysis.

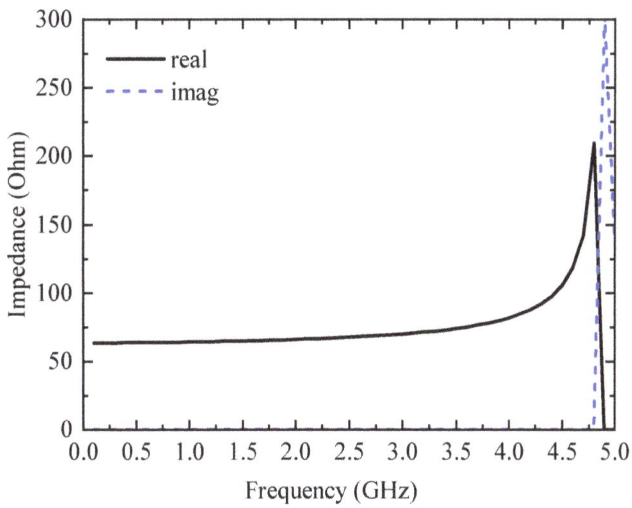

Figure 2.10. Analytical results for the characteristic impedance of the MML SWS.

couplers and match them to the SWS by employing impedance-matching techniques.

The real and imaginary parts of Z_0 calculated from (2.42) are presented in figure 2.10. It can be seen that the imaginary part of Z_0 is close to 0 Ω over most of the operating frequencies, while the real part varies from 66 Ω to about 82 Ω from 0 to 4 GHz. The imaginary part of Z_0 increases rapidly beyond 4.9 GHz, which represents the stopband of the SWS. The calculated results are similar to those obtained from the field theory analysis.

2.3 Application of MML SWS in TWTs

The methods used to obtain wave propagation properties such as the dispersion diagram, characteristic impedance, interaction impedance, etc. were described in the previous section. In this section, we present a comprehensive design procedure and performance of a TWT based on a practical Ka-band MML SWS. The dispersion characteristics of the SWS are studied first. We then introduce the design of input/output couplers for optimal wave transmission, based on the calculated characteristic impedance. Finally, the hot-test results of the TWT with a sheet beam, obtained using PIC simulation tools, are presented.

To be able to operate at millimeter-wave frequencies, the dimensions of the aforementioned SWS are scaled; the specific dimensional parameters are: $\varepsilon_r = 6.5$, loss tangent = 4e−4, $H_1 = 0.15$ mm, $l = 0.8$ mm, $w = 0.08$ mm, $s = 0.1$ mm, $H_2 = 0.4$ mm. The thickness of the copper microstrip line is 0.006 mm. Keeping in mind the impact of the surface roughness on the effective conductivity, the conductivity of the metal is set to 4×10^7 S m^{-1}.

2.3.1 Wave propagation properties

The dispersion characteristics of the SWS are obtained through simulations using both the CST MWS Eigenmode Solver and the extraction method based on the transmission matrix of three periods of the SWS. Detailed descriptions of both methods were included in section 1.7.2 of chapter 1. Figures 2.11(a) and (b) respectively show the phase constants and attenuation constants of the MML SWS obtained from both methods. The results from the two methods match very well.

Some discontinuities can be observed in the results obtained using the extraction method. These discontinuities are related to the finite number of cascaded unit cells and correspond to frequencies where the overall phase shift across the structure of N periods is a multiple of π. At these frequencies, the properties of the entire SWS circuit cannot be accurately captured, resulting in the appearance of band edges that do not exist physically. If the number of periods used for extraction is sufficiently large, the influence of these band edges can be minimized.

2.3.2 Input/output feed couplers

As the SWS consists of a meandering microstrip line, it is natural to use a microstrip line for the input/output couplers. To achieve optimal wave transmission and minimum reflection, it is important to ensure that the characteristic impedance of the input/output couplers matches that of the SWS.

Figure 2.12 shows the real and imaginary parts of the characteristic impedance Z_0 obtained using the extraction method. The imaginary part of Z_0 is close to zero, while the real part of Z_0 is about 63 Ω over most of the operating frequency range. The characteristic impedance of a uniform microstrip line with the same dimensions as those of the meander line is about 72 Ω, which is higher than that of the MML

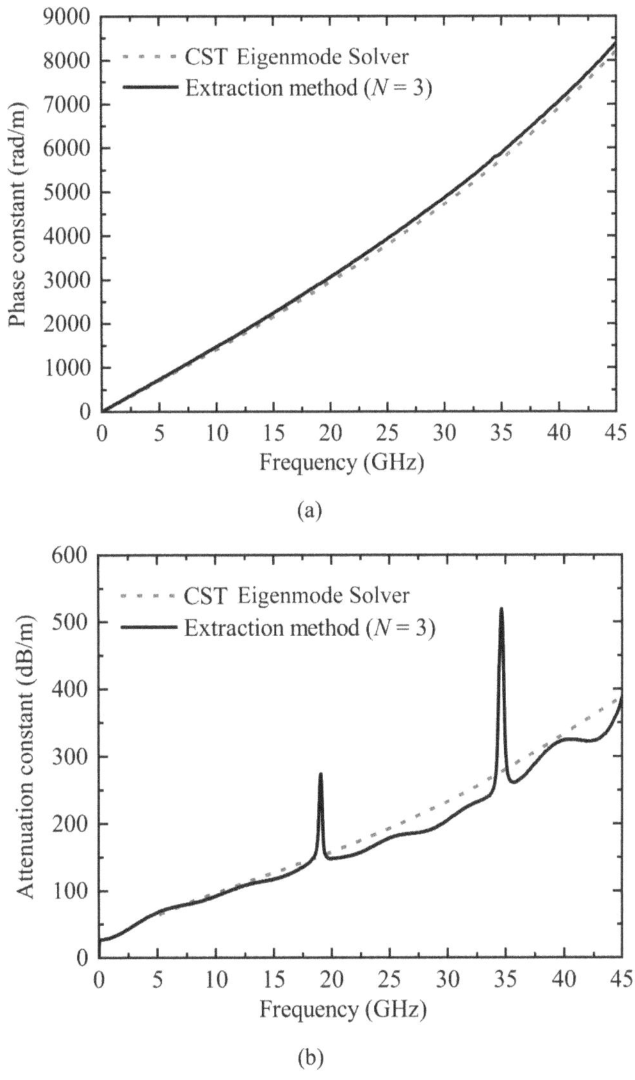

Figure 2.11. Simulated and extracted values of (a) the phase constant and (b) the attenuation constant of the MML SWS.

SWS. To ensure a good match, the SWS should be fed using a microstrip line with a characteristic impedance of 63 Ω.

As shown in figure 2.13(a), the proposed meander-line SWS of 100 periods is fed with a microstrip line with a width of w'. The corners in the right-angle bends in the feed are chamfered to eliminate the effect of the discontinuities arising from the bends. Figure 2.13(b) illustrates the simulated S-parameters for the proposed SWS. As shown, when $w' = 1.1$ mm, which corresponds to 63 Ω of the SWS, the reflection coefficient is below −25 dB over the entire frequency range from 0 to 45 GHz, which

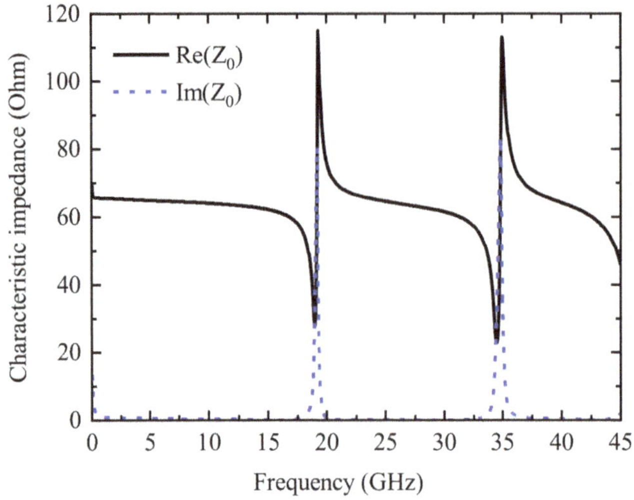

Figure 2.12. Extracted characteristic impedance of the MML SWS.

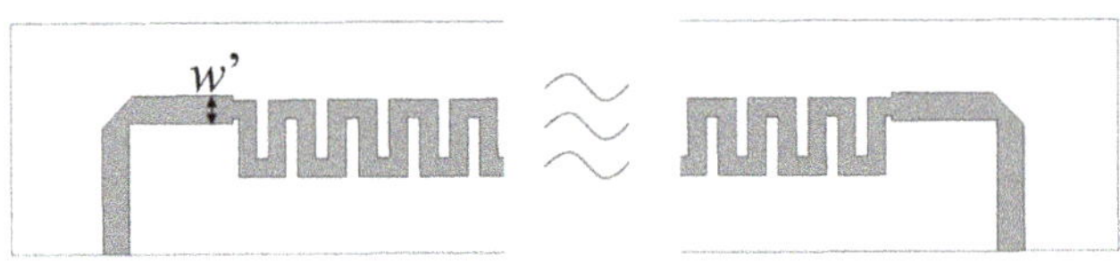

(a)

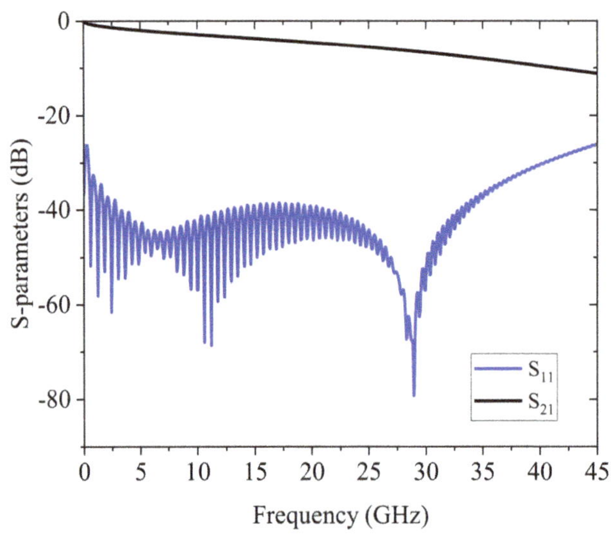

(b)

Figure 2.13. (a) Top view of the meander-line SWS with feeding structures. (b) *S*- parameters of the 100-period meander-line SWS.

is good enough for TWT applications. The attenuation increases with frequency. The transmission coefficient S_{21} is -10 dB at 41 GHz.

2.3.3 Particle-in-cell simulation results for a TWT based on MML SWS

In this subsection, PIC simulation results are presented for the MML SWS of 100 periods. As shown in figure 2.14, a rectangular sheet beam with a cross section of 0.4mm × 0.1 mm is assumed to be located 0.08 mm above the substrate.

The dispersion diagram and average interaction impedance over the beam cross section of the SWS are simulated using the CST MWS Eigenmode Solver. The average interaction impedance over the area of the electron-beam cross section can be calculated as follows [15]:

$$K_c = \frac{{E_{z,\,av}}^2}{2P\beta^2} \tag{2.43}$$

where $E_{z,av}$ is the average longitudinal electric field over the cross section of the electron beam and P is the power flow. The obtained dispersion diagram and average interaction impedance are presented in figure 2.15. The interaction impedance values decrease with frequency. A beam line of 5100 V intersects the phase constant curve at about 23 GHz. However, the best output power occurs at about 34 GHz, which is higher than 23 GHz. This is because the condition for wave amplification requires the speed of the beam to be slightly faster than the phase velocity of the electromagnetic wave propagating in the SWS.

The PIC simulation is carried out using CST Particle Studio. The beam voltage is 5100 V, and the beam current is 0.15 A. The electron beam is focused with a constant longitudinal magnetic field of 1.5 T. Figure 2.16 shows the output power and gain versus input power for the MML SWS with 100 periods at 33 GHz. For low levels of

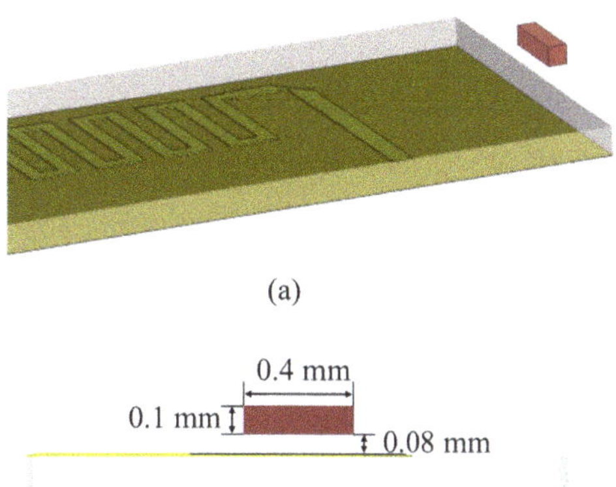

(a)

(b)

Figure 2.14. Sheet electron beam with a rectangular cross section above the MML SWS. (a) Perspective view and (b) cross-sectional view of the beam location.

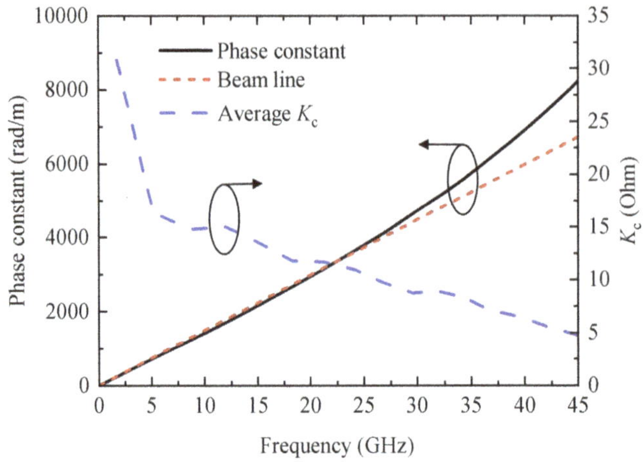

Figure 2.15. Simulated dispersion diagram and average interaction impedance.

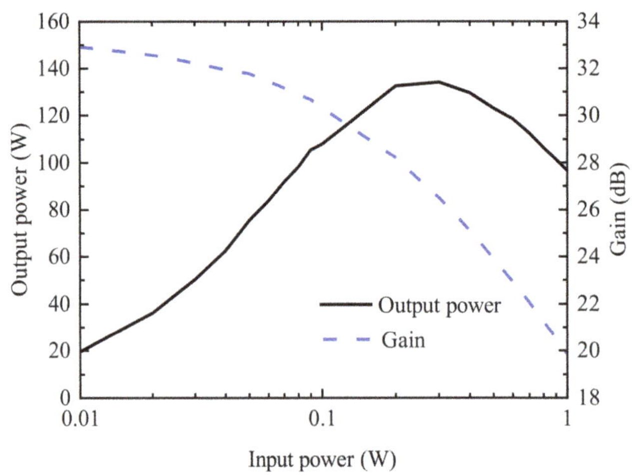

Figure 2.16. Output power and gain vs. input power for the MML SWS of 100 periods at 33 GHz.

input power, the output power increases linearly with the input power. The value of the linear gain is about 32 dB. As the input power increases, the gain begins to drop. A saturation power of 134 W is achieved when the input power is 0.3 W. The corresponding gain is 26.5 dB. Figure 2.17 shows the output power and linear gain versus frequency for an input power of 0.01 W. The maximum power of 20.3 W is achieved at 34 GHz, and the corresponding gain is 33 dB.

2.4 Summary

This chapter provides a comprehensive introduction to the MML SWS for TWT applications. We describe both the field analysis and the equivalent circuit analysis

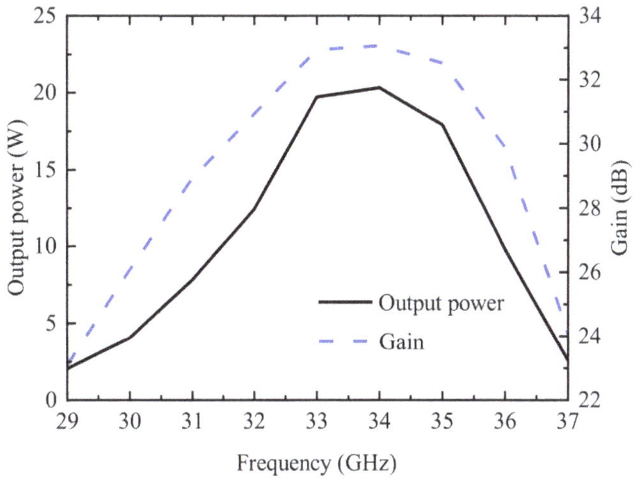

Figure 2.17. Output power and gain vs. frequency for the MML SWS of 100 periods for an input power of 0.01 W.

methods to obtain the propagation properties of the SWS, including the dispersion diagram, the attenuation constant, the characteristic impedance, and the interaction impedance. The analytical results show good agreement with the simulated or measured results, indicating the accuracy and reliability of the analytical models used. Furthermore, as an example, an MML SWS is designed for a Ka-band TWT. The hot-test parameters of the tube are presented. The results demonstrate the potential of the MML SWS to provide fairly high-power amplification at millimeter-wave frequencies in TWTs.

References

[1] Denlinger E J 1971 A frequency dependent solution for microstrip transmission lines *IEEE Trans. Microw. Theory Tech.* **19** 30–9

[2] Pozar D M 2017 *Microwave Engineering* 4th edn (New York: Wiley)

[3] Hammerstad E O 1975 Equations for microstrip circuit design *1975 5th European Microwave Conf.* (Piscataway, NJ: IEEE) 268–72

[4] Kobayashi M 1988 A dispersion formula satisfying recent requirements in microstrip CAD *IEEE Trans. Microw. Theory Tech.* **36** 1246–50

[5] Butcher P N 1957 The coupling impedance of tape structures *Proc. IEEE B: Radio Electron. Eng.* **104** 177–87

[6] Weiss J A 1974 Dispersion and field analysis of a microstrip meander-line slow-wave structure *IEEE Trans. Microw. Theory Tech.* **22** 1194–201

[7] Worm S B and Pregla R 1984 Hybrid-mode analysis of arbitrarily shaped planar microwave structures by the method of lines *IEEE Trans. Microw. Theory Tech.* **32** 191–6

[8] Glandorf F-J and Wolff I 1988 A spectral-domain analysis of periodically nonuniform coupled microstrip lines *IEEE Trans. Microw. Theory Tech.* **36** 522–8

[9] Zhang D G, Yung E K N and Ding H Y 1996 Dispersion characteristics of a novel shielded periodic meander line *Microw. Opt. Technol. Lett.* **12** 1–5

[10] Wen Z *et al* 2020 Theory, simulation, and analysis of the high-frequency characteristics for a meander-line slow-wave structure based on field-matching methods with dyadic green's function *IEEE Trans. Electron Devices* **67** 697–703
[11] Bryant T G and Weiss J A 1968 Parameters of microstrip transmission lines and of coupled pairs of microstrip lines *IEEE Trans. Microw. Theory Tech.* **16** 1021–7
[12] Carter R G 2018 *Microwave and RF Vacuum Electronic Power Sources* (Cambridge: Cambridge University Press)
[13] Courtney W E 1975 Printed-Circuit RF-Keyed Crossed-Field Amplifier *Tech. Note FSD-TR-75-321* (Lincoln Laboratory: Massachusetts Institute of Technology)
[14] Garg R and Bahl I J 1978 Microstrip discontinuities *Int. J. Electron.* **45** 81–7
[15] Pierce J R 1950 *Traveling Wave Tubes* (New York: Van Nostrand)

Chapter 3

Planar helix slow-wave structure

3.1 Introduction

As mentioned in chapter 1, the circular helix is the most widely used slow-wave structure (SWS) in traveling-wave tubes (TWTs), especially in the fields of communications and electronic countermeasures (ECMs). The circular helix has the widest bandwidth and low dispersion among all SWSs. Its interaction impedance is also unmatched by many other SWSs. However, as the frequency of operation increases, the circular helix suffers from limitations such as difficult fabrication and low capacity for thermal dissipation.

In this context, a pair of parallel unidirectionally conducting (UC) screens has been found to have propagation properties similar to those of the circular helix and is therefore called a 'flattened sheath helix' [1] or a 'planar helix' [2]. Since the pair of UC screens can be fabricated using photolithographic techniques, the planar helix offers easier fabrication and a higher frequency of operation compared to the conventional circular helix. In this chapter, we describe the electromagnetic guiding properties of a pair of UC screens and briefly consider the effect of the periodicity of the conductors comprising the screens. We then consider the potential of the planar helix for application in TWTs [3], including the effect of features such as a metal shield as well as the dielectric substrates on which the screens may be printed [4].

3.2 Propagation properties of a pair of UC screens (flattened sheath helix)

The structure considered here consists of a pair of parallel screens conducting in different but symmetric directions and having different dielectric media in the sandwiched and outer regions (figure 3.1) [1]. This structure is a generalization of the structure considered by Arora [5], in which the screens are immersed in air. The symmetry of the structure ensures that the direction of phase propagation and that

doi:10.1088/978-0-7503-5764-7ch3

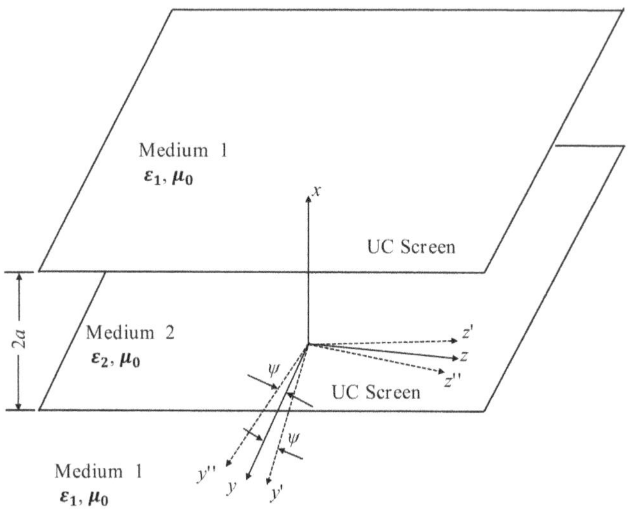

Figure 3.1. The fattened sheath helix. z: direction of propagation; y', y'': directions of conduction of the top and bottom UC screens, respectively; ψ: helix angle. © [1977] IEEE. Reprinted, with permission, from [1].

of the Poynting vector averaged over the cross section (xy-plane) coincide. The structure is assumed to be of infinite extent in the transverse (y) direction, thus facilitating an exact analysis. These results should be useful as a first-order approximation for a structure that is limited in the transverse direction. To simplify the analysis further, the screens with discrete conductors are replaced by UC screens that are infinitely thin and are assumed to be perfectly conducting, but only in the direction of the conductors. This concept is similar to the 'sheath-helix' approximation, in which the discrete turns of a helix are replaced by a uniform sheath that is assumed to conduct only in the direction of the helix windings.

Similar to the circular helix, the pair of UC screens is an SWS, i.e. the phase velocity of wave propagation on the structure is less than the free-space velocity of light, and it supports a 'surface wave.' Barlow and Brown [6] suggest the definition of a surface wave as 'one that propagates along an interface between two different media without radiation.' A characteristic feature of the surface waves is the decay of the fields away from the guiding surface. It is an implication of this characteristic that the surface waves on lossless structures have a phase velocity less than the free-space velocity of light.

3.2.1 Field expressions and characteristic equations

In figure 3.1, the screens are situated at $x = \pm a$. Medium 1, which constitutes the region outside the screens, has a permittivity of ε_1, while medium 2, the sandwiched region, has a permittivity of ε_2. The screens at the top and bottom conduct in the y' and y'' directions, which make angles of ψ and $-\psi$, respectively, with the y-axis. Correspondingly, they are perfectly insulating in the directions z' and z''. For surface wave propagation, the fields must decay exponentially in the regions $|x| > a$, whereas in the region $|x| < a$, they may have either a hyperbolic or a trigonometric dependence on x. Further, the following boundary conditions must be satisfied at $x = \pm a$:

$$\begin{matrix} E_{y'} \\ E_{y''} \end{matrix} = E_y \cos\psi \pm E_z \sin\psi = 0 \quad \text{on} \quad x = \pm a \tag{3.1}$$

$$\begin{matrix} E_{z'} \\ E_{z''} \end{matrix} = \mp E_y \sin\psi + E_z \cos\psi \text{ is continuous across } x = \pm a \tag{3.2}$$

$$\begin{matrix} H_{y'} \\ H_{y''} \end{matrix} = H_y \cos\psi \pm H_z \sin\psi \text{ is continuous across } x = \pm a. \tag{3.3}$$

In (3.1) and (3.3), the upper and lower symbols go together. Taken together, conditions (3.1) and (3.2) imply that E_y and E_z are continuous across $x = \pm a$.

In the following, a time dependence of the form $\exp(j\omega t)$ and propagation in the positive z-direction are assumed. It is also assumed that the field quantities have no variation in the y-direction, i.e. $\partial/\partial y = 0$. The structure supports hybrid modes, i.e. both E_z and H_z are simultaneously present. The symmetry of the structure suggests that the general solution can be decomposed into transverse symmetric (even) and transverse antisymmetric (odd) solutions. The term 'transverse symmetric' implies that the transverse components E_x, E_y, H_x, and H_y are symmetric with respect to x, while the longitudinal components E_z and H_z are antisymmetric. Similarly, the transverse antisymmetric solution implies that the transverse components are antisymmetric and the longitudinal components are symmetric. Furthermore, due to symmetry, one may consider only the upper half of the structure.

The longitudinal components of the transverse antisymmetric modes for the region $x \geqslant 0$ can be written as

$$E_{z1} = A_1 e^{-u_1(x-a)} e^{-j\beta z} \tag{3.4a}$$

$$H_{z1} = B_1 e^{-u_1(x-a)} e^{-j\beta z} \tag{3.4b}$$

$$E_{z2} = A_2 \cosh(u_2 x) e^{-j\beta z} \tag{3.4c}$$

$$H_{z2} = B_2 \cosh(u_2 x) e^{-j\beta z} \tag{3.4d}$$

where the subscripts 1 and 2 represent the regions $|x| > a$ and $|x| < a$, respectively. A_1, A_2, B_1, and B_2 are the amplitude coefficients. The propagation constant β and the transverse decay coefficients u_1 and u_2 are related by the equations:

$${u_1}^2 = \beta^2 - {k_1}^2, \quad {u_2}^2 = \beta^2 - {k_2}^2 \tag{3.5a}$$

where

$${k_1}^2 = \omega^2 \mu_0 \varepsilon_1, \quad {k_2}^2 = \omega^2 \mu_0 \varepsilon_2. \tag{3.5b}$$

Expressions (3.4) are solutions of the Helmholtz equations for E_z and H_z for a uniform dielectric medium:

$$\nabla^2 E_z + \omega^2 \mu\varepsilon E_z = 0 \tag{3.6}$$

$$\nabla^2 H_z + \omega^2 \mu\varepsilon H_z = 0. \tag{3.7}$$

The transverse field components can be obtained from Maxwell's equations as follows [7]:

$$E_{xn} = \frac{j}{{u_n}^2}\left(\beta\frac{\partial E_{zn}}{\partial x} + \omega\mu\frac{\partial H_{zn}}{\partial y}\right) \tag{3.8a}$$

$$E_{yn} = \frac{-j}{{u_n}^2}\left(-\beta\frac{\partial E_{zn}}{\partial y} + \omega\mu\frac{\partial H_{zn}}{\partial x}\right) \tag{3.8b}$$

$$H_{xn} = \frac{-j}{{u_n}^2}\left(\omega\varepsilon\frac{\partial E_{zn}}{\partial y} - \beta\frac{\partial H_{zn}}{\partial x}\right) \tag{3.8c}$$

$$H_{yn} = \frac{j}{{u_n}^2}\left(\omega\varepsilon\frac{\partial E_{zn}}{\partial x} + \beta\frac{\partial H_{zn}}{\partial y}\right) \tag{3.8d}$$

where the subscript n represents region 1 or 2. All $\partial/\partial y$ terms can be ignored, since there is no variation in the y-direction. After applying the boundary conditions, the characteristic equation for the transverse antisymmetric modes is obtained:

$$\frac{k_1^2/u_1 + (k_2^2/u_2)\tanh u_2 a}{u_1 + u_2 \coth u_2 a} = \tan^2\psi. \tag{3.9}$$

For the transverse symmetric case, the 'cosh' function in (3.4c) and (3.4d) is replaced by the 'sinh' function. The transverse antisymmetric mode should be chosen for TWT applications, since, in this mode, the E_z component is nonzero at $x = 0$.

3.2.2 Dispersion characteristics

The characteristic equations for the present structure are transcendental equations. The dispersion characteristics of the structure (i.e. propagation constant vs. frequency curves) for given a, ψ, and dielectric properties of the media can be obtained numerically. For decaying fields outside the screens, u_1 must be real and positive. On the other hand, u_2 may be real or imaginary. Special cases of interest are (i) the normal helix, in which the inner medium is dielectric and the outer medium is air, and (ii) the inverted helix, in which the two media are interchanged.

(i) Normal helix

For the case of a normal helix ($\varepsilon_1 = \varepsilon_0$, $\varepsilon_2 = \varepsilon_r\varepsilon_0$, $k_1 = k_0$, $u_1 = u_0$), βa is plotted versus $k_0 a$ in Figures 3.2 and 3.3 for $\varepsilon_r = 2.56$ (polystyrene) and $\psi = 30^\circ$ and 60°, respectively. Like modes are numbered identically in figures 3.2 and 3.3. It is found

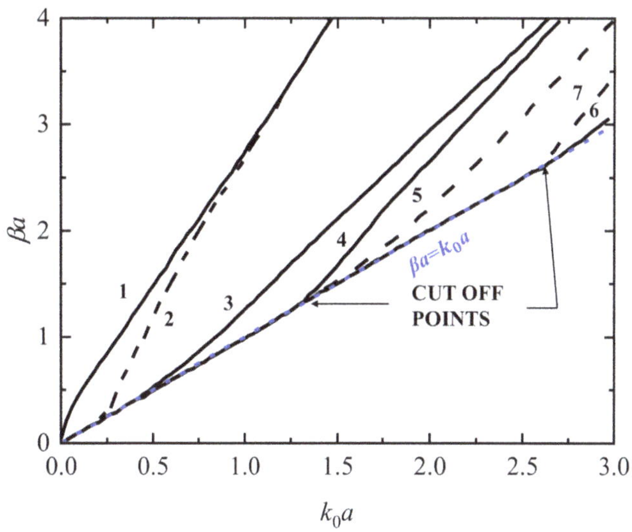

Figure 3.2. Dispersion characteristics of a normal helix for $\varepsilon_r = 2.56$ (polystyrene) and $\psi = 30°$. Continuous lines represent symmetric modes and dashed lines antisymmetric modes. © [1977] IEEE. Reprinted, with permission, from [1].

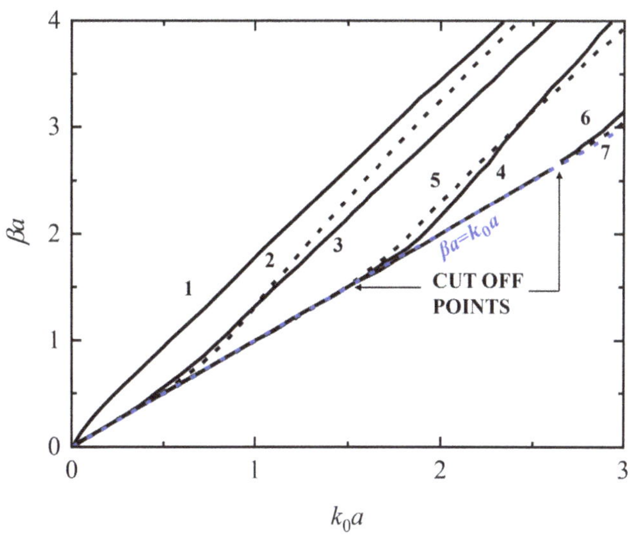

Figure 3.3. Dispersion characteristics of a normal helix for $\varepsilon_r = 2.56$ (polystyrene) and $\psi = 60°$. Continuous lines represent symmetric modes and dashed lines antisymmetric modes. © [1977] IEEE. Reprinted, with permission, from [1].

that there are two symmetric modes (1 and 3) and one antisymmetric mode (2) which propagate down to zero frequency. Modes 1 and 2 correspond to the two modes that exist when medium 2 is also air [5]; these are called the 'helix modes.' Higher-order modes of both symmetric and antisymmetric types have low-frequency cutoffs

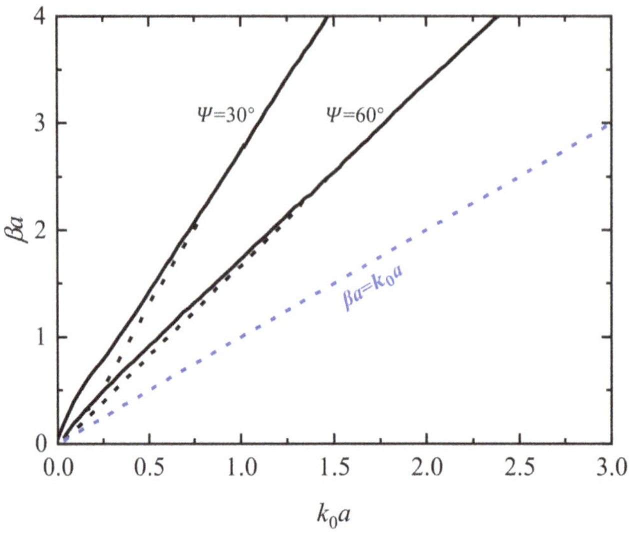

Figure 3.4. Dispersion characteristics of an inverted helix for $\varepsilon_r = 2.56$. Continuous lines represent symmetric modes and dashed lines antisymmetric modes. © [1977] IEEE. Reprinted, with permission, from [1].

identical to those of a dielectric slab [8]. Mode 3 and the higher-order modes are attributed to the presence of the dielectric between the screens.

(ii) Inverted helix

For the case of an inverted helix ($\varepsilon_2 = \varepsilon_0,\ \varepsilon_1 = \varepsilon_r\varepsilon_0,\ k_2 = k_0,\ u_2 = u_0$), βa is plotted versus $k_0 a$ in figure 3.4 for $\varepsilon_r = 2.56$ (polystyrene) and $\psi = 30°$ and $60°$, respectively. In this case, the characteristic equation for each type of mode, symmetric or antisymmetric, has only one solution. The modes that arise due to the presence of the dielectric between the screens in the case of the normal helix are absent in this case, and only the helix modes exist.

In the cases of both normal and inverted helices, consistent with the slow-wave nature of the structure, $\beta > k_0$; in other words, the phase velocity of wave propagation on the structure, $v_p = \omega/\beta$, is less than the free-space velocity of light. The helix modes are appreciably slower than the other modes, and the phase velocity decreases as the angle ψ decreases. In fact, in the absence of any dielectric, $v_p \propto \sin\psi$ at sufficiently high frequencies, as in the case of the circular helix. Also, in both the transverse symmetric and transverse antisymmetric cases, there is a single mode that has no cutoff frequency. By a proper choice of excitation, symmetric modes may be eliminated [9], and the guide dimensions may be chosen in such a way as to ensure single-mode operation.

The field patterns and attenuation due to conductor and dielectric losses have also been studied [10]. The helix modes suffer relatively greater conductor loss than the dielectric modes. All modes, in general, are elliptically polarized. Therefore, a pair of UC screens with ferrite slabs can be used to obtain nonreciprocal propagation characteristics similar to those of a circular helix [11].

3.3 Infinitely wide planar helix with periodicity

Section 3.2 described the results of the study of a pair of UC screens, or a planar helix, based on the sheath-helix approximation. In practice, the conductors on the screens are periodic and have a finite conductor width. The effects of these features should be incorporated into the analysis for more accurate results. To this end, the 'tape-helix' model has been employed in the case of the circular helix [12]; this analysis reveals spatial harmonics and a forbidden region in the dispersion characteristics of the circular helix, similar to other periodic structures.

The tape-helix model considers the helix to be a perfectly conducting tape with finite width and infinitesimal thickness. The current is assumed to flow only on the tape. When the width of the tape is much smaller than the period of the helix, one can further assume that the current flows only along the direction of the conductors. The electric field intensity vanishes on the perfectly conducting tape. In the following, the tape-helix model is used for the analysis of the planar helix, including its periodicity.

The structure considered for analysis is shown in figure 3.5 [2]. It consists of a pair of arrays of parallel, straight, and perfectly conducting strips, which are assumed to be infinitesimal in thickness. The arrays are separated by a distance of $2a$ in the x-direction. Strips in the top array are oriented at an angle of ψ, while those in the bottom array are oriented at an angle of $-\psi$ with respect to the y-axis. The directions parallel to the strips in the top and bottom arrays are y' and y'', respectively. The centerlines of the strips intersect the axes at $y = mp_y$ and

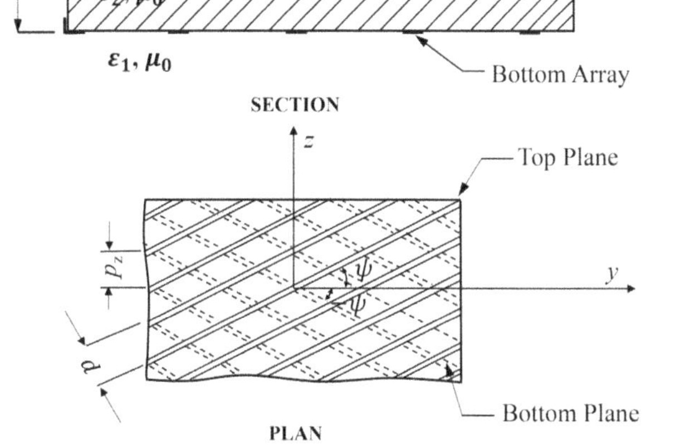

Figure 3.5. Cross section of the planar helix; z is the direction of propagation. © [1979] IEEE. Reprinted, with permission, from [2].

$z = np_z$, where m and n are integers, and p_y and p_z are the periodic spacings between the strips measured in the y and z-directions, respectively. The spacings p_y and p_z are related by the equation $p_z/p_y = \tan\psi$. Similarly, the strip width w is related to w_y and w_z such that $w_y = w/\sin\psi$ and $w_z = w/\cos\psi$. In general, the permittivities of medium 1 and medium 2 are considered to be different.

3.3.1 Symmetry properties

The planar helix remains unchanged under any one or any combination of the following transformations:

$$(x, y, z) \quad \rightarrow \quad (-x, -y, z) \tag{3.10a}$$

$$(x, y, z) \quad \rightarrow \quad \left(x, y \pm \frac{p_y}{2}, z \pm \frac{p_z}{2}\right) \tag{3.10b}$$

$$(x, y, z) \quad \rightarrow \quad (x, y, z \pm p_z) \tag{3.10c}$$

$$(x, y, z) \quad \rightarrow \quad (x, y \pm p_y, z). \tag{3.10d}$$

As in the previous section, the structure admits two independent solutions: transverse symmetric and transverse antisymmetric. Also, only one half of the structure need be considered for a solution.

3.3.2 Field expressions and boundary conditions

The invariant transformation (3.10b) is shared in common with cross-wound twin helices described by Chodorow and Chu [13]. Therefore, the forms of the exponent in the field expressions used there should also be applicable here. Moreover, the field expressions should reflect the periodicity of the structure. In the following, a time dependence of the form $\exp(j\omega t)$ and propagation in the positive z-direction are assumed. Based on these considerations, field expressions for the transverse symmetric case for regions 1 and 2 are:

$$E_{z1} = \sum_{m=-\infty}^{+\infty}\sum_{n=-\infty}^{+\infty}\left[A_{mn}\cos\left(\frac{2\pi ny}{p_y}\right) + A'_{mn}\sin\left(\frac{2\pi ny}{p_y}\right)\right]e^{-u_{mn}(x-a)}e^{-j\beta_{mn}z} \tag{3.11a}$$

$$H_{z1} = \sum_{m=-\infty}^{+\infty}\sum_{n=-\infty}^{+\infty}\left[B_{mn}\cos\left(\frac{2\pi ny}{p_y}\right) + B'_{mn}\sin\left(\frac{2\pi ny}{p_y}\right)\right]e^{-u_{mn}(x-a)}e^{-j\beta_{mn}z} \tag{3.11b}$$

$$E_{z2} = \sum_{m=-\infty}^{+\infty}\sum_{n=-\infty}^{+\infty}\left[C_{mn}\sinh(k_{mn}x)\cos\left(\frac{2\pi ny}{p_y}\right) + C'_{mn}\cosh(k_{mn}x)\sin\left(\frac{2\pi ny}{p_y}\right)\right]e^{-j\beta_{mn}z} \tag{3.11c}$$

$$H_{z2} = \sum_{m=-\infty}^{+\infty} \sum_{n=-\infty}^{+\infty} \left[D_{mn} \sinh(k_{mn}x) \cos\left(\frac{2\pi ny}{p_y}\right) + D'_{mn} \cosh(k_{mn}x) \sin\left(\frac{2\pi ny}{p_y}\right) \right] e^{-j\beta_{mn}z}. \quad (3.11\text{d})$$

In (3.11), u_{mn}, restricted to be positive and real for fields decaying away from the structure, are given by:

$$u_{mn}^2 = (2\pi n/p_y)^2 + (\beta_{mn})^2 - k_1^2 \quad (3.12\text{a})$$

$$k_{mn}^2 = (2\pi n/p_y)^2 + (\beta_{mn})^2 - k_2^2 \quad (3.12\text{b})$$

$$\beta_{mn} = \beta_{00} + (2\pi/p_z)(2m + n) \quad (3.12\text{c})$$

$$k_1{}^2 = \omega^2\mu_0\varepsilon_1, \quad k_2{}^2 = \omega^2\mu_0\varepsilon_2. \quad (3.12\text{d})$$

In the transverse antisymmetric case, cosh replaces sinh and vice versa. The transverse field components in the two regions are obtained using Maxwell's equations. The coefficients A_{mn} etc. are related by the following boundary conditions: at $x = a$, the tangential components of the electric and magnetic field are continuous, and

$$H_{y1} - H_{y2} = J_z \quad (3.13\text{a})$$

$$H_{z2} - H_{z1} = J_y. \quad (3.13\text{b})$$

Furthermore, for narrow strips, one can assume that $J = J_{y'}$, so that:

$$J_z = J_{y'} \sin\psi \quad \textit{and} \quad J_y = J_{y'} \cos\psi. \quad (3.14)$$

Analogous to (3.11), $J_{y'}$ can be expressed in the form:

$$J_{y'} = \sum_{m=-\infty}^{+\infty} \sum_{n=-\infty}^{+\infty} \left[J_{mn} \cos\left(\frac{2\pi ny}{p_y}\right) + J'_{mn} \sin\left(\frac{2\pi ny}{p_y}\right) \right] e^{-j\beta_{mn}z} \quad (3.15)$$

By applying the boundary conditions, one evaluates the coefficients A_{mn}, A'_{mn} etc. in terms of J_{mn} and J'_{mn}.

3.3.3 Characteristic equation and numerical results

To obtain the characteristic equations, the variational technique introduced in [13] can be used. Furthermore, for an approximate solution, one may assume $J_{y'}$ to have a single term. In this 'zeroth-order' approximation, the characteristic equation reduces to a singly infinite series with $m = 0$. Some of the details and the characteristic equation for the case when $\varepsilon_1 = \varepsilon_2 = 1$ are given in [2]. More details are available in [10].

For a numerical solution of the characteristic equations for the transverse symmetric and transverse antisymmetric cases, we assume the following values for various parameters: $\psi = 10°$, $w_z/p_z = 0.1$, and $p_z/2\pi a = 1$. The infinite series in the

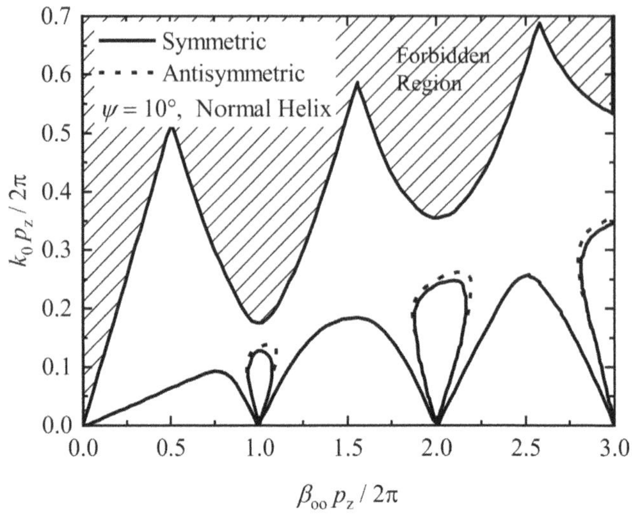

Figure 3.6. Fundamental space harmonic for the normal helix case: $\varepsilon_{r1} = 1$, $\varepsilon_{r2} = 2.56$, $\psi = 10°$. Dashed curves represent the transverse antisymmetric solution where it is visibly different from the transverse symmetric one. © [1979] IEEE. Reprinted, with permission, from [2].

characteristic equations are found to converge sufficiently fast to allow us to truncate the series after five terms. Figure 3.6 gives the dispersion curves for the 'normal helix' case when region 2 is filled with polystyrene ($\varepsilon_r = 2.56$) and region 1 is air. The forbidden region, in which no solutions corresponding to the fields decaying away from the arrays are possible, is clearly marked. For the sake of clarity, only β_{00} is plotted. Wherever the transverse antisymmetric solution is different from the transverse symmetric one, it is shown by dashed lines.

Dispersion curves for the 'inverted helix' case, in which region 1 consists of dielectric (polystyrene) and region 2 consists of air, are shown in figure 3.7. The dashed curves represent the antisymmetric solution where it is different from the symmetric one. Thus, in both 'normal helix' and 'inverted helix' cases, analysis that includes the periodicity and finite width of the conductors in the pair of arrays that constitute the planar helix provides more accurate dispersion characteristics. In particular, it reveals the space harmonics and the forbidden region in the dispersion characteristics.

3.4 Application of the planar helix in TWTs

It is observed in section 3.2 that the helix modes on the infinitely wide planar helix are appreciably slower than the other modes, and in the absence of any dielectric, $v_p \propto \sin\psi$ at sufficiently high frequencies. This behavior is very similar to that observed by Pierce in the case of the circular helix [14]. In view of this similarity in the electromagnetic guiding properties of the planar helix and the circular helix, it is natural to expect the application of the planar helix to be similar to that of the circular helix in general, and in particular, in TWTs. In this section, we first describe the

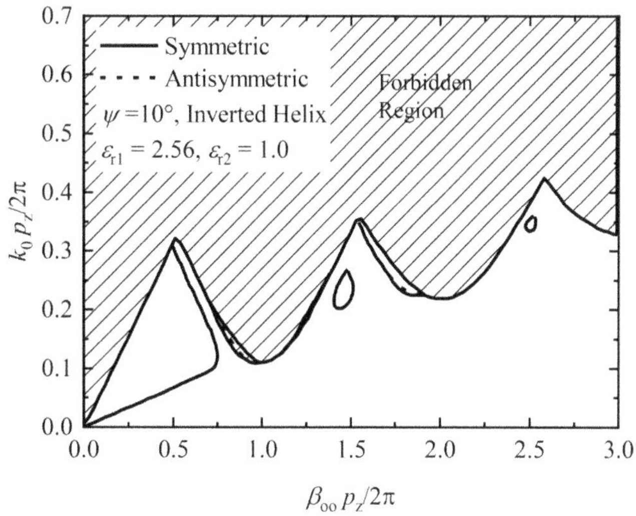

Figure 3.7. Forbidden region for $\varepsilon_{r1} = 2.6$ and $\psi = 10°$ and dispersion characteristics for the inverted helix case: $\varepsilon_{r1} = 2.6$, $\varepsilon_{r2} = 1$, $\psi = 10°$. Dashed curves represent the transverse antisymmetric solution where it is different from the symmetric one. © [1979] IEEE. Reprinted, with permission, from [2].

dispersion characteristics of the planar helix in the presence of an electron beam [3]; these indicate amplification for one of the forward waves. This is followed by considering a general configuration in which the UC screens constituting the planar helix are printed on dielectric substrates and shielded by metal plates [4]. As in section 3.2, to keep the analysis simple, the planar helix is assumed to be infinitely wide, and the sheath-helix approximation is used.

3.4.1 Basic configuration including an electron beam

The structure considered here is similar to that in figure 3.1 (section 3.2) and is shown in figure 3.8 [3]. A sheet electron beam of thickness b is located symmetrically between the UC screens. The rest of the regions, both inside and outside the screens, are considered to be vacuum. The beam current density is constant over the beam cross section, and the current has only a z-component—an assumption very closely realized in practice by means of a focusing magnetic field of appropriate strength. To facilitate an exact analysis, the screens and the beam are considered to be of infinite extent in the y-direction, so that fields are constant in the y-direction and $\partial/\partial y = 0$.

The boundary conditions require both the transverse electric (TE) and transverse magnetic (TM) mode types to be present. Assuming that the fields vary as exp ($j\omega t - \Gamma z$), the field equations for both types of modes can be written as follows:

For the TE modes:

$$\Gamma E_y + j\omega\mu_0 H_x = 0 \tag{3.16a}$$

$$\Gamma H_x + \partial H_z/\partial x + j\omega\varepsilon_0 E_y = 0 \tag{3.16b}$$

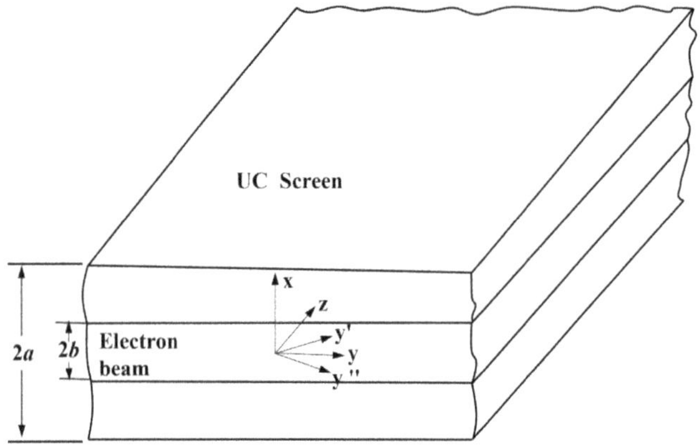

Figure 3.8. The planar TWT. z: direction of propagation; y' and y'': directions of conduction of the top and bottom UC screens, respectively; ψ: helix angle. © [1983] IEEE. Reprinted, with permission, from [3].

$$\partial E_y/\partial x + j\omega\mu_0 H_z = 0. \tag{3.16c}$$

For the TM modes:

$$\Gamma H_y - j\omega\varepsilon_0 E_x = 0 \tag{3.17a}$$

$$-\Gamma E_x - \partial E_z/\partial x + j\omega\mu_0 H_y = 0 \tag{3.17b}$$

$$\partial H_y/\partial x - j\omega\varepsilon_0 E_z = J_z \tag{3.17c}$$

where $\Gamma = \alpha + j\beta$ is the complex propagation constant in the z-direction and J_z is the current density. The TE modes are unaffected by the presence of the beam, and the Helmholtz equation for H_z simplifies to

$$\frac{\partial^2 H_z}{\partial x^2} - u^2 H_z = 0 \tag{3.18a}$$

where

$$u^2 = -\Gamma^2 - k_0^2 \quad \text{and} \quad k_0^2 = \omega^2\mu_0\varepsilon_0. \tag{3.18b}$$

However, the TM modes are affected by the current in the beam, and the Helmholtz equation for E_z becomes:

$$\frac{\partial^2 E_z}{\partial x^2} - u^2 E_z - \frac{u^2}{j\omega\varepsilon_0} J_z = 0. \tag{3.19}$$

Using the continuity and force equations for the charges, J_z can be expressed in terms of E_z [14] as

$$J_z = -\left[\frac{j\omega_p^2}{\omega}\frac{\beta_e^2\varepsilon_0}{(\beta_e + j\Gamma)^2}\right]E_z. \tag{3.20}$$

In (3.20), $\omega_p^2 = \rho_0 e/(m\varepsilon_0)$ and $\beta_e = \omega/v_0$, where ρ_0 is the average charge density, e/m is the charge-to-mass ratio of the electron, and v_0 is the average electron velocity. Using (3.20), (3.19) becomes

$$\frac{\partial^2 E_z}{\partial x^2} - p^2 E_z = 0 \tag{3.21}$$

where

$$p^2 = -(\Gamma^2 + k_0^2)\left[1 - \left(\frac{\omega_p}{\omega}\right)^2 \left(\frac{\beta_e}{\beta_e + j\Gamma}\right)^2\right]. \tag{3.22}$$

3.4.2 Field expressions and characteristic equation for the basic configuration

The cross section of the structure under consideration is shown in figure 3.9. Due to its symmetry, as mentioned in section 3.2, the structure supports transverse symmetric and transverse antisymmetric modes. For application in TWTs, the transverse antisymmetric modes are considered, since these have a symmetric variation of the longitudinal field components E_z and H_z about the $x = 0$ plane. Also, due to symmetry, only half of the structure, i.e. $x \geqslant 0$ needs to be considered.

For the transverse antisymmetric modes, the longitudinal field components satisfying (3.18a) and (3.21) can be expressed as

$$E_z = \begin{cases} Ae^{-u(x-a)} & x > a \\ B_1 e^{u(x-b)} + B_2 e^{-u(x-b)} & b < x < a \\ C\cosh(px) & 0 < x < b \end{cases} \tag{3.23}$$

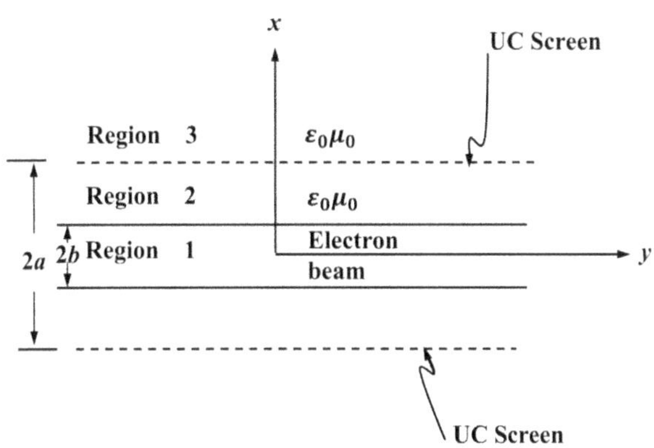

Figure 3.9. Configuration considered for theoretical analysis. The structure is of infinite extent in the y- and z-directions. The top and bottom UC screens conduct in directions that make angles of ψ and $-\psi$, respectively, with the y-axis. © [1983] IEEE. Reprinted, with permission, from [3].

$$H_Z = \begin{cases} Me^{-u(x-a)} & x > a \\ N\cosh(ux) & 0 < x < a \end{cases}. \tag{3.24}$$

In (3.23) and (3.24), the factor of $\exp(-\Gamma z)$ is dropped. Also, A_1, B_1, B_2, etc. are constant coefficients which are related by the following boundary conditions:

(i) The electric field components are zero and the magnetic field components are continuous along y' *at* $x = a$.
(ii) The tangential components of the electric field are continuous at $x = a$.
(iii) The tangential components of the electric and magnetic fields are continuous at $x = b$.

The boundary conditions yield the characteristic equation:

$$[1 + \coth(ua)]\tan^2\psi = \frac{k_0^2}{u^2}\left[1 + \frac{(p/u)\tanh pb + \tanh u(a-b)}{1 + (p/u)\tanh pb\tanh u(a-b)}\right]. \tag{3.25}$$

3.4.3 Modes of propagation and gain

The numerical solution of the characteristic equation (3.25) gives the modes that can propagate on the structure. In the presence of an electron beam, there are three forward-wave modes. Figures 3.10(a) and (b) show the real and imaginary parts of the propagation constant, respectively, as functions of beam voltage for the set of parameters mentioned in the figure caption. When the difference between the phase velocity of the wave and the velocity of the electrons is large, the three propagation constants are purely imaginary (β_1, β_2, and β_3), with no attenuation or amplification. When the difference between the phase velocity of the wave and the velocity of the electrons is small, two of the modes (β_2 and β_3) have complex propagation constants with the same value of the imaginary parts but equal and opposite real parts. The mode with the negative real part increases as the wave advances along the structure —resulting in amplification. With an increase in beam current, the electron velocity range over which amplification can occur increases, and so does the maximum attainable amplification.

The order of magnitude of gain can be calculated following [15]. All higher-order modes are neglected, and the amplitudes of the three forward waves at the input are taken to be equal. Therefore, the amplitude gain in the TWT for the growing wave is:

$$\frac{E_0}{E_i} = \frac{1}{3}e^{\alpha l} \tag{3.26}$$

where E_i is the total electric field at the input, α is the growth constant for the growing wave, and l is the tube length. Hence, the power gain G in decibels is:

$$\mathrm{G} = 20\log 0.333 + 20\alpha l\log e = -9.54 + 8.686\left(\frac{\alpha a}{\beta a}\right)\beta l = -9.54 + 54.575\left(\frac{\alpha a}{\beta a}\right)N \tag{3.27}$$

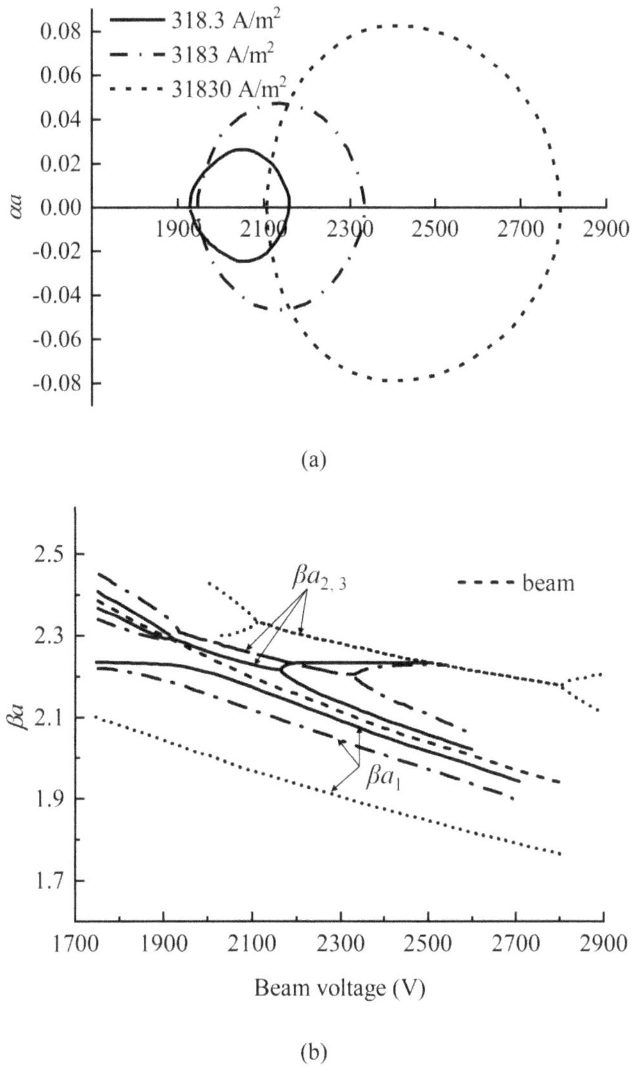

Figure 3.10. (a) Attenuation constant and (b) phase constant versus beam voltage for $b/a = 0.2$, $a = 5$ mm, frequency = 1.875 GHz, $\psi = 5°$ for three different values of beam current density, $J_0 = 318.3$ A m^{-2}, $J_0 = 3183$ A m^{-2}, and $J_0 = 31\,830$ A m^{-2}. © [1983] IEEE. Reprinted, with permission, from [3].

where N is the tube length in terms of guide wavelengths. When the values of αa and βa from figure 3.10 for a beam current density of 31 830 A m^{-2} and a beam voltage of 2400 V are substituted into (3.27), the value of G obtained is 12 dB for a tube ten wavelengths long. For comparison, the gain obtained in the case of a circular helix TWT for a corresponding set of dimensional and beam parameters is 11.5 dB [16]; the diameter of the circular helix is taken to be equal to the spacing between the UC screens of the planar helix.

3.4.4 General configuration

In a circular helix TWT, the helix is generally supported by ceramic rods inside an evacuated glass or metal tube envelope. The presence of the dielectric rods and metal shield changes the dispersion characteristics and reduces the interaction impedance [17]. In some cases, the dispersion characteristics can be 'shaped' to achieve a wider operating bandwidth by adjusting the physical dimensions of the support rods and metal shield and by selecting a suitable dielectric material for the rods. Such dispersion shaping usually comes at the expense of the maximum gain of the TWT.

In the case of the planar helix TWT, the UC screens constituting the planar helix are likely to be printed on dielectric substrates. The effects of such dielectric loading and shielding with metal plates have been studied in detail [4, 18]. The general configuration for this analysis is shown in figure 3.11. The UC screens are printed on a pair of dielectric substrates of thickness t_2, and another pair of dielectric slabs of thickness t_3 sandwiches the UC screens. Two metal plates, separated by a distance of $2d$, are located symmetrically on either side of the UC screens.

Beginning with the general configuration, different fabrication possibilities are considered. The dispersion characteristics and interaction impedance are studied for two commonly used ceramic substrate materials, alumina and beryllia. The details of the derivation of the characteristic equations and interaction impedance, as well as plots of the dispersion characteristics and the interaction impedance for different fabrication methods, are reported in [4, 18]. In general, it is observed that the phase velocity and interaction impedance reduce due to dielectric loading and metal shields. On the other hand, the phase velocity vs. frequency curve flattens, increasing the potential operating bandwidth of the structure.

A detailed study of the planar helix with transverse confinement by straight-edge connections (known as the PH-SEC), which is a more practical structure, is described in the next chapter.

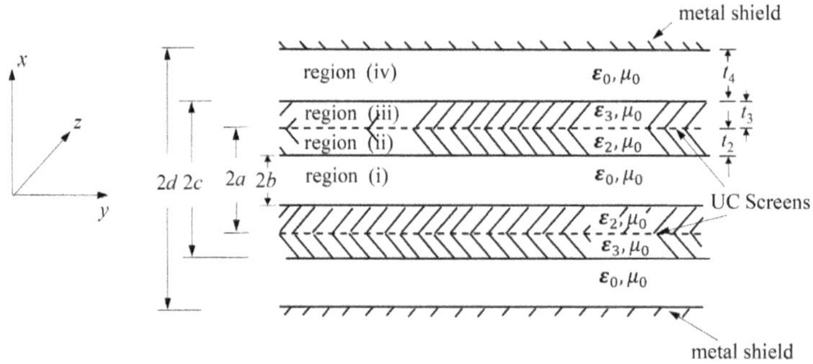

Figure 3.11. General configuration considered for analysis in the presence of dielectric substrates and metal shield [18].

3.5 Summary

This chapter describes studies of the planar helix SWS, which consists of a pair of UC screens. The planar helix, considered to be infinitely wide, permits an exact analysis. The dispersion characteristics obtained from an analysis based on the sheath-helix approximation reveal a close similarity with those of the circular helix. The results of an analysis based on the tape-helix approximation are also presented. These results highlight the spatial harmonics and forbidden regions in the dispersion characteristics that result from the periodicity and finite width of the conductors in the planar helix. Next, the analysis of the planar helix is extended to include an electron beam inside the pair of UC screens, yielding a mode of propagation that corresponds to the amplification of the propagating wave. The estimated gain for this mode compares quite well with the gain of a circular helix TWT for a corresponding set of dimensional and beam parameters. Briefly mentioned at the end of the chapter is a general configuration which includes a multilayer dielectric environment and a metal shield for the planar helix.

References

[1] Arora R K, Bhat B and Aditya S 1977 Guided waves on a flattened sheath helix *IEEE Trans. Microw. Theory Tech.* **25** 71–2

[2] Aditya S and Arora R K 1979 Guided waves on a planar helix *IEEE Trans. Microw. Theory Tech.* **27** 860–3

[3] Chadha D, Aditya S and Arora R K 1983 Field theory of planar helix traveling-wave tube *IEEE Trans. Microw. Theory Tech.* **31** 73–6

[4] Chadha D, Aditya S and Arora R K 1984 Study of planar-helix slow-wave structure for application to TWTs *IEEE Proc. H Microwaves Antennas Propag.* **131** 14–20

[5] Arora R 1966 Surface waves on a pair of parallel undirectionally conducting screens *IEEE Trans. Antennas Propag.* **14** 795–7

[6] Barlow H M and Brown J 1961 *Radio Surface Waves* (London: Oxford University Press)

[7] Jordan E C and Balmain K G 1969 *Electromagnetic Waves and Radiating Systems* (New Delhi: Prentice-Hall of India)

[8] Collin R E 1960 *Field Theory of Guided Waves* (New York: McGraw-Hill)

[9] Arora R K 1967 Field of a line source situated parallel to a surface-wave structure comprising a pair of unidirectionally conducting screens *Can. J. Phys.* **45** 2145–72

[10] Aditya S 1979 *Studies on a Planar Helical Slow-Wave Structure* (Indian Institute of Technology)

[11] Aditya S and Arora R K 1979 Nonreciprocal dispersion characteristics of a planar helix on magnetized ferrite slabs *IEEE Trans. Microw. Theory Tech.* **27** 864–8

[12] Sensiper S 1955 Electromagnetic wave propagation on helical structures (a review and survey of recent progress) *Proc. IRE* **43** 149–61

[13] Chodorow M and Chu E L 1955 Cross-wound twin helices for traveling-wave tubes *J. Appl. Phys.* **26** 33–43

[14] Pierce J R *Travelling Wave Tubes* (Princeton, NJ: D. Van Nostrand Co., Inc)

[15] Collin R E 1992 *Foundations for Microwave Engineering* (New York: McGraw-Hill)

[16] Chu L J and Jackson J D 1948 Field theory of traveling-wave tubes *Proc. IRE* **36** 853–63

[17] Paik S F 1969 Design formulas for helix dispersion shaping *IEEE Trans. Electron Devices* **16** 1010–4

[18] Chadha D 1983 *Investigations on a Planar Helical Slow-Wave Structure with Particular Reference to Travelling-Wave Tube Applications* (Indian Institute of Technology)

IOP Publishing

Planar Slow-Wave Structures: Applications in Traveling-Wave Tubes

Chen Zhao and Sheel Aditya

Chapter 4

Planar helix with straight-edge connections (PH-SEC) slow-wave structure

4.1 Introduction

As mentioned in chapters 1 and 3, although the circular helix is very widely used as a slow-wave structure (SWS) in traveling-wave tubes (TWTs), it faces fabrication and other challenges when one tries to use it for operation at millimeter-wave frequencies (30–300 GHz) and higher. To address these challenges, a pair of unidirectionally conducting (UC) screens was suggested; as described in chapter 3, this structure has guiding properties similar to those of the circular helix [1–4]. Hence, this structure is called the planar helix. To keep the analysis simple, the planar helix was analyzed using the approximation of infinite width. To obtain a practically useful structure, Cier Siang Chua of the authors' research group suggested the addition of straight connections between the individual conductors of the top and bottom screens, yielding a planar helix with straight-edge connections (PH-SEC) [5].

The PH-SEC offers several advantages compared to the conventional circular helix. It is suitable for printed circuit fabrication as well as microfabrication due to its planar configuration and straight-edge connections. Printed circuit TWTs have the potential to be fabricated at very low cost compared to conventional TWTs, since relatively fewer machined parts are required. Another advantage of the planar geometry is the possibility of using a sheet electron beam. Very significantly, the PH-SEC does not have an inherent limitation on its bandwidth, since its guiding properties resemble those of a circular helix.

In this chapter, we describe the dispersion characteristics and interaction impedance properties of the PH-SEC obtained using field-theory analyses and simulations. Beginning with a PH-SEC immersed in free space [5], a more general configuration is also considered. The general configuration includes practical modifications such as a beam tunnel, a metal shield, and multilayer dielectric

doi:10.1088/978-0-7503-5764-7ch4

substrates [6]. We describe the scaled fabrication of two special cases of the general configuration of the PH-SEC using printed circuit techniques and provide measurement results. While the studies reported in [5, 6] were carried out using the sheath-helix approximation, the effects of periodicity considered in [7] are also described in this chapter. The chapter ends with a description of hot-test parameters (gain, power, efficiency) obtained using particle-in-cell (PIC) simulation for a TWT amplifier based on a PH-SEC configuration operating around 4.75 GHz [8].

4.2 Field theory analysis of the PH-SEC

In connection with confining the pair of UC screens (planar helix) in the transverse direction, it was suggested in [9] that for one of the modes, the edges of the UC screens could be connected, making the structure similar to a rectangular tape helix, as shown in figure 4.1. In such a rectangular helix, the connections between the individual conductors of the top and bottom screens require inclined conductors; this may result in fabrication difficulties if printed circuit or microfabrication techniques are used. On the other hand, in the PH-SEC, as shown in figure 4.2, the individual connections of the top and bottom screens are straight, i.e., $\psi_2 = 0°$. Such connections can easily be realized, e.g. as vias, with the additional advantages mentioned earlier.

The generation of accurate simulation results for phase velocity and interaction impedance for the PH-SEC can be time-consuming. On the other hand, analytical methods can provide quick estimations of these properties; this can be very helpful in

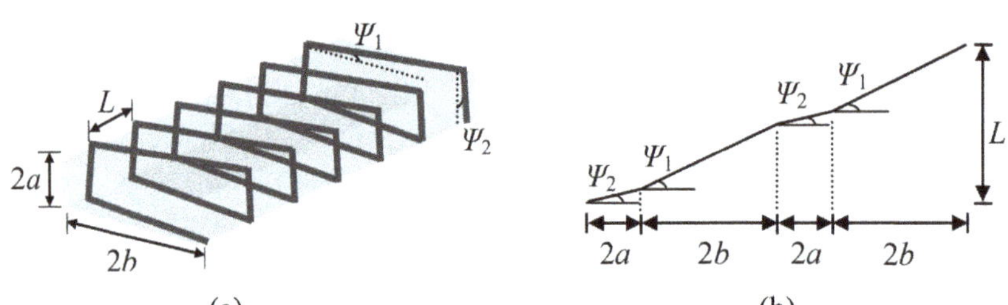

Figure 4.1. (a) Perspective view of a generalized rectangular helix. (b) Unfolded view of a rectangular helix turn. © [2009] IEEE. Reprinted, with permission, from [5].

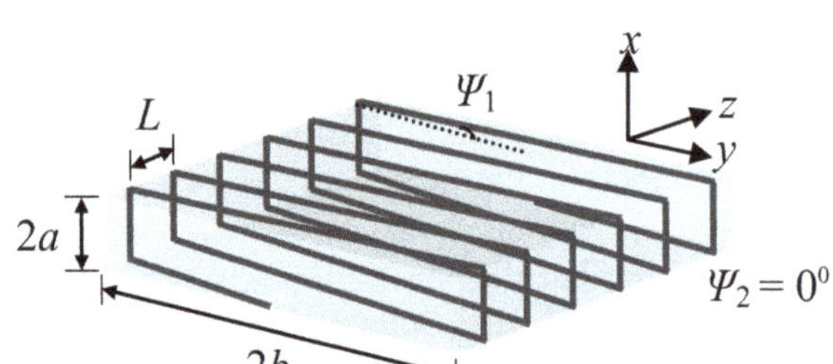

Figure 4.2. Perspective view of a PH-SEC. © [2009] IEEE. Reprinted, with permission, from [5].

the initial selection of structural parameters for the design of a particular TWT. Moreover, analytical methods can provide better insight into the physics of the structure.

4.2.1 PH-SEC immersed in free space

For the rectangular helix shown in figure 4.1, dispersion characteristics were obtained by using the sheath-helix approximation, assuming average boundary conditions on the four sides, and ignoring the fields in the corner regions outside the helix [10]. Even the approximate approach [10] becomes tedious when applied to multilayer structures which are likely to arise in practice. Our research group has calculated the dispersion characteristics of the PH-SEC (figure 4.2) using the effective dielectric constant (EDC) method [11], which is a simple and fast way of estimating the propagation constant. It has been shown that the EDC method works quite well for a range of dimensional parameters. The EDC method can also simplify the analysis of multilayer structures.

Parts of this section have been reprinted, with permission, from [5]. © [2009] IEEE.

4.2.1.1 Effective dielectric constant method

The EDC method has been extensively applied to open waveguides of rectangular cross section [11]. The results of the method are accurate in the frequency range that is far from the cutoff. In this method, the original three-dimensional (3D) waveguide is replaced by two related two-dimensional (2D) structures, each of which is easy to analyze.

The geometry of the PH-SEC is shown in figure 4.2. L is the period of the structure; the dimensions along the x- and y-directions are $2a$ and $2b$, respectively. The structures in figures 4.1 and 4.2 are assumed to be immersed in free space. Figure 4.3(a) shows the rectangular cross section of the structure. Corresponding to immersion in free space, ε_{r1} equals 1. Let us consider the case $b/a \geqslant 1$ first. In the sheath-helix approximation, one has UC screens at $x = \pm a$ and straight-edge connections at $y = \pm b$. Using the EDC method, the structure is analyzed in two steps. In the first step, we consider the x-dependent profile only. This corresponds to a pair of UC screens separated by a distance of $2a$ and infinite in the y-direction, as shown in figure 4.3(b). Thus, the 'slowing-down' effect of the helical structure is taken into account in this step. In the second step, we consider the y-dependent profile only, accounting for the transverse confinement in the y-direction. As shown in figure 4.3(c), this corresponds to a symmetric dielectric slab of thickness $2b$ that has a dielectric constant of $\varepsilon_{\text{eff}}{}'$, which is obtained in the first step. For the case $b/a < 1$, the two steps are reversed; the y-dependent profile is considered first, taking into account the 'slowing-down' effect; the x-dependent profile is considered next, accounting for the confinement in the x-direction. The details of the analysis are given below, assuming $b/a \geqslant 1$.

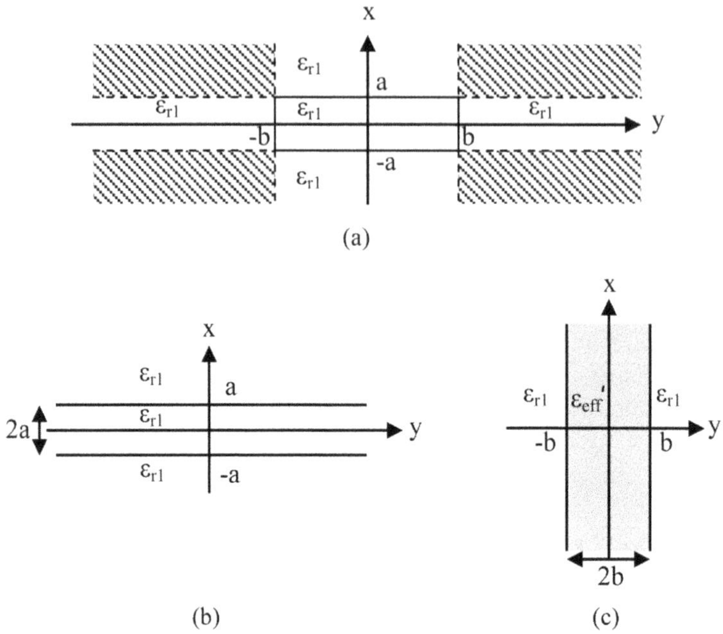

Figure 4.3. Model for applying the EDC method. (a) Rectangular cross section of the proposed structure. (b) x-dependent profile. (c) y-dependent profile. © [2009] IEEE. Reprinted, with permission, from [5].

a. **The *x*-dependent profile**
The geometry of the x-dependent profile, shown in figure 4.3(b), corresponds to a pair of UC screens immersed in air; the screens conduct in directions making angles $\pm\psi_1$ with respect to the y-axis. The symmetry of the structure implies that the solutions can be decomposed into transverse symmetric (even) and transverse antisymmetric (odd) cases [12], meaning that E_x, E_y, H_x, and H_y are symmetric or antisymmetric, respectively, with respect to the x-axis. To capture the slowing-down effect of the helical conductor in the original structure, we modify the pitch angle of the UC screens, ψ_1, to an effective pitch angle, ψ_{eff}, which is simply the arctangent of the ratio of the period L and the perimeter of the cross section $4a + 4b$. The hybrid mode field solution and characteristic equations for the structure in figure 4.3(b) were derived in [12]. Choosing the transverse antisymmetric mode, the characteristic equation is

$$\frac{k^2}{u^2}\cot^2\psi_{\text{eff}} = \coth(ua) \tag{4.1}$$

where $k = \omega^2\mu_0\varepsilon_0\varepsilon_{r1}$. Here, u is the decay coefficient in the x-direction and is related to the phase constant β' for this step by

$$u^2 = \beta'^2 - k^2. \tag{4.2}$$

β' is numerically obtained from (4.1) and (4.2). Then, $\varepsilon_{\text{eff}}'$ for the second step is calculated from $\varepsilon_{\text{eff}}' = (\beta'/k)^2$. One may consider the transverse symmetric mode in a similar manner.

b. **The *y*-dependent profile**
The slowing-down effect of the helical conductor in the original structure is considered in the x-dependent profile. The y-dependent profile (figure 4.3(c)) is simply an infinite dielectric slab immersed in free space, with a dielectric constant $\varepsilon_{\text{eff}}'$ that is calculated from the x-dependent profile. The dielectric slab supports transverse magnetic (TM) and transverse electric (TE) modes, which can be even or odd with respect to y. To support the transverse antisymmetric mode of the original waveguide, and keeping in mind that E_x is odd w.r.t. x in the previous step, here we consider modes with an even variation of E_x w.r.t. y. Thus, we choose the TE-even mode for this step; this is the lowest-order mode for a dielectric slab. The characteristic equation for the TE-even mode is [13]:

$$\frac{v_1}{v_2} = \tan(v_2 b) \tag{4.3}$$

where v_1 and v_2 are the y-direction decay coefficients for the air and substrate regions, respectively, and are related to the overall β in the z-direction by

$$\beta^2 = {v_1}^2 + {k_1}^2 = {k_{\text{eff}}}'^2 - {v_2}^2 \tag{4.4}$$

In (4.4), ${k_1}^2 = \omega^2\mu_0\varepsilon_0\varepsilon_{r1}$ and ${k_{\text{eff}}}'^2 = \omega^2\mu_0\varepsilon_0\varepsilon_{\text{eff}}'$ are the wave numbers in the free-space and dielectric regions, respectively. The dispersion characteristics of the original structure can be obtained by solving the characteristic equations (4.1) and (4.3) numerically.

c. **Calculated and simulation results**
Figure 4.4 shows the normalized phase velocities for the rectangular helix and the PH-SEC, calculated using the EDC method. Also included are simulation results obtained using CST Microwave Studio (MWS) Eigenmode Solver for the full 3D models shown in figures 4.1 and 4.2. For the rectangular helix in figure 4.1(a) ($\psi_1 = \psi_2 = 2.86°$) with $a = 0.06$ mm and $b = 0.24$ mm ($b/a = 4$), it can be seen that the normalized phase velocity approaches 0.052 at high frequencies. This closely matches the result presented in figure 4.2(a) in [10] for the same dimensions. Figure 4.4 also shows the results for the structure with straight-edge connections shown in figure 4.1(c) ($\psi_2 = 0°$). In this case, for $a = 0.06$ mm and $b = 0.24$ mm, the normalized phase velocity approaches a lower value of 0.04 at high frequencies. Further, for $a = 0.24$ mm and $b = 0.06$ mm, the normalized phase velocity approaches an even lower value of 0.01. Thus, the proposed structure with straight-edge connections can offer a phase velocity which is significantly lower than that of a rectangular helix with the same cross-sectional area. Also, except at low frequencies, the results of the EDC method match the simulation results very well. The inaccuracy of the EDC

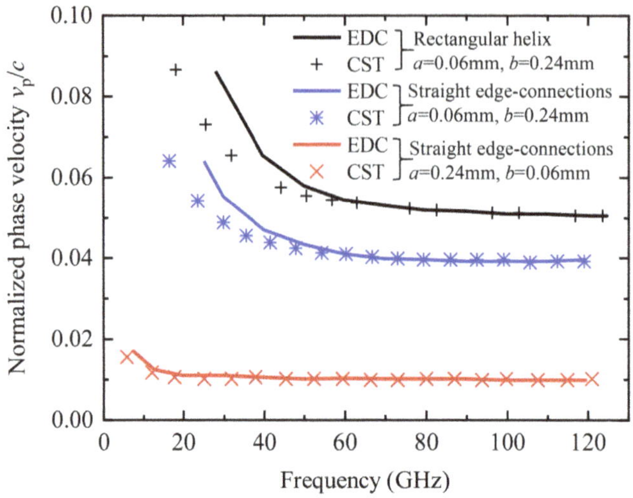

Figure 4.4. Comparison between the normalized phase velocity of the rectangular helix and that of the PH-SEC. © [2009] IEEE. Reprinted, with permission, from [5].

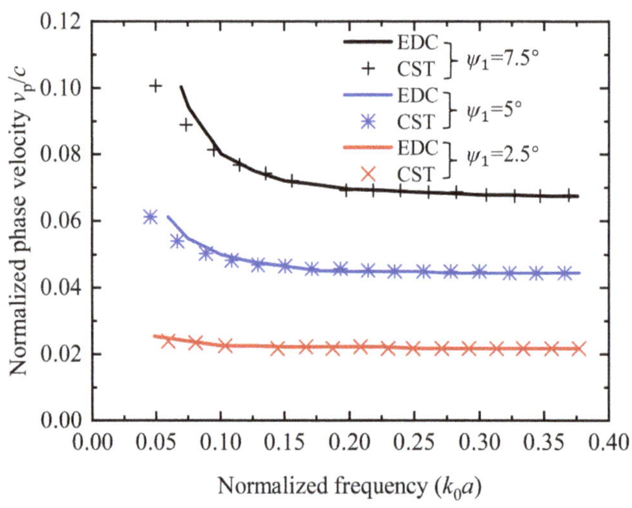

Figure 4.5. Normalized phase velocity of the PH-SEC for $\psi_1 = 2.5°$, 5° and 7.5° with $a = b = 1.2$ mm. © [2009] IEEE. Reprinted, with permission, from [5].

method at low frequencies arises because the fields in the corner regions are neglected [11]; furthermore, a large aspect ratio also contributes to this inaccuracy, as explained later in figure 4.6.

Figure 4.5 shows the normalized phase velocity for straight-edge connections for different values of ψ_1, with $a = b = 1.2$ mm. As expected, when the pitch angle decreases, the phase velocity also decreases. Again, except at

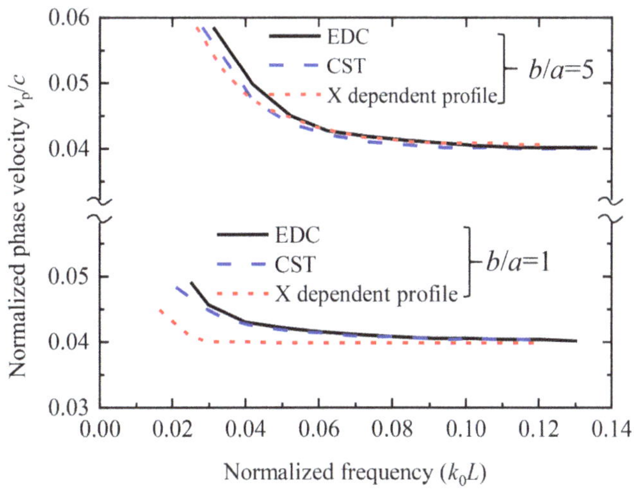

Figure 4.6. Normalized phase velocity of the PH-SEC for $b/a = 1$ and 5 with fixed L and cross section perimeter. © [2009] IEEE. Reprinted, with permission, from [5].

low frequencies, the match between the calculated and simulation results is very good. Finally, figure 4.6 shows the effect of variations in b/a for straight-edge connections, with the period L and cross section perimeter fixed at 0.48 and 12 mm, respectively. As b/a increases, the simulation results more closely follow the results of the x-dependent profile (infinite planar helix), and the accuracy of the EDC method deteriorates. This indicates that for large width-to-height ratios, it may be enough to consider just the x-dependent profile; in such cases, the y-dependent profile does not represent the original structure correctly.

4.2.2 PH-SEC in the presence of multilayer dielectric substrates

Parts of this section have been reprinted, with permission, from [8]. © [2011] IEEE.

This subsection describes a PH-SEC combined with multilayer substrates, a vacuum tunnel within the helical structure, and a metal shield. Such modifications are required when the planar helix structure is used in TWTs. Based on the modified effective dielectric constant (MEDC) method (which is described next), results for phase velocity and interaction impedance are presented. The effects of variations in dimensional and material parameters are studied in detail with a view to obtaining a flat phase velocity vs. frequency curve. The design of a wideband coplanar waveguide (CPW) feed for one of the configurations, together with measured results, is also presented. These results confirm the ease of fabrication, low loss, and wideband potential of the PH-SEC. The contents of this subsection follow the description in [6, 14].

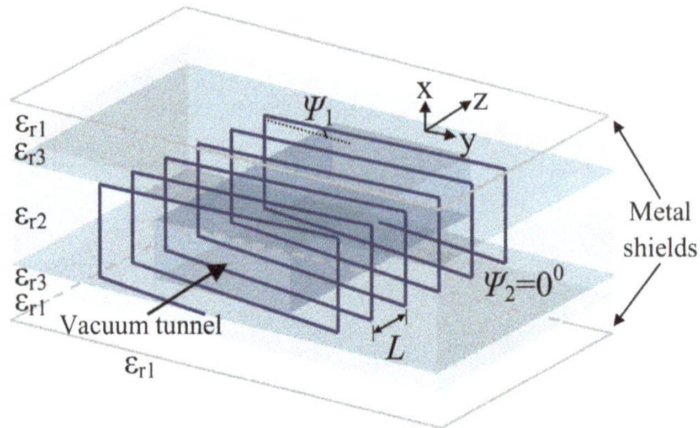

Figure 4.7. General configuration of a shielded PH-SEC combined with multilayer dielectric substrates and a vacuum tunnel. © [2010] IEEE. Reprinted, with permission, from [6].

The general configuration of a shielded PH-SEC in the presence of multilayer dielectric substrates is shown in figure 4.7. Figure 4.8(a) shows the cross-sectional view of the general configuration corresponding to figure 4.7. The straight-edge connections ($\psi_2 = 0°$) are embedded in the middle substrate layer, which has a dielectric constant of ε_{r2} and a thickness of $2a$ in the x-direction. Corresponding to the sheath-helix approximation, it is assumed that a pair of UC screens is printed on the top and bottom of the middle substrate layer; the screens conduct in the $\pm\psi_1$ directions with respect to the y-axis and have a width of $2b$ in the y-direction. In order to accommodate a sheet electron beam, a vacuum tunnel ($\varepsilon_{r1} = 1$) is centered within the helical structure with dimensions $2c$ and $2d$ along the x- and y-directions, respectively. L is the period of the structure. A substrate layer and a vacuum layer with dielectric constants of ε_{r3} and ε_{r1} and thicknesses of t_3 and t_4, respectively, are stacked between the UC screens and metal shields. The metal shields and the dielectric substrates are assumed to be infinite in the y-direction. The separation between the vacuum tunnel and the top and bottom UC screens is t_2; the distance between the edges of the vacuum tunnel and the straight-edge connections is g.

The general configuration described above can lead to a number of simpler configurations. Five such configurations are shown in table 4.1: (A) an unshielded normal helix, (B) an unshielded normal helix with a vacuum tunnel, (C) a shielded normal helix with a vacuum tunnel, (D) a shielded inverted helix with a vacuum tunnel and (E) a shielded helix with a vacuum tunnel.

4.2.2.1 Problems with the EDC method

There are two alternative ways in which the 3D structure of figure 4.7 can be decomposed into 2D structures. These two alternative ways are denoted by EDC-case I and EDC-case II. In EDC-case I, the planar helix and dielectric slab are infinite in the y- and x-directions, respectively, as depicted in figures 4.8(b) and (c); in EDC-case II, the planar helix and dielectric slab are infinite in the x- and y-directions, respectively, as depicted in figures 4.8(d) and (e). Depending on the aspect

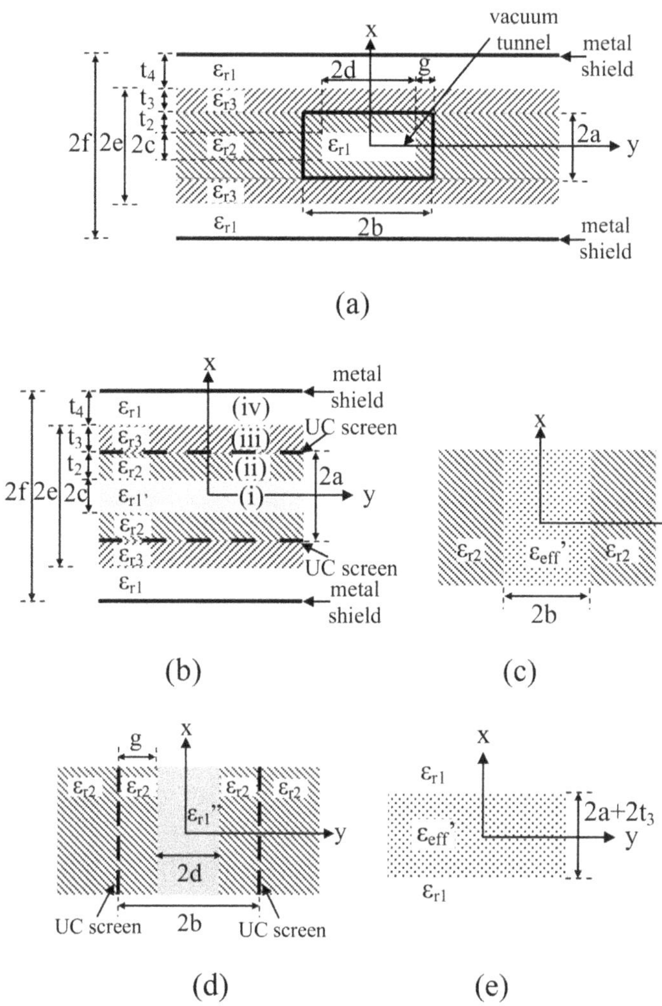

Figure 4.8. Model for applying the MEDC method. (a) Cross section of the general shielded helix configuration; (b) planar helix infinite in the *y*-direction; (c) dielectric slab infinite in the *x*-direction; (d) planar helix infinite in the *x*-direction; and (e) dielectric slab infinite in the *y*-direction.

ratio of the original 3D structure, one case may yield more accurate results than the other [13]. Figure 4.9 shows a comparison of the simulated phase velocity of the full 3D structure with the results calculated using EDC-case I and EDC-case II for an unshielded normal helix (table 4.1: configuration A) for $b/a = 1$ and 3 with the period L fixed at 0.315 mm and the cross-sectional perimeter $4a + 4b$ fixed at 4.8 mm. The characteristic equation for this configuration is given in the following subsections. Simulation results have been obtained by using CST MWS. It can be noted that EDC-case I overestimates the phase velocity, whereas EDC-case II underestimates it, even in the region far from the cutoff.

Table 4.1. Configurations for PH-SEC. © [2010] IEEE. Reprinted, with permission, from [6].

Configuration	Cross-sectional view
A. Unshielded normal helix	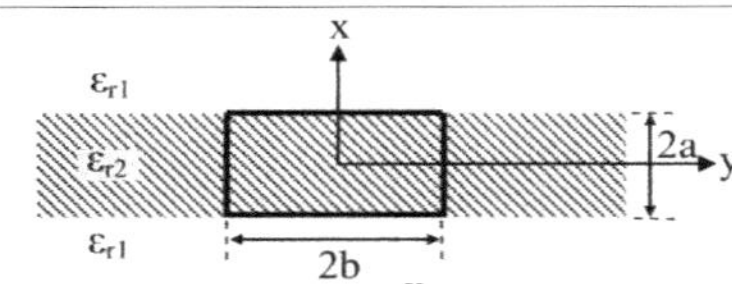
B. Unshielded normal helix with a vacuum tunnel	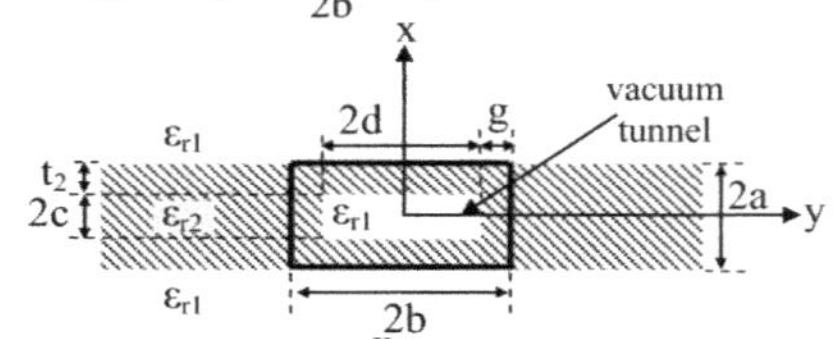
C. Shielded normal helix with a vacuum tunnel	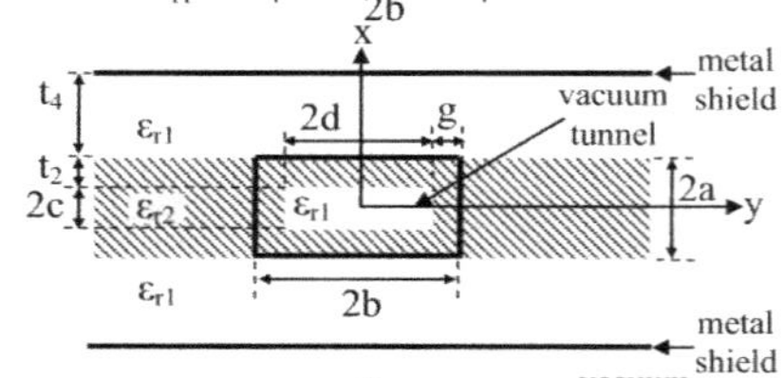
D. Shielded inverted helix with a vacuum tunnel	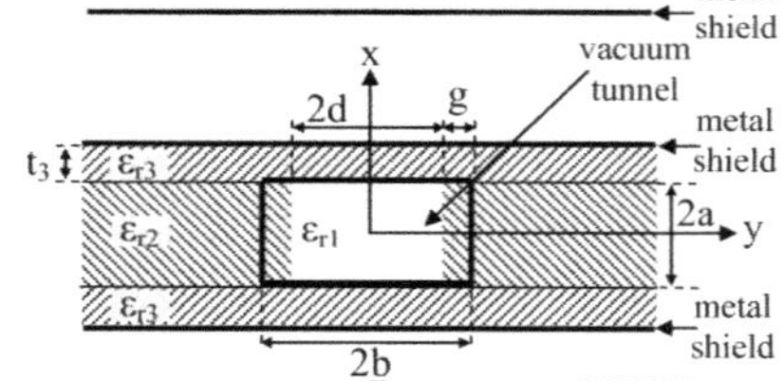
E. Shielded helix with a vacuum tunnel	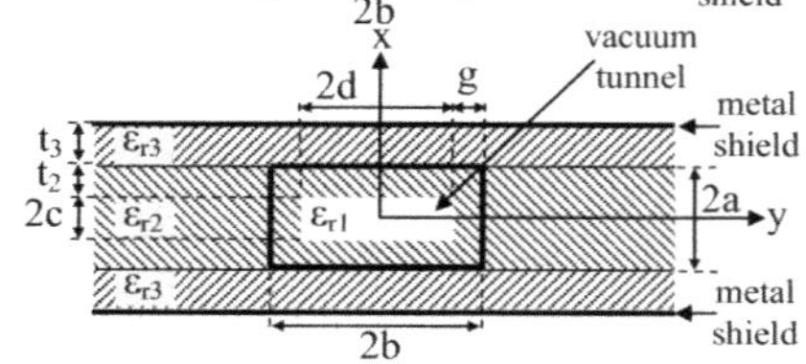

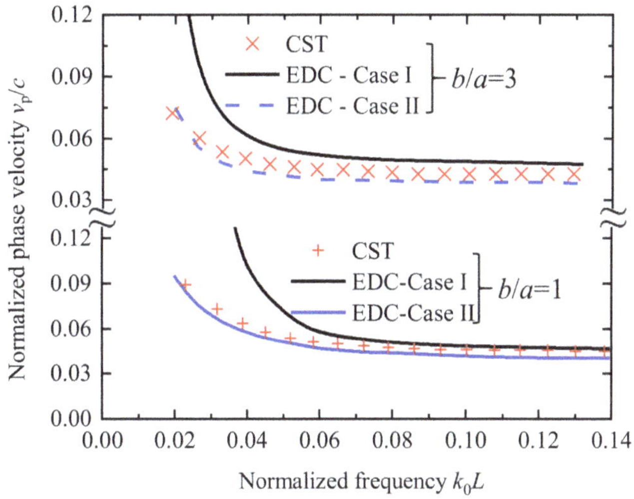

Figure 4.9. Normalized phase velocities calculated using the conventional EDC method [5] for the unshielded normal helix (table 4.1: A) for $b/a = 1$ and 3 with fixed L and cross-sectional perimeter ($\varepsilon_{r1} = 1$, $\varepsilon_{r2} = 3.02$). Simulation results for the 3D structure are also included. © [2010] IEEE. Reprinted, with permission, from [6].

4.2.2.2 Modified effective dielectric constant method

A combination of EDC-case I and EDC-case II, the so-called 'dual EDC method,' was reported in [15] for the analysis of rectangular dielectric waveguides. In the dual EDC method, a combination of the possible applications of the conventional EDC method is used to cancel the errors produced in each case, and hence a more accurate approximation is achieved. However, the dual EDC method developed for the rectangular dielectric waveguides cannot be directly applied to the present structure. The infinite planar helix supports hybrid modes only, unlike the infinite (2D) dielectric waveguides, which support the TE and TM modes. In addition, in the case of a PH-SEC, the first step—an infinite planar helix—plays a dominant role in the solution. In view of the errors observed for both EDC-case I and EDC-case II in the previous subsection, the MEDC method combines the results for both cases of the conventional EDC method by using a weighted average of the results in each case. The weights for the 2D waveguides are based on the aspect ratio of the original 3D structure. This is simpler than the dual EDC method and is directly related to the geometry of the problem. The propagation constant values obtained from the 2D planar helices that are infinite in the y- and x-directions, depicted in figures 4.8(b) and (d), respectively, are weight averaged and form the $\varepsilon_{\text{eff}}'$ for the subsequent 2D dielectric slabs that are infinite in the x- and y-directions, depicted in figures 4.8(c) and (e), respectively. The propagation constant values obtained from the 2D dielectric slabs are weight averaged again to yield the propagation constant of the original 3D structure.

The details of the analysis based on the MEDC method are given below. It is assumed that $b/a \geqslant 1$. The description given below is for the PH-SEC (i.e., $\psi_2 = 0$ in figure 4.7). However, the MEDC method is also expected to be applicable to a rectangular helix embedded in multilayer dielectric substrates (i.e., $\psi_2 = \psi_1 \neq 0$ in figure 4.7).

a. **Case I (Step 1)—planar helix infinite in the *y*-direction**

The general configuration shown in figure 4.8(a) is analyzed using the MEDC method. The 3D structure is decomposed into two 2D structures. Following the approach used in EDC-case I, the first 2D structure, a planar helix which is infinite in the y-direction, is shown in figure 4.8(b). It includes a pair of UC screens separated by a distance of $2a$ and conducting in directions $\pm\psi_1$ with respect to the y-axis. The dielectric constant in the middle region, ε_{r1}', represents the vacuum tunnel as well as the substrate surrounding the vacuum tunnel in the y-direction in the original 3D structure. Hence, ε_{r1}' is obtained by taking the weighted average of the dielectric constant in each region along the y-direction:

$$\varepsilon_{r1}' = \frac{\varepsilon_{r1} \times d + \varepsilon_{r2} \times g}{d + g} \tag{4.5}$$

The other details of this 2D structure follow from the description of the general configuration given at the beginning of this section 4.2.2.2. The multilayer 2D structure in figure 4.8(b) is similar to the infinite planar helix structure reported in [16], with the difference that in the present case $\varepsilon_{r1}' \neq \varepsilon_{r1}$. The longitudinal field components of the transverse antisymmetric mode for this configuration, as well as the application of boundary conditions using the sheath-helix model, are shown in the appendix of this chapter. The characteristic equation for the transverse antisymmetric mode is:

$$\frac{\dfrac{\varepsilon_{r1}'}{u_1'}\dfrac{1 + (u_1'\varepsilon_{r2}/u_2'\varepsilon_{r1}')T_1T_2}{T_1 + (u_2'\varepsilon_{r1}'/u_1'\varepsilon_{r2})T_2} + \dfrac{\varepsilon_{r1}}{u_4'}\dfrac{1 + (u_4'\varepsilon_{r3}/u_3'\varepsilon_{r1})T_3T_4}{T_4 + (u_3'\varepsilon_{r1}/u_4'\varepsilon_{r3})T_3}}{\dfrac{u_1'T_1 + u_1'(u_2'/u_1')T_2}{1 + (u_1'/u_2')T_1T_2} + \dfrac{u_4' + u_4'(u_3'/u_4')T_3T_4}{T_4 + (u_4'/u_3')T_3}} = \frac{1}{k_0^2}\tan^2\psi_{\text{eff}} \tag{4.6}$$

where $T_1 = \coth(u_1'c)$, $T_2 = \tanh(u_2't_2)$, $T_3 = \tanh(u_3't_3)$, and $T_4 = \tanh(u_4't_4)$. u_1', u_2',u_3', and u_4' are decay coefficients in the x-direction. The phase constant for this structure, which has finite dimensions in the x-direction, is called $\beta_{\text{x\_helix}}'$. The decay coefficients are related to the phase constant $\beta_{\text{x\_helix}}'$ as follows:

$$\begin{aligned} u_1'^2 &= \beta_{\text{x\_helix}}'^2 - k_1'^2,\ u_2'^2 = \beta_{\text{x\_helix}}'^2 - k_2^2, \\ u_3'^2 &= \beta_{\text{x\_helix}}'^2 - k_3^2 \text{ and } u_4'^2 = \beta_{\text{x\_helix}}'^2 - k_0^2 \end{aligned} \tag{4.7}$$

where

$$k_1'^2 = \omega^2\mu_0\varepsilon_0\varepsilon_{r1}',\ k_2^2 = \omega^2\mu_0\varepsilon_0\varepsilon_{r2},\ k_3^2 = \omega^2\mu_0\varepsilon_0\varepsilon_{r3}, \text{ and } k_0^2 = \omega^2\mu_0\varepsilon_0. \tag{4.8}$$

To capture the slowing-down effect of the original 3D structure, the pitch angle of the UC screens, ψ_1, is changed to ψ_{eff}, which is simply the arctangent of the ratio of the period L and the perimeter of the cross section $4a + 4b$. One may consider the transverse symmetric mode in a similar manner.

The characteristic equation for the 2D planar helix that is infinite in the y-direction can easily be deduced from equation (4.6) for each configuration listed in table 4.1. For example, setting $t_3 = 0$ and $t_4 = \infty$ leads to the characteristic equation for the unshielded normal helix with a vacuum tunnel. The numerical solution of (4.6), which is the transcendental characteristic equation for the planar helix that is finite in the x-direction, yields the phase constant $\beta_{\text{x\_helix}}'$.

b. **Case II (step 1)—planar helix infinite in the *x*-direction**

Following EDC-case II, the first 2D structure, a planar helix that is infinite in the x-direction, is shown in figure 4.8(d). It includes a pair of UC screens separated by a distance of $2b$. In a manner similar to that used in the previous subsection, the dielectric constant in the middle region, ε_{r1}'', is obtained from:

$$\varepsilon_{r1}'' = \frac{\varepsilon_{r1} \times c + \varepsilon_{r2} \times t_2}{c + t_2}. \tag{4.9}$$

The other details of this 2D structure follow from the description of the general configuration given at the beginning of this section 4.2.2.2. The characteristic equation for this case can be obtained by setting $c = d$, $t_2 = g$, $t_3 = \infty$, $t_4 = 0$, and $\varepsilon_{r3} = \varepsilon_{r2}$ in figure 4.8(b). Hence, the characteristic equation for the transverse antisymmetric mode is

$$\frac{[\varepsilon_{r1}'' + \varepsilon_{r1}''(u_1''\varepsilon_{r2}/u_2''\varepsilon_{r1}'')T_1T_2]/[u_1''T_1 + u_1''(u_2''\varepsilon_{r1}''/u_1''\varepsilon_{r2})T_2] + \varepsilon_{r2}/u_2''}{[u_1''T_1 + u_1''(u_2''/u_1'')T_2]/[1 + (u_1''/u_2'')T_1T_2] + u_2''} = \frac{1}{k_0{}^2}\tan^2\Psi_{\text{eff}} \tag{4.10}$$

where $T_1 = \coth(u_1''d)$ and $T_2 = \tanh(u_2''g)$. u_1'' and u_2'' are decay coefficients in the y-direction. The phase constant for this structure, which has finite dimensions in the y-direction, is called $\beta_{\text{y\_helix}}'$. The decay coefficients are related to the phase constant $\beta_{\text{y\_helix}}'$ as follows:

$$u_1''^2 = \beta_{\text{y\_helix}}'^2 - k_1''^2 \text{ and } u_2''^2 = \beta_{\text{y\_helix}}'^2 - k_2^2 \tag{4.11}$$

where

$$k_1''^2 = \omega^2\mu_0\varepsilon_0\varepsilon_{r1}''. \tag{4.12}$$

For the case of an unshielded normal helix (table 4.1: A), the characteristic equation for the 2D planar helix that is infinite in the x-direction is obtained by setting $d = 0$ and $g = b$ in (4.10); for configurations B to E listed in table 4.1, equation (4.10) can be used directly. The numerical solution of (4.10) yields the phase constant $\beta_{\text{y\_helix}}'$.

The phase constant values obtained from (4.6) and (4.10) for the 2D planar helices are weight averaged based on the aspect ratio of the original 3D structure:

$$\beta'_{\text{helix}} = \frac{\beta'_{\text{x\_helix}} \times b + \beta'_{\text{y\_helix}} \times a}{a + b}. \tag{4.13}$$

Through several comparisons of the results from the MEDC method applied to configurations with different aspect ratios, we find that for $b/a \geqslant 3$, β_{helix}' calculated using (4.13) is generally a good approximation for the phase constant of the 3D structure. For smaller aspect ratios, e.g. $1 \leqslant b/a < 3$, the subsequent step involving the infinite dielectric slabs is required for a more accurate approximation. The EDC for the subsequent infinite dielectric slabs is simply:

$$\varepsilon_{\text{eff}}' = \left(\frac{\beta_{\text{helix}}'}{k_0}\right)^2. \tag{4.14}$$

c. **Case I (step 2)—dielectric slab infinite in the *x*-direction**
The slowing-down effect of the helix in the original waveguide has been taken into account in the first step of Case I and Case II using the 2D infinite planar helix structures. The subsequent 2D structure for Case I, a dielectric slab that is infinite in the x-direction, is depicted in figure 4.8(c). The dielectric slab of width $2b$, immersed in a medium with dielectric constant ε_{r2}, is considered to have a dielectric constant of $\varepsilon_{\text{eff}}'$, which is calculated from (4.14). The dielectric slab supports the TM and TE modes, which can be even or odd with respect to y. To support the transverse antisymmetric mode of the original waveguide, and keeping in mind that E_x is odd w.r.t. x in the planar helix which is infinite in the y-direction, we consider modes with an even variation of E_x w.r.t. y. Thus, we choose the TE-even mode for this step; this is the lowest-order mode for a dielectric slab. The characteristic equation for the TE-even mode is [17]:

$$\frac{v_1}{v_2} = \tan(v_2 b) \tag{4.15}$$

where v_1 and v_2 are decay coefficients in the y-direction for the regions with dielectric constant ε_{r2} and $\varepsilon_{\text{eff}}'$, respectively; these are related to the phase constant, called $\beta_{\text{y\_sub}}$ in this case, as follows:

$$\beta_{\text{y\_sub}}^2 = v_1^2 + k_2^2 = k_{\text{eff}}'^2 - v_2^2. \tag{4.16}$$

In (4.16), $k_{\text{eff}}'^2 = \omega^2 \mu_0 \varepsilon_0 \varepsilon_{\text{eff}}'$ is the wave number in the slab region. The same characteristic equation applies to this step for all the configurations listed in table 4.1. The solution of (4.15) yields the phase constant $\beta_{\text{y\_sub}}$.

d. **Case II (step 2)—dielectric slab infinite in the *y*-direction**
The 2D dielectric slab for Case II is depicted in figure 4.8(e). The dielectric slab, infinite in the y-direction and immersed in free space, is considered to have a dielectric constant of $\varepsilon_{\text{eff}}'$; the slab thickness is generally taken to be $2a + 2t_3$, since the effect of the dielectric layer ε_{r3} has not yet been considered in this case. The slab width is changed to $2a$ for configurations A to C in table 4.1. The characteristic equation for the TE-even mode is:

$$\frac{w_1}{w_2} = \tan[w_2(a + t_3)] \tag{4.17}$$

where w_1 and w_2 are the decay coefficients in the x-direction for the air and substrate regions, respectively; they are related to the phase constant, called β_{x_sub} in this case, by

$$\beta_{x\_sub}^2 = w_1{}^2 + k_0{}^2 = k_{eff}'^2 - w_2{}^2. \tag{4.18}$$

The solution of (4.17) yields the phase constant β_{x_sub}. In the last step, the phase constant values for the 2D infinite dielectric slabs obtained from (4.15) and (4.17) are weight averaged according to the aspect ratio as follows:

$$\beta = \frac{\beta_{y\_sub} \times b + \beta_{x\_sub} \times a}{a + b} \tag{4.19}$$

to give β, which is the final approximation of the phase constant of the original 3D structure.

4.2.2.3 *Interaction impedance*

As mentioned in chapter 1, the value of the interaction impedance, K_c, is crucial for the application of an SWS in a TWT. Following [18], the average interaction impedance is written as:

$$K_c = \frac{E_{z,\,av}^2}{2P\beta^2} \tag{4.20}$$

where $E_{z,av}$ is the average value of the axial electric field of the transverse antisymmetric mode, calculated over the cross section of the sheet beam, and P is the total power propagating along the z-direction. In general, we assume a solid and uniform sheet electron beam whose height and width are half those of the vacuum tunnel and centered within the structure. For configuration A in table 4.1, a hypothetical sheet electron beam whose height and width are half those of the helix is centered within the helix. To evaluate (4.20), we use the β value of the 3D structure calculated from (4.13) or (4.19) depending on the aspect ratio. The calculation of the power P is simplified, as in [16], by using the approximation of a 2D planar helix that is infinite in the y-direction, as shown in figure 4.8(b). For the general configuration, the total power in the transverse antisymmetric mode can be obtained from

$$P = 2(P_i + P_{ii} + P_{iii} + P_{iv})(2b) \tag{4.21}$$

where P_i, P_{ii}, P_{iii}, and P_{iv} are the power flows per unit width for the corresponding regions in the upper half of the structure in figure 4.8(b). The width of the 2D structure in the y-direction is taken to be $2b$. By applying Poynting's theorem, the power flow in each region can be obtained using the following expression:

$$P_k = \frac{1}{2}\frac{\beta}{\omega\mu_0}\int_x\left(\left(\frac{\mu_0}{\varepsilon_k}\right)^2|H_y|^2 + |E_y|^2\right)dx \tag{4.22}$$

where k is the index for region i to iv. The relations between the constant coefficients A, B_1, B_2, C_1, C_2, D, etc. can be derived from the boundary conditions given in the appendix of this chapter. The power flow for the other configurations can be derived in a similar manner.

4.2.2.4 Results and discussion

Figures 4.10(a) and (b) show the normalized phase velocity and average interaction impedance for the unshielded normal helix (table 4.1: A) calculated using the methods described in sections 4.2.2.2 and 4.2.2.3, respectively. Simulation results for the actual 3D structure with the actual helical conductor, obtained using CST, are also included for comparison. The period and the cross-sectional perimeter are fixed at 0.315 and 4.8 mm, respectively, to maintain a constant value of ψ_{eff}. The curve labeled 'MEDC' represents the results calculated using both steps 1 and 2 in the MEDC method, while 'MEDC: β_{helix}'' represents the results calculated using only step 1, i.e. the infinite planar helix. As can be seen in figure 4.10(a), for $b/a = 1$, both steps of the MEDC method are required to get an accurate estimate of the phase velocity. However, for $b/a = 3$, step 1 itself gives accurate results; step 2 actually causes a deterioration of results, since for such aspect ratios, this step does not represent the original structure correctly; this was also observed in [5]. Except at low frequencies, the results of the MEDC method match the simulation results very well. The inaccuracy of the EDC method at low frequencies is well-known and arises because the fields in the corner regions are neglected [11]. The results in figure 4.10(a) indicate that a lower and flatter phase velocity curve can be obtained by lowering the aspect ratio of the helix. This is attributed to a stronger effect of ε_{r2} for a lower aspect ratio, i.e. a larger value of a for a given value of b. The interaction impedance results in figure 4.10(b) match the simulation results very well; this shows that the method of calculation proposed in section 4.2.2.3, i.e. the use of an accurate phase constant together with a 2D infinite planar helix approximation, works satisfactorily.

Figures 4.11(a) and (b) show the normalized phase velocity and interaction impedance of the unshielded normal helix (table 4.1: A) for different values of ψ_1, with $a = b = 0.6$ mm. As expected, when the pitch angle decreases, the phase velocity also decreases. Again, except at low frequencies, the match between the calculated and simulated results is very good. Furthermore, as the pitch angle increases, the actual structure deviates more from the sheath-helix model, and the accuracy of the MEDC method reduces.

A comparison of unshielded and shielded normal helix configurations, both with a vacuum tunnel (table 4.1: B and C), is shown in figures 4.12(a) and (b). For the shielded configuration, we consider two values of the metal shield separation from the planar helix, $t_4 = 0.3$ and 0.5 mm. As can be seen in figure 4.12(a), the metal shield can reduce the phase velocity within lower frequency ranges and hence produce a flatter phase velocity curve compared to that of the unshielded configuration. The phase velocity curve can be further flattened by reducing the shield separation t_4. These results show that the presence of the metal shield causes a higher concentration of the field in the dielectric region ε_{r2}, even at low frequencies.

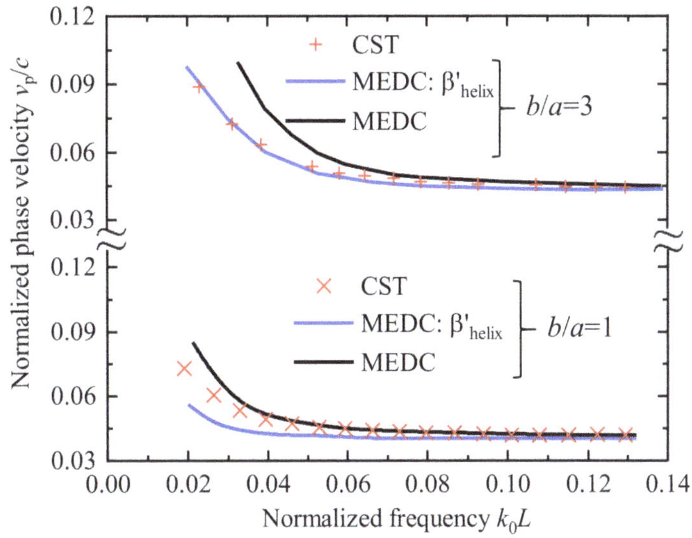

(a)

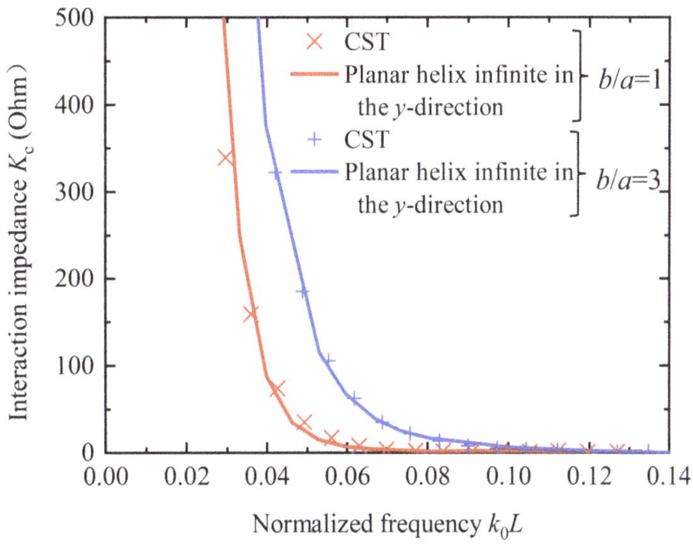

(b)

Figure 4.10. (a) Normalized phase velocity and (b) average interaction impedance of the unshielded normal helix (table 4.1: A) for $b/a = 1$ and 3 with fixed L and cross-sectional perimeter ($\varepsilon_{r1} = 1$, $\varepsilon_{r2} = 3.02$). © [2011] IEEE. Reprinted, with permission, from [8].

Figure 4.12(b) shows that the metal shield also reduces the variation of interaction impedance over the bandwidth.

Figures 4.13(a) and (b) show the normalized phase velocities and interaction impedances of the shielded normal helix, shielded inverted helix, and shielded helix

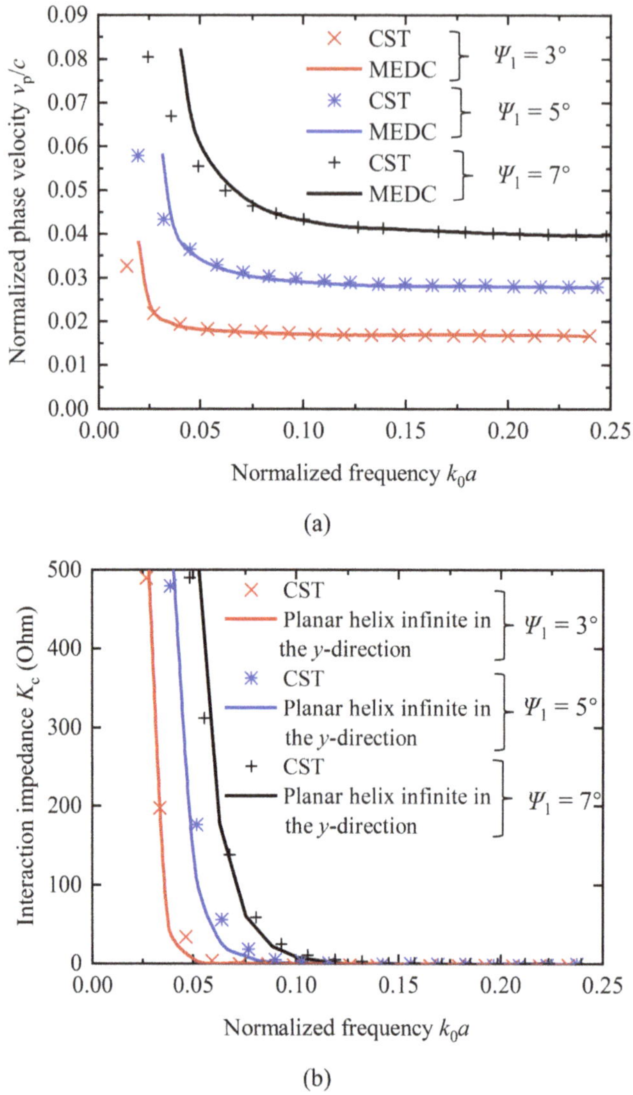

Figure 4.11. (a) Normalized phase velocity and (b) interaction impedance of the unshielded normal helix (table 4.1: A) for $\psi_1 = 3°$, $5°$, and $7°$ with $a = b = 0.6$ mm, $\varepsilon_{r1} = 1$, and $\varepsilon_{r2} = 3.02$. © [2011] IEEE. Reprinted, with permission, from [8].

configurations, all with a vacuum tunnel (table 4.1: C, D, and E). Among these configurations, the shielded normal helix with a vacuum tunnel has the widest bandwidth; however, it also has the largest variation in phase velocity and interaction impedance over the bandwidth. The shielded inverted helix and shielded helix with a vacuum tunnel can produce much flatter phase velocity curves due to the dielectric loading outside the UC screens. Also, for the latter configuration, a larger amount of dielectric material surrounding the helix structure causes a lower phase

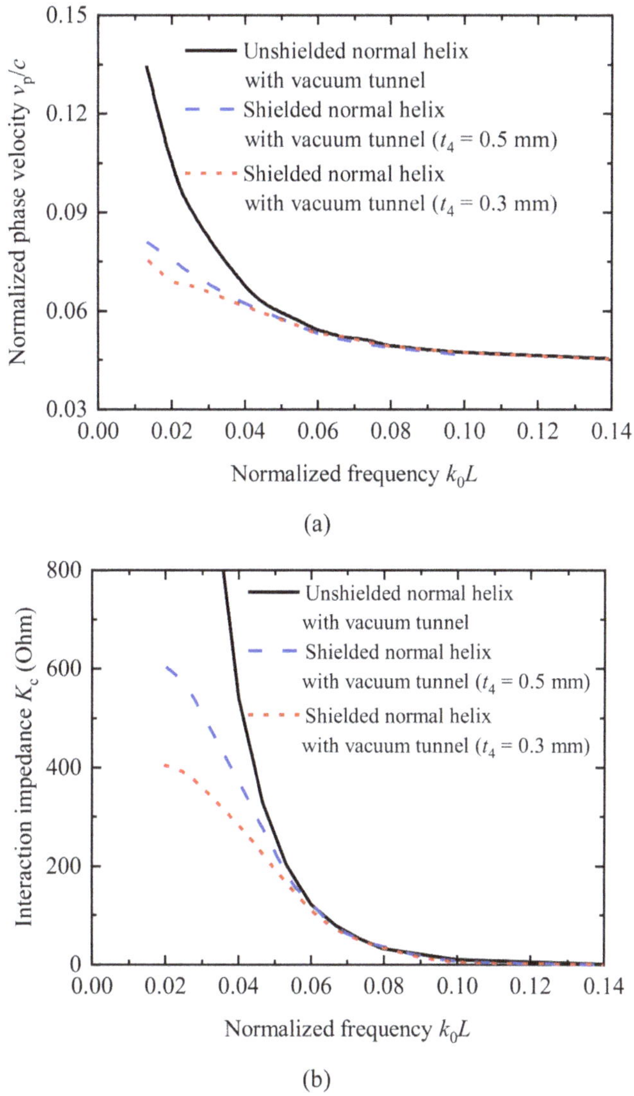

Figure 4.12. Comparison between the unshielded normal helix with a vacuum tunnel and the shielded normal helix with a vacuum tunnel: (a) normalized phase velocity and (b) interaction impedance (table 4.1: B and C). © [2011] IEEE. Reprinted, with permission, from [8].

velocity. The interaction impedance of the shielded helix with a vacuum tunnel is very similar to that of the shielded inverted helix with a vacuum tunnel.

Figures 4.14(a) and (b) show the effect of varying the dielectric constants ε_{r2} and ε_{r3} for the shielded helix with a vacuum tunnel (table 4.1: E). As expected, a higher dielectric constant surrounding the helix, e.g. $\varepsilon_{r2} = \varepsilon_{r3} = 6.15$, reduces the phase velocity, interaction impedance, and the bandwidth. Figure 4.14(a), shows the results for this configuration. The flattest phase velocity curve occurs at $\varepsilon_{r2} = 3.02$

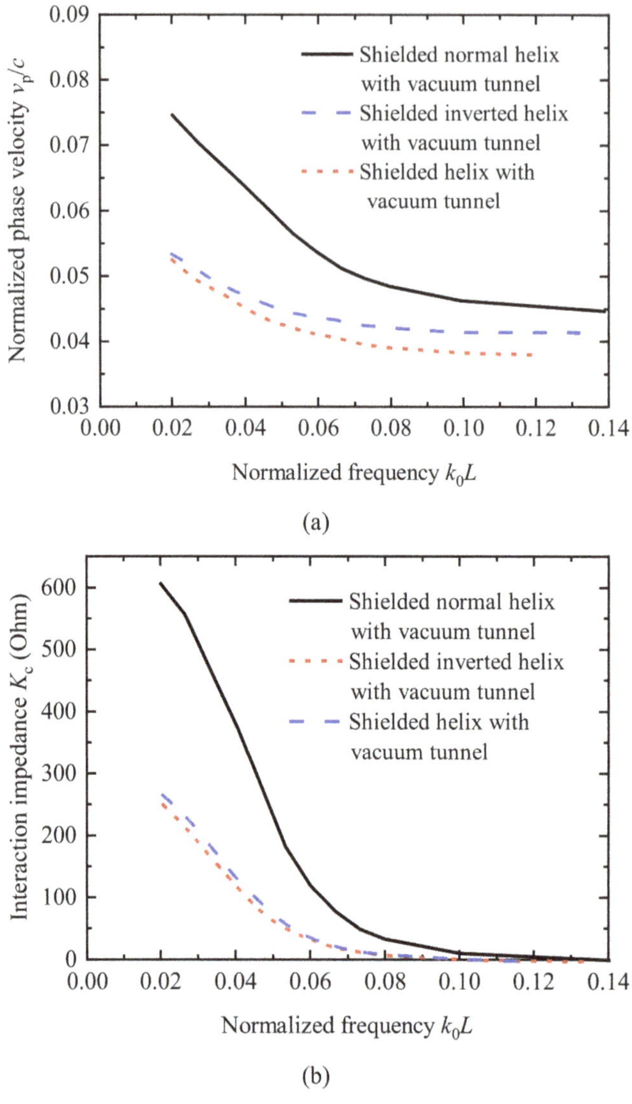

Figure 4.13. Comparison between the shielded normal helix with a vacuum tunnel, the shielded inverted helix with a vacuum tunnel, and the shielded helix with a vacuum tunnel: (a) normalized phase velocity and (b) interaction impedance (table 4.1: C, D, and E). © [2011] IEEE. Reprinted, with permission, from [8].

and $\varepsilon_{r3} = 6.15$. For this case, the bandwidth with a ±5% variation is 85.7% (for a k_0L range of 0.04 to 0.1). It should be possible to achieve an even flatter phase velocity curve by optimizing parameters such as the aspect ratio, the metal shield distance, and the dielectric constant of the material surrounding the helix. The results in figure 4.14(b) also indicate that the value of the interaction impedance remains significant while the phase velocity curve becomes flatter.

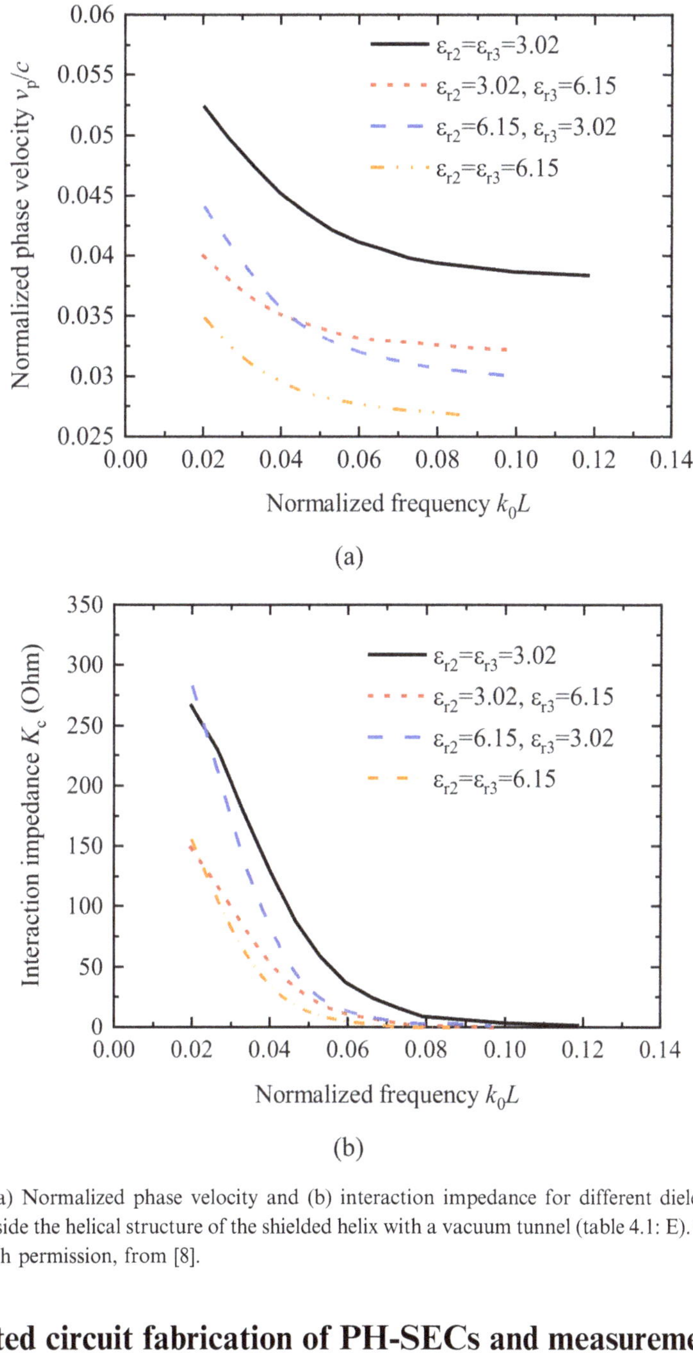

Figure 4.14. (a) Normalized phase velocity and (b) interaction impedance for different dielectric constants inside and outside the helical structure of the shielded helix with a vacuum tunnel (table 4.1: E). © [2011] IEEE. Reprinted, with permission, from [8].

4.3 Printed circuit fabrication of PH-SECs and measurement results

The analytical and simulated results for the cold-test parameters of the PH-SECs immersed in air and multilayer dielectric substrates were presented in the previous sections. The realization of the PH-SEC SWSs using printed circuit fabrication is described in this section. The design of a wideband CPW feed for the unshielded normal helix structure (table 4.1: A), together with measured results, is presented to confirm the ease of fabrication, low loss, and wideband potential of the PH-SEC. The measured S-parameters and phase velocity of the fabricated structures over the frequency range 0.5–10 GHz are compared with simulation results obtained using CST MWS [6].

Figure 4.15(a) shows part of the schematic of a CPW feed integrated with an unshielded normal helix structure. A photograph of the fabricated structure is shown in figure 4.15(b). The helical structure is printed on a Roger RO3203 substrate with a dielectric constant of $\varepsilon_{r2} = 3.02$. The environment surrounding the fabricated structure is air ($\varepsilon_{r1} \approx 1$). The top and bottom helical conductor strips have a width of 1 mm in the z-direction. The vias that act as straight-edge connections are realized by plated-through hole (PTH) technology with a 0.5 mm via diameter. A CPW section starts from a via end of the helix structure, followed by a tapered section to join a 50 Ω CPW section, which is connected to a standard SubMiniature version A (SMA) connector. The length of the tapered CPW section and the impedance of the initial CPW section are important parameters used to achieve wideband impedance matching. In addition, a reduction in 'b' and an increase in 'L' for the first turn can improve the matching by lowering the helix impedance. The CPW ground planes should extend for a few turns of the helix for better matching at low frequency. The detailed dimensions of the designed CPW feed are given in the caption of figure 4.15 (a). In order to provide same-side feeds and to accommodate an electron gun and collector, the CPW feeds at both ends of the helix structure are at right angles to the

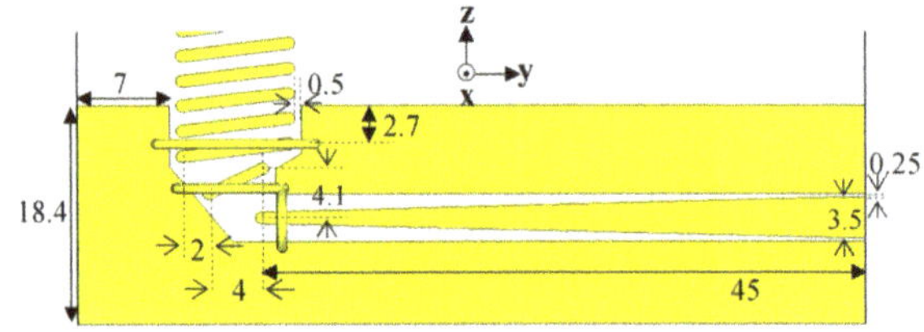

(a)

(b)

Figure 4.15. (a) Schematic of the proposed CPW feed. (b) Photo of the fabricated structure integrated with 25 periods of unshielded normal helix (table 4.1: A: $a = 0.762$ mm, $b = 4$ mm and $L = 2.1$ mm). All dimensions are in millimeters. © [2010] IEEE. Reprinted, with permission, from [6].

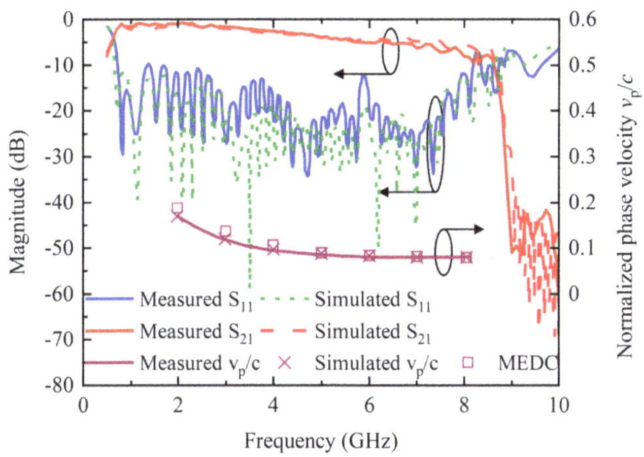

Figure 4.16. S-parameters and normalized phase velocities of the fabricated structure shown in figure 4.15. © [2010] IEEE. Reprinted, with permission, from [6].

helix structure. Air bridges are added to ensure that the CPW ground planes are at the same potential.

Figure 4.16 presents the measured and simulated S-parameters of the complete structure shown in figure 4.15(b) over a frequency range from 0.5 to 10 GHz. The simulation results were obtained using the CST MWS Transient solver and include dielectric and conductor losses. The measured S-parameters match the simulated ones very well. The frequency range over which S_{11} is below -10 dB extends from 1 to 8 GHz, i.e. three octaves. We measured the attenuations of two fabricated structures that were identical except for the number of helical turns (25 and 30). The results, obtained using the difference in the magnitude of S_{21}, are 0.02, 0.05, and 0.1 dB mm^{-1} for 4, 5, and 6 GHz, respectively. The attenuation values compare well with the typical attenuation for circular helices [19]. The attenuation increases as the frequency increases, mainly due to the dielectric loss in the substrate and the conductor loss in the PTHs.

Figure 4.16 also shows the measured phase velocity values. The phase velocity is calculated from the measured phase constant, which is obtained from the per-unit-length phase difference in the phase values of S_{21} for the two fabricated structures. The measured phase velocity values match well with the simulated values as well as the calculated values obtained using the MEDC method.

To accommodate an electron beam, a tunnel is needed inside the SWS. This leads to the unshielded normal helix with an air tunnel (table 4.1: B). This structure was also fabricated, and measurements were carried out over a frequency range of 0.5–7 GHz [14]. Once again, the measured S-parameters and phase velocity results match the simulated values very well.

More design and fabrication results are described in detail in chapter 8. These include the fabrication of a C/X-band PH-SEC using printed circuit techniques and the microfabrication of Ka-band and W-band PH-SEC SWSs.

4.4 Effect of periodicity on the PH-SEC

The results of the study of the PH-SEC in free space, as well as in the multilayer dielectric environment described in sections 4.2 and 4.3, are based on the sheath-helix approximation. To account for the effects of periodicity and width of the conductors in the infinitely wide planar helix, the 'tape-helix' model was employed in section 3.3. The same model is employed in this section for a more accurate analysis of the PH-SEC. The tape-helix model considers the helix to be a perfectly conductive tape with finite width and infinitesimal thickness. The tape width is considered to be much smaller than the pitch. The description in this section follows [7, 20].

Although the analysis presented here considers the PH-SEC to be situated in free space and ignores the effects of the supporting dielectric material and metal shielding, it does provide a simple and fast means to investigate the effects of conductor (tape) width and space harmonics in the phase velocity and interaction impedance of PH-SEC.

The analysis in section 3.3, as described below, considered different space harmonic distributions in the y- and z-directions, leading to a complex derivation with a double infinite summation for field expressions. More recently, a tape-helix analysis of the rectangular helix (figure 4.1(a)) has been reported in [21]. This approach makes use of the modified Marcatili's method and average power flow matching at boundaries. This tape-helix analysis considers the confinement of the rectangular helix in both x- and y-directions in a single step, which leads to a complex analysis and a rather complicated characteristic equation. Compared to [2], the approach described in [7] reduced the complexity of the derivation by assuming that the fields have the same space harmonic distribution in both y- and z-directions. The complexity was further reduced by obtaining the characteristic equation by simply equating the electric field parallel to the strips to zero, whereas [2] uses the rather complicated variational technique for deriving the characteristic equation. Compared to [21], the analysis is simplified by using the EDC method described in section 4.2.1.1. In this method, the analysis is carried out in two steps; the original three-dimensional (3D) waveguide is replaced by two related 2D structures, each of which is easy to analyze.

Parts of this section have been reprinted, with permission, from [7]. © [2018] IEEE.

4.4.1 Infinitely wide periodic planar helix

In the application of the EDC method, the first of the two 2D structures is the infinitely wide structure corresponding to the PH-SEC shown in figure 4.2. This structure takes into account the x-dependent profile of the PH-SEC; it captures the 'slowing-down' effect of the planar helix and yields a propagation constant denoted by β_1. The confinement in the y-direction is considered in the next step.

As shown in figure 4.17, the infinitely wide periodic planar helix can be modeled as a pair of arrays of parallel, straight, and perfectly conductive strips. The strips in the top and bottom arrays are oriented at angles of $+\psi$ and $-\psi$ with respect to the y-axis, respectively. The top and bottom arrays are separated by a distance $2a$ in the

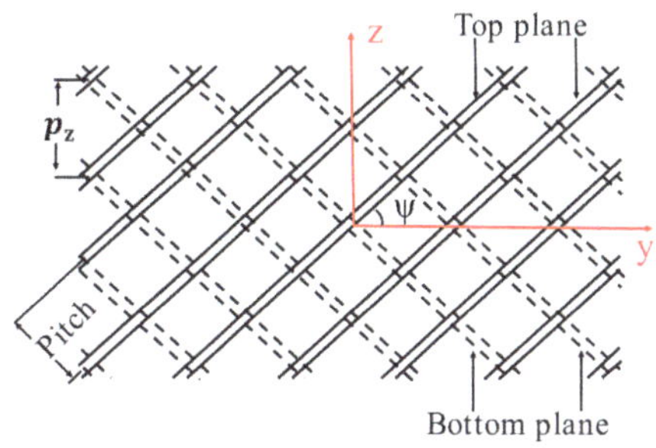

(a)

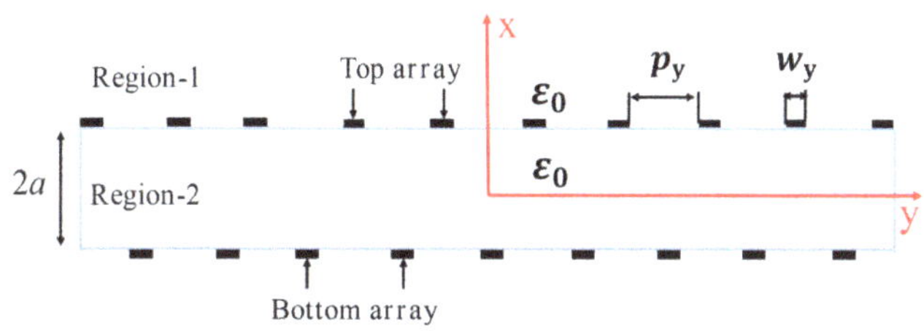

(b)

Figure 4.17. Plan view (a) and cross-sectional view (b) of an infinitely wide periodic planar helix. © [2018] IEEE. Reprinted, with permission, from [7].

x-direction and are infinitely wide in the y-direction. The thickness of the strips is assumed to be infinitesimally small. The periodic spacings between the strips are p_y and p_z, and the strip widths are w_y and w_z along the y-and z-directions, respectively. Wave propagation is assumed to take place in the z-direction.

The only difference with respect to figure 3.5 is that here, both regions 1 and 2 are considered to be free space. The invariant transformations, or symmetry properties, expressed in (3.10) remain the same, and the structure supports two independent solutions: transverse symmetric and transverse antisymmetric. For the operation of TWTs, we are interested in the transverse antisymmetric solution, since this solution corresponds to longitudinal symmetric fields, i.e. non-zero values of E_z in the region around the z-axis.

Floquet theory and Helmholtz's equation are used to write the general expressions for longitudinal fields in regions 1 and 2. The expressions for the longitudinal electric and magnetic fields for the transverse antisymmetric (longitudinal symmetric) mode are:

$$E_{z1} = \sum_{n=-\infty}^{\infty} A_n \exp\{-u_n(x-a)\}\exp\left\{j\frac{2\pi n y}{p_y}\right\}\exp\{-j\beta_n z\} \tag{4.23}$$

$$E_{z2} = \sum_{n=-\infty}^{\infty} B_n \cosh(u_n x)\exp\left\{j\frac{2\pi ny}{p_y}\right\}\exp\{-j\beta_n z\} \tag{4.24}$$

$$H_{z1} = \sum_{n=-\infty}^{\infty} C_n \exp\{-u_n(x-a)\}\exp\left\{j\frac{2\pi ny}{p_y}\right\}\exp\{-j\beta_n z\} \tag{4.25}$$

$$H_{z2} = \sum_{n=-\infty}^{\infty} D_n \cosh(u_n x)\exp\left\{j\frac{2\pi ny}{p_y}\right\}\exp\{-j\beta_n z\} \tag{4.26}$$

In (4.23)–(4.26), A_n, B_n, C_n, and D_n are the unknown amplitude constants for the nth space harmonic; β_n is the propagation constant of the nth space harmonic, which is related to the propagation constant of the fundamental space harmonic β_0 as $\beta_n = \beta_0 + (2\pi n/p_z)$, and u_n is the decay constant in the x-direction, obtained from the following equation:

$$u_n^2 = \left(\frac{2\pi n}{p_y}\right)^2 + \beta_n^2 - k_0^2. \tag{4.27}$$

In (4.27), $k_0 = \omega(\mu_0\varepsilon_0)^{1/2}$ is the wave number in vacuum at the angular frequency ω.

Due to symmetry, we need only consider the structure for $x \geqslant 0$. The continuity of the tangential electric field and the discontinuity of the tangential magnetic field at $x = a$ lead to the following boundary conditions:

$$E_{z1,\,n}(x = a) = E_{z2,\,n}(x = a) \tag{4.28a}$$

$$E_{y1,\,n}(x = a) = E_{y2,\,n}(x = a) \tag{4.28b}$$

$$[H_{z2,\,n}(x = a) - H_{z1,\,n}(x = a)] = J_{y,\,n} \tag{4.28c}$$

$$[H_{y1,\,n}(x = a) - H_{y2,\,n}(x = a)] = J_{z,\,n} \tag{4.28d}$$

where $J_{y,\,n}$ and $J_{z,\,n}$ are the surface current densities corresponding to the nth space harmonic on the $x = a$ interface in the y- and z-directions, respectively. For a strip width much smaller than the pitch, the current in the strips can be considered to be flowing only along the strips. We can then express $J_{y,\,n}$ and $J_{z,\,n}$ in terms of $J_{\|n}$, the current flowing parallel to the strips corresponding to the nth space harmonic:

$$J_{z,\,n} = J_{\|n} \sin\psi \tag{4.29a}$$

$$J_{y,\,n} = J_{\|n} \cos\psi. \tag{4.29b}$$

Similar to the tape-helix analysis of the circular helix in [22], $J_{\|n}$ can be obtained by finding the Fourier components of the total current on the strips. The total current is assumed to have a constant magnitude J along the strip width while flowing parallel to the strips. The current is also assumed to have a phase variation according to $e^{-j\beta_0 z}$. The expression for $J_{\|n}$ is:

$$J_{\|n} = \frac{Jw_z}{p_z}\text{sinc}\left(\frac{\beta_n w_z}{2}\right). \tag{4.30}$$

The unknown amplitude constants obtained by applying the boundary conditions are:

$$A_n = \frac{-k_c^2 \cosh(u_n a)(j\omega\mu_0)u_n J_{\|n}}{u_n^2 k^2 e^{u_n a}}\left[\sin\psi + \frac{1}{k_c^2}\left(\frac{2\pi n}{p_y}\right)\beta_n \cos\psi\right] \tag{4.31a}$$

$$B_n = \frac{A_n}{\cosh(u_n a)} \tag{4.31b}$$

$$C_n = \frac{-J_{\|n}\cos\psi\sinh(u_n a)}{e^{u_n a}} \tag{4.31c}$$

$$D_n = \frac{-C_n}{\sinh(u_n a)} \tag{4.31d}$$

where $k_c^2 = k_0^2 - \beta_n^2$.

The characteristic equation for determining β_1 is obtained using the condition $E_{1\|}(x = a) = 0$ at the center of the strips [22]. $E_{1\|}$ is the component of the electric field intensity parallel to the strips and is given by:

$$E_{1,\|} = \sum_{n=-\infty}^{+\infty} E_{1,\|n} = \sum_{n=-\infty}^{+\infty}\left(E_{y1,n}\cos\psi + E_{z1,n}\sin\psi\right). \tag{4.32}$$

The following is the resulting characteristic equation for the infinitely wide periodic planar helix:

$$\sum_{n=-\infty}^{+\infty}\left\{\frac{\dfrac{(j\omega\mu_0)u_n\cos^2\psi}{k_c^2 e^{u_n a}}\left(\dfrac{Jw_z}{p_z}\text{sinc}\left(\dfrac{\beta_n w_z}{2}\right)\right)}{\left[\sinh(u_n a) - \dfrac{\cosh(u_n a)}{u_n^2 k_o^2}\left(\dfrac{2\pi n}{p_y}\beta_n + k_c^2\tan\psi\right)^2\right]}\right\} = 0 \tag{4.33}$$

Although (4.33) involves the sum of an infinite number of space harmonics, it converges with an error of less than 0.5% when the first seven space harmonics are considered.

Similar to section 4.2.1.1, the pitch angle ψ in (4.33) needs to be modified to correctly capture the slowing-down effect of the original PH-SEC [5]. The modified pitch angle ψ_{eff} is obtained as:

$$\psi_{\text{eff}} = \tan^{-1}\left(\frac{\text{pitch}}{4a + 4b}\right) \tag{4.34}$$

4.4.2 Effect of transverse confinement

Next, we use the EDC method to include the effect of the transverse confinement of the planar helix in the y-direction by straight-edge connections. The second 2D structure, which takes into account the y-dependent profile, is considered to be a dielectric slab of thickness $2b$ in the y-direction, infinite in the x-direction, and situated in free space. The dielectric constant of the slab is taken to be $\varepsilon'_{r,\,\text{eff}}$, which is calculated from the propagation constant β_1 obtained for the first structure:

$$\varepsilon'_{r,\,eff} = \left(\frac{\beta_1}{k_0}\right)^2. \tag{4.35}$$

The infinite dielectric slab in this step can support TM and TE modes, which can be even or odd with respect to y. An even variation of E_x with respect to y is considered here to support the transverse antisymmetric mode of the original SWS [5]. Thus, the characteristic equation for the TE-even mode of the dielectric waveguide is used here. The relations used to determine the propagation constant β_2 are:

$$\frac{v_1}{v_2} = \tan(v_2 b) \tag{4.36}$$

$$\beta_2^2 = v_1^2 + k_1^2 = k_{\text{eff}}'^2 - v_2^2 \tag{4.37}$$

where $k_1^2 = \omega^2\mu_0\varepsilon_0$ and $k_{\text{eff}}'^2 = \omega^2\mu_0\varepsilon_0\varepsilon'_{r,\,\text{eff}}$ are the wave numbers in the free-space and dielectric regions, respectively, while v_1 and v_2 are the y-direction decay coefficients for the free-space and dielectric regions, respectively.

The overall propagation constant is obtained by taking the weighted average of the propagation constants β_1 and β_2 using:

$$\beta = \frac{b\beta_1 + a\beta_2}{a + b} \tag{4.38}$$

4.4.3 Interaction impedance

Following the definition of interaction impedance given in chapter 1, the on-axis interaction impedance for the fundamental space harmonic in a SWS is written as:

$$K_c = \frac{\left|E_{z,\,0}^2(0)\right|}{2P\beta_0^2} \tag{4.39}$$

where $|E_{z,\,0}(0)|$ is the magnitude of the longitudinal component of the electric field at the center of the SWS for the fundamental space harmonic and P is the total RF power propagating through the SWS. The total RF power is the sum of the powers contributed by all space harmonics. We consider the sum of the power contributed by the first seven space harmonics to be the total power, since the power contributed by the higher-order harmonics is negligibly small. The power flow per unit width corresponding to the nth space harmonic for an infinitely wide planar helix, shown in figure 4.17, is given by:

$$P_{uw,\,n} = \frac{1}{2}Re[\int_x (E_{x,\,n}H^*_{y,\,n} - E_{y,\,n}H^*_{x,\,n})dx] \tag{4.40}$$

The total power flow in the PH-SEC is obtained by multiplying $P_{uw,\,n}$ by $2b$, i.e. the width in the y-direction:

$$P_n = P_{uw,\,n} \times 2b. \tag{4.41}$$

The interaction impedance of the PH-SEC is obtained by substituting the results of the phase constant of the PH-SEC from (4.38), the longitudinal field expressions for the infinitely wide planar helix given in (4.23)–(4.26), and the transverse field components obtained using (3.8) into (4.39).

4.4.4 Calculated and simulated results

The propagation constant of the fundamental space harmonic (β_0) is obtained using the EDC method described in the previous subsections. The results for the normalized phase velocity and the interaction impedance for the fundamental space harmonic of the PH-SEC, calculated based on the analysis, are compared with the simulation results from the CST MWS Eigenmode Solver in figures 4.18–4.20 for different values of the aspect ratio a/b. In each of these figures, the perimeter $4(a + b)$ of the PH-SEC is fixed at 9.6 mm. The results are given for three different pitch angles, $\psi = 2.5°$, 5°, and 7.5°, and the strip width is fixed at 20% of the pitch, which is different for each pitch angle.

Figure 4.18 shows the results for $a/b = 0.25$ and $a = 0.48$ mm, figure 4.19 shows the results for $a/b = 1$ and $a = 1.2$ mm, and figure 4.20 shows the results for $a/b = 4$ and $a = 1.92$ mm. For these different aspect ratios and pitch angles, the normalized phase velocity obtained using analysis shows a good match with simulation results, except at relatively low frequencies. It is well-known that the EDC method becomes less accurate at lower frequencies [11]. As expected, when the pitch angle decreases, the phase velocity also decreases. The interaction impedance obtained using analysis also shows a good match with simulation results. The plots show that generally higher values of the interaction impedance can be obtained from a PH-SEC by increasing the pitch angle.

Figure 4.18 also includes the normalized phase velocity and interaction impedance of the PH-SEC obtained from the sheath-helix approximation using the EDC [5]. The figure shows that the results of the tape-helix approximation using the EDC provide a more accurate estimate compared to those produced using the sheath-helix

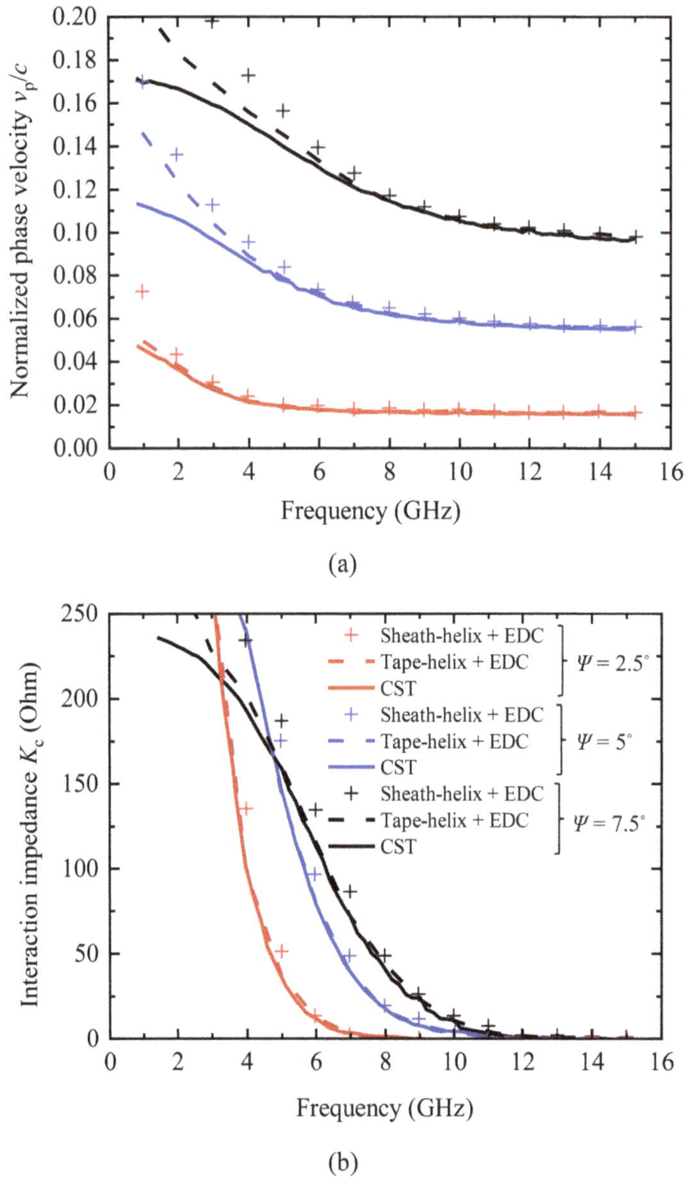

Figure 4.18. Comparison between simulated and analytical results for a PH-SEC for three different pitch angles ($\psi = 2.5°$, 5°, 7.5°) with $a/b = 0.25$, $a = 0.48$ mm, and $w/p_z = 0.2$. (a) Normalized phase velocity, (b) interaction impedance. © [2018] IEEE. Reprinted, with permission, from [7].

approximation and the EDC. From the results shown in figure 4.18, we observe that for $\psi = 7.5°$, the tape-helix model provides values of phase velocity higher by 20% at the lower-frequency end and higher by 4% at the higher-frequency end compared to the sheath-helix model. The interaction impedance also shows a similar percentage increase.

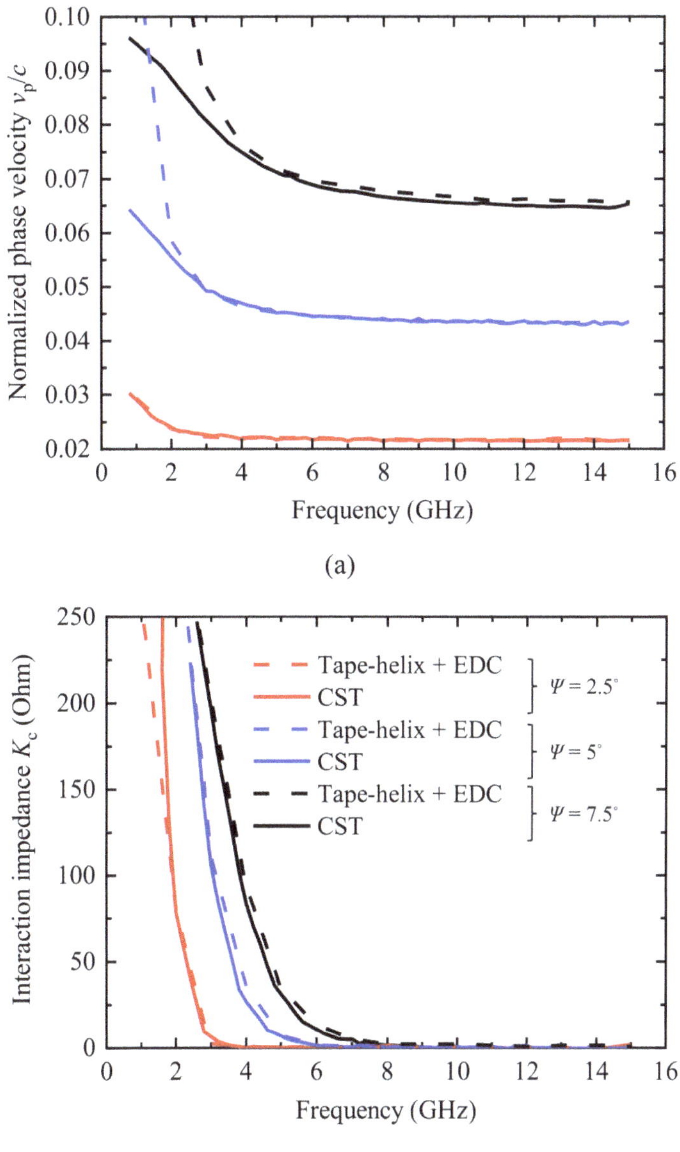

Figure 4.19. Comparison between simulated and analytical results for a PH-SEC for three different pitch angles ($\psi = 2.5°$, $5°$, $7.5°$) with $a/b = 1$, $a = 1.2$ mm, and $w/p_z = 0.2$. (a) Normalized phase velocity, (b) interaction impedance. © [2018] IEEE. Reprinted, with permission, from [7].

To enable comparison of the calculated and simulated results in the W-band frequency range, figure 4.21 shows the normalized phase velocity and interaction impedance of a PH-SEC with $a/b = 1$ and $a = 0.15$ mm for three different pitch angles ($\psi = 2.5°$, $5°$, $7.5°$). The strip width of the PH-SEC is fixed at 20% of the pitch. Similar to the previous results, the normalized phase velocity obtained using

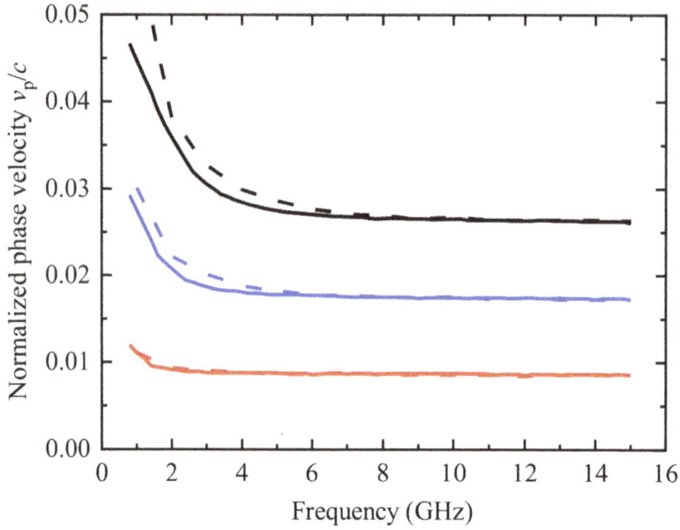

(a)

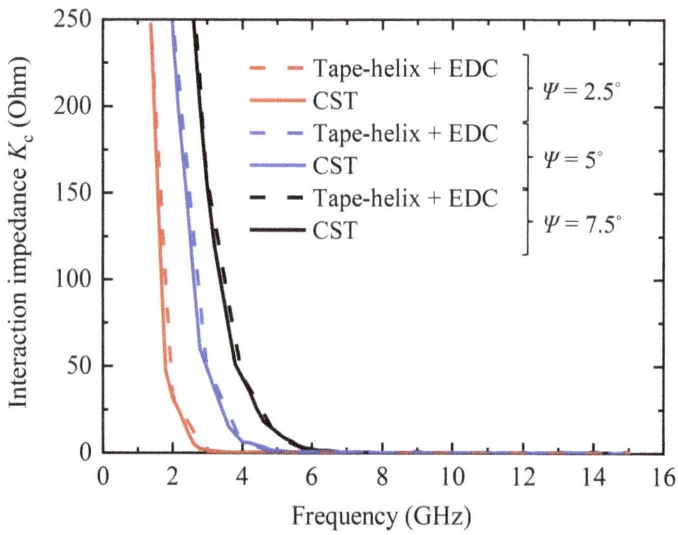

(b)

Figure 4.20. Comparison between simulated and analytical results for a PH-SEC for three different pitch angles (ψ = 2.5°, 5°, 7.5°) with a/b = 4, a = 1.92 mm, and w/p_z = 0.2. (a) Normalized phase velocity, (b) interaction impedance. © [2018] IEEE. Reprinted, with permission, from [7].

analysis also shows a good match with the simulation results, except at relatively low frequencies. Moreover, the interaction impedance results obtained from simulation and analysis also show a reasonably good match.

The dispersion diagram for a PH-SEC with $a = b = p_z = 1.2$ mm and a strip width of 0.24 mm is given in figure 4.22. This diagram is obtained from the numerical

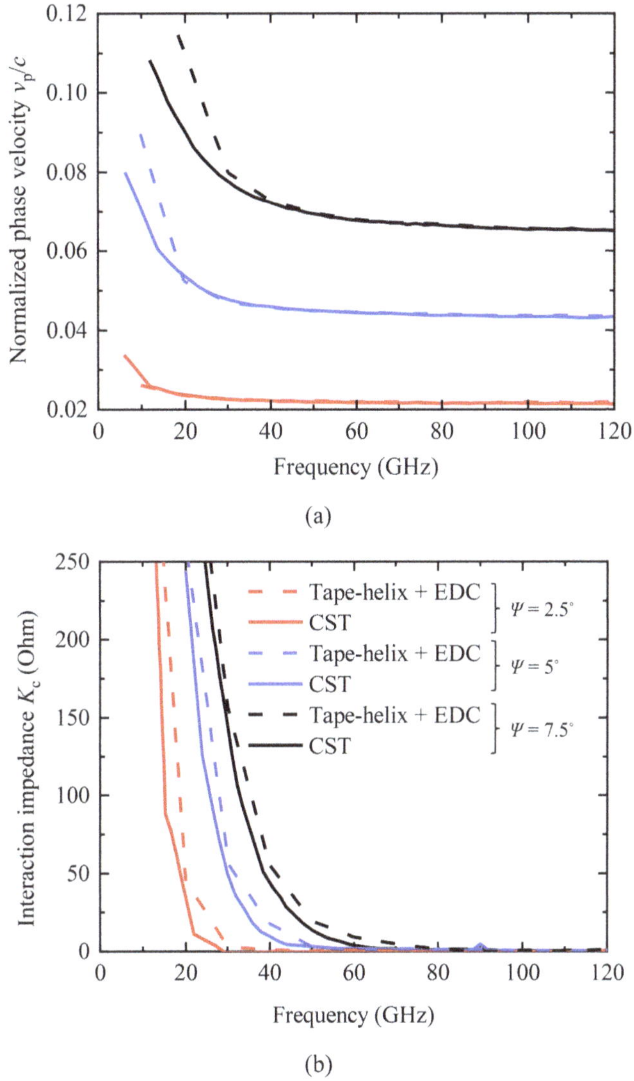

Figure 4.21. Comparison between simulated and analytical results for a PH-SEC for three different pitch angles ($\psi = 2.5°$, $5°$, $7.5°$) with $a/b = 1$, $a = 0.15$ mm, and $w/p_z = 0.2$. (a) Normalized phase velocity, (b) Interaction impedance. © [2018] IEEE. Reprinted, with permission, from [7].

solution of the analytical equations for frequencies extending beyond the pi frequency. The diagram shows the existence of passbands and stopbands in the PH-SEC; in this respect, the PH-SEC is like other periodic SWSs.

The calculated results based on our analysis are also compared with those in [2] and [21]. In both cases, the results are shown to be more accurate over a wider frequency range [20]. The analysis is also extended to a PH-SEC placed inside a metal enclosure [20]. Once again, the results from the analysis agree very well with

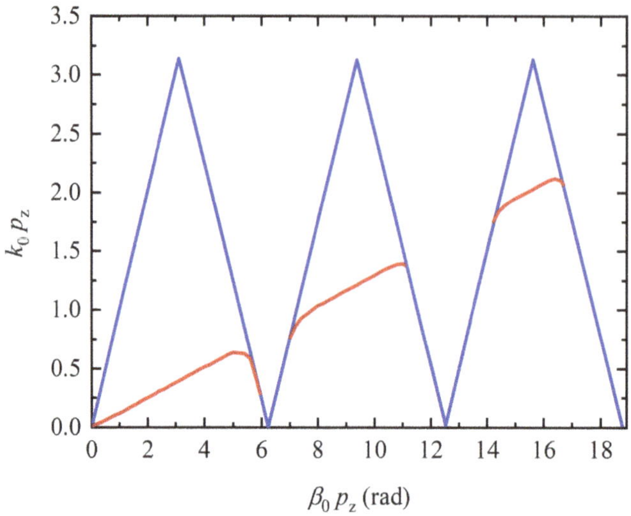

Figure 4.22. Dispersion diagram of a PH-SEC with $a/b = 1$, $a = 1.2$ mm, $w/p_z = 0.2$, and $a/p_z = 1$. © [2018] IEEE. Reprinted, with permission, from [7].

the simulation results at frequencies far from the cutoff region. The main effects of the metal enclosure are that the PH-SEC becomes less dispersive and that the interaction impedance decreases as the size of the metal enclosure decreases.

4.4.5 Validation through experiment

Since the above analysis considers a PH-SEC immersed in a homogeneous medium, for an experimental validation of the analysis, a structure was designed and fabricated that substantially emulates a PH-SEC immersed in a homogeneous dielectric medium [20]. An RO3203 substrate with a thickness of 1.524 mm, a dielectric constant of 3.02, and a loss tangent of 0.0016 is used for the fabrication; $a = 0.762$ mm, $b = 3$ mm, pitch = 2.1 mm, and strip width = 0.6 mm. PTHs with a radius of 0.25 mm are used for the straight-edge connections of the PH-SEC. The straight-edge connections and the horizontal strips of the PH-SEC are connected together using circular patches with a radius of 0.4 mm.

Two SWSs, identical apart from their lengths, were designed and fabricated. The first SWS has 25 periods of PH-SEC, while the second has 30 periods. Figure 4.23(a) shows the fabricated prototype of the SWS with 30 periods of PH-SEC. As shown in the figure, CPWs are used to connect the SWS with the SMA connectors. The width of the CPW is tapered to achieve wideband matching. In addition, different pitch lengths are used for two periods at both ends of the SWS to achieve better matching. As shown in figure 4.23(b), two layers of RO3203 substrate are attached above and below the SWS using nylon screws to approximate an SWS immersed in a homogeneous dielectric medium.

The S-parameters of the SWS are obtained from the transient solver in CST MWS. An open-boundary condition (available in this solver) is applied all around

(a)

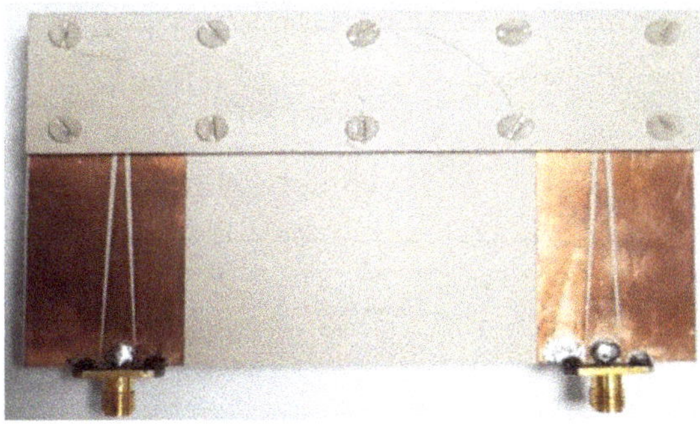

(b)

Figure 4.23. Fabricated prototype of the SWS with 30 periods. (a) The SWS before assembly of the top and bottom dielectric sheets; (b) the final SWS. Reprinted with permission from [20].

the SWS, since the structure is in free space. The width of the CPW feed line is gradually tapered to match the 50-Ohm measurement system. Waveguide ports in the simulator are used to excite the CPW feed lines. Losses in the SWS are considered by specifying the conductivity of the metal and the loss tangent of the dielectric material.

The measured S-parameters of the SWS are compared with those from simulations. As shown in figure 4.24(a), the simulated and measured S_{21} match very well over the frequency range of 2–5 GHz, which is the targeted operational range of the fabricated SWS. The measured value of S_{21} is slightly lower than the simulated one. This is attributed to the fabrication tolerance and the soldering of SMA connectors in the fabricated SWS. Also, the simulations do not include the small losses due to the SMA connectors. Another reason for the mismatch is the difference between simulated straight-edge connections and fabricated PTHs. Between 2 and 5 GHz, the simulated S_{11} values are better than -15 dB, while the measured S_{11} values are better than -12.5 dB.

Regarding the deviation between the measured and simulated S-parameter values above 5.5 GHz, it should be noted that the stopband of the periodic structure starts around this frequency (see figure 4.22). Significant changes occur in the values of

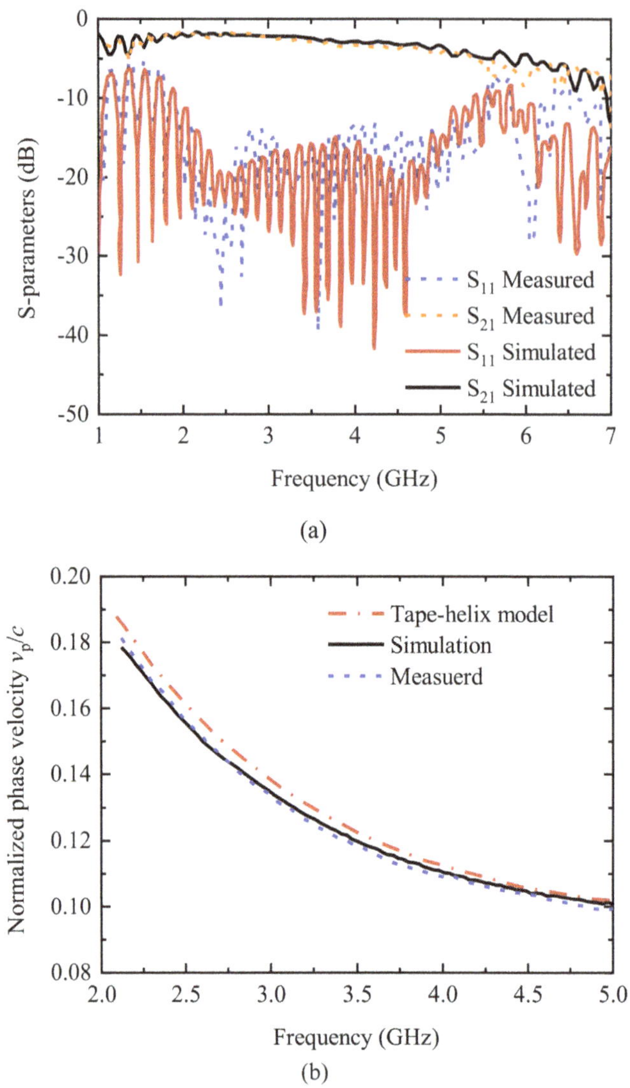

Figure 4.24. (a) Comparison between the measured and simulated S-parameters of the SWS with 30 periods. (b) Comparison between the normalized phase velocities of the PH-SEC. Reprinted with permission from [20].

S-parameters around the stopband. Due to fabrication tolerances, there is usually a slight mismatch between the stopband of the fabricated structure and the simulated one, causing the measured and simulated S-parameter values to appear to have more deviation around the stopband.

As mentioned in section 4.3, the phase velocity is estimated from the measured phase constant for the fabricated structures. This value of the phase velocity is compared with the results from tape-helix analysis and simulation in figure 4.24(b). As shown in the figure, the measured phase velocity values match well with those from the analysis. In addition, the measured values of phase velocity closely match the simulation results. The slight deviation between the phase velocity from the analysis and the measured phase velocity arises from the finite thicknesses of the top and bottom dielectric substrates in the fabricated SWSs, whereas the analysis considers a homogeneous medium extending to infinity.

The abovementioned results show that the analysis presented here can provide more accurate results than the previous sheath-helix and tape-helix analyses for the PH-SEC and the rectangular helix reported in the research literature. Also, our analysis approach is applicable to a wide range of aspect ratios and pitch angles, as well as for frequencies extending to hundreds of gigahertz.

4.5 Particle-in-cell simulation results for a TWT based on a PH-SEC

As shown in section 4.3 and chapter 8, the PH-SEC is amenable to fabrication using printed circuit as well as microfabrication technology. It is also suitable for the incorporation of a sheet electron beam. Sheet beam devices offer advantages such as higher beam current capacity, decreased beam voltage, and reduced power density compared to circular-cylindrical beam geometries. In this section, we present the PIC simulation of a PH-SEC with a sheet beam, as shown in figure 4.25 [8]. The sheet beam, assumed to be elliptical in cross section, passes through a vacuum tunnel that is centrally located inside the PH-SEC. Table 4.2 lists the main dimensional parameters.

4.5.1 Cold-test results

Figure 4.26 shows the simulated normalized phase velocity, the on-axis interaction impedance, and the average interaction impedance over the beam cross section obtained using the CST MWS Eigenmode Solver. The maximum thickness of the elliptical sheet beam is assumed to be half the height of the vacuum tunnel in order to minimize the dielectric charging effect. The 2392 V beam line intersects the operating mode at 4.2 GHz. The average interaction impedance of the PH-SEC is generally higher than the on-axis values.

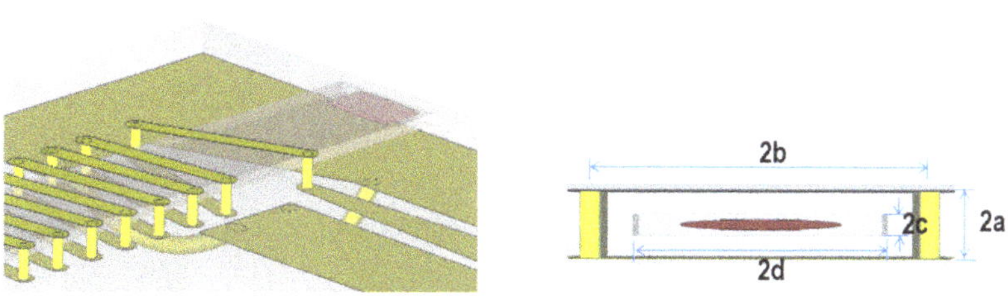

(a) (b)

Figure 4.25. Sheet electron beam with an elliptical cross section inside the PH-SEC SWS. (a) Transparent dielectric showing the sheet beam location; (b) cross-sectional view. © [2011] IEEE. Reprinted, with permission, from [8].

Table 4.2. Main circuit parameters (dimensions in millimeters).

Parameter	Value
2*a*	1.524
2*b*	8
2*c*	0.508
2*d*	6
Dielectric constant	3.02
Helix period	2.116
Via diameter	0.5
Beam height	0.25
Beam width	4

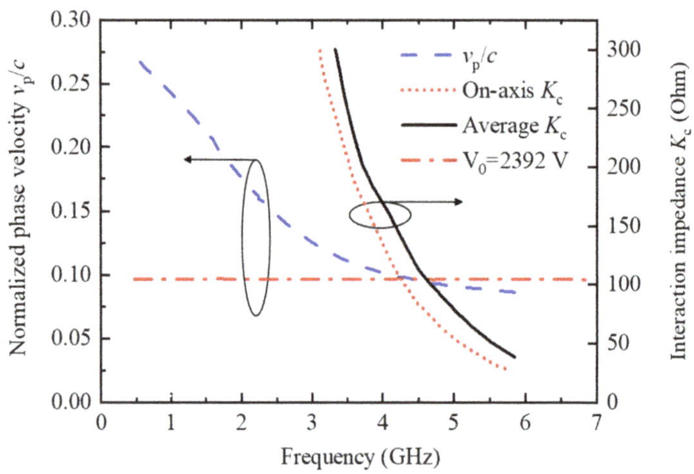

Figure 4.26. Simulated normalized phase velocity, on-axis, and average interaction impedance. © [2011] IEEE. Reprinted, with permission, from [8].

4.5.2 PIC simulation results

The beam current is set to 32 mA, and the input power is assumed to be 0.1 W. A 45-turn PH-SEC with a CPW feed is simulated in the CST PIC Solver, including the dielectric and conductor losses. As shown in figure 4.25, the elliptical emission surface and the input port for the RF signal are located at the $-z$ end. The beam propagates in the $+z$-direction with a homogeneous axial magnetic field of 0.5 T. Figure 4.27 shows the particle trajectory plot at 5 GHz. The colors indicate alternating high- and low-energy particles due to velocity modulation. Figure 4.28 shows the simulated small-signal gain over the frequency range of 4.2–5.3 GHz. The simulation includes conductor and dielectric losses. The gain (loss) values obtained without the electron beam (zero beam current) are a good match for the simulated

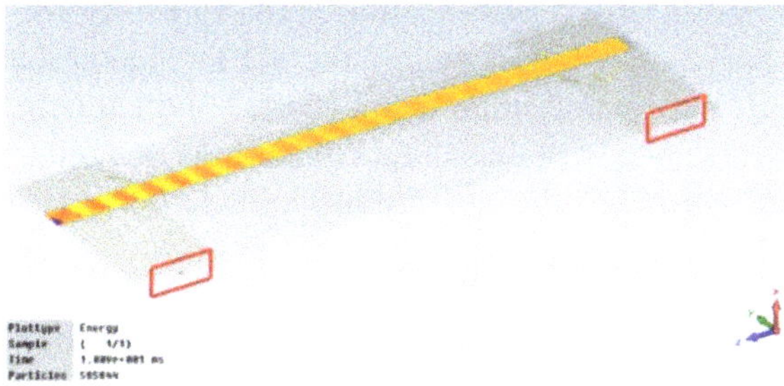

Figure 4.27. Particle trajectory plot captured at 5 GHz. © [2011] IEEE. Reprinted, with permission, from [8].

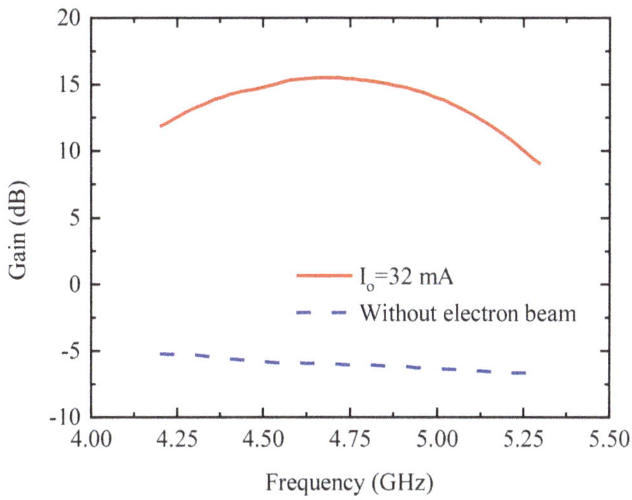

Figure 4.28. Simulated small-signal gain for a PH-SEC with 45 periods. © [2011] IEEE. Reprinted, with permission, from [8].

S_{21} of the cold structure. The maximum gain is 15 dB at 4.75 GHz for 32 mA of beam current.

More PIC simulation results for TWTs incorporating PH-SEC SWS are described in chapter 8, along with the design and fabrication of various PH-SEC SWSs. These results cover the C/X-, Ka-, and W-band frequencies.

4.6 Summary

This chapter describes the PH-SEC, which is a transversely confined and practical version of the infinitely wide planar helix SWS. Included is a field-theory-based analysis of a PH-SEC immersed in free space; the analysis is simplified by using the sheath-helix approximation and by making use of the EDC method; the latter provides accurate results at frequencies sufficiently greater than the cutoff frequencies. A similar analysis is then carried out for a PH-SEC immersed in a multilayer dielectric environment that includes a beam tunnel as well as a metal enclosure; this part of the analysis uses a simple modification of the EDC method called the MEDC method. Analysis based on field theory and the MEDC method is then extended to consider the periodic nature and finite width of the conductors in the PH-SEC. For this part of the work, the tape-helix approximation is used, which reveals the passband–stopband nature of the dispersion characteristics as well as the spatial harmonics. In most of these cases, the interaction impedance is also calculated using field expressions. To estimate the performance of a TWT based on a PH-SEC, PIC simulations are also carried out for a TWT operating around 4.75 GHz.

The results of the abovementioned analyses, namely dispersion characteristics and interaction impedance, calculated over a wide range of frequencies, compare quite well with simulation results obtained from CST. Furthermore, to demonstrate the feasibility of fabrication, two PH-SEC SWSs operating at relatively low frequencies are fabricated using printed circuit technology. The measured S-parameters and phase velocity results for these structures compare quite well with the CST simulation results.

The analytical, simulation, and measured results presented in this chapter show that the PH-SEC SWS has very good potential for application in TWTs.

Appendix A

Fields are assumed to have a variation of the form $e^{(j\omega t - j\beta z)}$. The longitudinal field components for the transverse antisymmetric mode of the general configuration in figure 4.8(b) are written as:

$$E_z = \begin{cases} A \sinh[u_4'(f - x)] & e \leqslant |x| \leqslant f \\ B_1 e^{u_3'(x-a)} + B_2 e^{-u_3'(x-a)} & a \leqslant |x| \leqslant e \\ C_1 e^{u_2'(x-c)} + C_2 e^{-u_2'(x-c)} & c \leqslant |x| \leqslant a \\ D \cosh(u_1'x) & |x| \leqslant c \end{cases} \tag{A.1}$$

$$H_z = \begin{cases} M \cosh[u_4'(f - x)] & e \leqslant |x| \leqslant f \\ N_1 e^{u_3'(x-a)} + N_2 e^{-u_3'(x-a)} & a \leqslant |x| \leqslant e \\ R_1 e^{u_2'(x-c)} + R_2 e^{-u_2'(x-c)} & c \leqslant |x| \leqslant a \\ Q \cosh(u_1'x) & |x| \leqslant c \end{cases}. \tag{A.2}$$

For the sheath-helix model, the boundary conditions are as follows:

(i) continuity of E_z, H_z, E_y and H_y at $x = \pm c$;
(ii) continuity of E_z and E_y at $x = \pm a$;
(iii) the tangential electric field is zero and the tangential magnetic field is continuous at $x = \pm a$ along the direction of conduction;
(iv) continuity of E_z, H_z, E_y, and H_y at $x = \pm e$.

References

[1] Arora R K, Bhat B and Aditya S 1977 Guided waves on a flattened sheath helix *IEEE Trans. Microw. Theory Tech.* **25** 71–2

[2] Aditya S and Arora R K 1979 Guided waves on a planar helix *IEEE Trans. Microw. Theory Tech.* **27** 860–3
[3] Aditya S and Arora R K 1979 Nonreciprocal dispersion characteristics of a planar helix on magnetized ferrite slabs *IEEE Trans. Microw. Theory Tech.* **27** 864–8
[4] Chadha D, Aditya S and Arora R K 1983 Field theory of planar helix traveling-wave tube *IEEE Trans. Microw. Theory Tech.* **31** 73–6
[5] Chua C, Aditya S and Shen Z 2009 Effective dielectric constant method for a planar helix with straight-edge connections *IEEE Electron Device Lett.* **30** 1215–7
[6] Chua C, Aditya S and Shen Z 2010 Planar helix with straight-edge connections in the presence of multilayer dielectric substrates *IEEE Trans. Electron Devices* **57** 3451–9
[7] A. Kumar M M and Sheel A 2018 Simplified tape-helix analysis of the planar helix slow wave structure with straight-edge connections *IEEE Trans. Electron Devices* **65** 2280–6
[8] Chua C, Aditya S, Shen Z, Tang M and Tsai J 2011 Planar helix with straight-edge connections and a sheet electron beam for traveling-wave tube applications *2011 IEEE Int. Vacuum Electronics Conf. (IVEC)* 519–20
[9] Fink H J and Whinnery J R 1982 Slow waves guided by parallel plane tape guides *IEEE Trans. Microw. Theory Tech.* **30** 2020–3
[10] Fu C, Wei Y, Wang W and Gong Y 2008 Dispersion characteristics of a rectangular helix slow-wave structure *IEEE Trans. Electron Devices* **55** 3582–9
[11] Knox R M and Toulios P P 1970 Integrated circuits for the millimeter through optical frequency range *Proc. Symp. on Submillimeter Waves* (Brooklyn: Polytechnic Press) 497–516
[12] Arora R 1966 Surface waves on a pair of parallel undirectionally conducting screens *IEEE Trans. Antennas Propag.* **14** 795–7
[13] Chen C L 2007 *Foundations for Guided-Wave Optics* (New York: Wiley)
[14] Chua C S 2012 *Studies on Planar Helical Slow-Wave Structures for Travelling-Wave Tube Applications* (Singapore: Nanyang Technological University)
[15] Chiang K S 1986 Dual effective-index method for the analysis of rectangular dielectric waveguides *Appl. Opt.* **25** 2169
[16] Chadha D, Aditya S and Arora R K 1984 Study of planar-helix slow-wave structure for application to TWTs *IEE Proceedings, Part H - Microwaves, Optics and Antennas* **131** 14–20
[17] Harrington R F 1961 *Time-Harmonic Electromagnetic Fields* (New York: McGraw-Hill)
[18] Pierce J R 1950 *Travelling Wave Tubes* (Princeton, NJ: D. Van Nostrand)
[19] Gilmour A S 1986 *Microwave Tubes* (Boston, MA: Artech House)
[20] Kumar M M N A 2019 *Application of Planar Helix Slow-Wave Structure in Backward-Wave Oscillators* (Singapore: Nanyang Technological University)
[21] Wei W, Wei Y, Wang W, Zhang M, Gong H and Gong Y 2015 Dispersion equations of a rectangular tape helix slow-wave structure *IEEE Trans. Microw. Theory Tech.* **63** 1445–56
[22] Sensiper S 1955 Electromagnetic wave propagation on helical structures (a review and survey of recent progress) *Proc. IRE* **43** 149–61

Chapter 5

Dispersion control in planar slow-wave structures

5.1 Introduction

For slow-wave structures (SWSs), a relatively flat dispersion curve is important for wideband operation [1], and negative dispersion can help to reduce the in-band harmonic content of a wideband TWT [2]. Therefore, dispersion-shaping techniques for SWSs have been studied by many researchers. The dispersion and the interaction impedance often act as a pair of opposing parameters. When dispersion is low, the value of interaction impedance K_c tends to be low as well. One of the challenges in dispersion shaping is preventing the interaction impedance from becoming too low, since that would reduce the gain and efficiency of the TWT.

Many studies have examined the dispersion shaping of circular helices, including those by Paik [3], Galuppi *et al* [4], Kumar *et al* [5], Ghosh *et al* [6], as well as more recent ones [7–9]. The dispersion-shaping techniques that have been studied for the circular helix include bringing the metal shield closer to the helix, reducing the dielectric constant of the rods that support the helix inside the shield, incorporating metal vanes of different shapes in the shield, and coating the dielectric support rods with metal.

Printed circuit techniques are important for miniaturization as well as low-cost mass production. Microfabrication techniques become important at high frequencies of operation, where the dimensions become small; this will become an important issue as future microwave systems operate at higher frequencies [10, 11]. The microstrip meander-line (MML) SWS and the planar helix with straight-edge connections (PH-SEC) are two kinds of popular planar SWSs that are quite compatible with printed circuit or microfabrication techniques. The dispersion control or dispersion shaping of the microfabricated structures is important for

doi:10.1088/978-0-7503-5764-7ch5

enhancing the operating bandwidth. This chapter focuses on the dispersion control techniques used with these two kinds of SWSs.

5.2 Dispersion control in meander-line SWSs

The dispersion characteristics of the MML SWS, which is composed of meandering microstrip lines, can be attributed to two main factors. First, the dispersion properties of the MML SWS are affected by the dispersion of the microstrip line itself. For low-frequency applications, a quasi-static approximation of the microstrip line is commonly employed for analysis. In this approximation, the electromagnetic wave propagating on the microstrip line is treated as a pure transverse electromagnetic (TEM) wave, and the phase velocity of the MML SWS does not vary significantly with frequency. However, as the operating frequency becomes higher, the dispersion of the microstrip line becomes increasingly significant. Second, as has been introduced in section 2.2, there are couplings between adjacent metal strips that greatly contribute to the dispersive characteristics of the MML SWS. The dispersion properties of both microstrip lines and coupled microstrip lines have been studied in the research literature [12–15]. It has been proven that the dispersion properties of the coupled microstrip lines are related to parameters including the strip width, the distance between strips, the thickness of the dielectric substrate, etc. Therefore, dispersion control or dispersion shaping of the MML SWS can be achieved by paying attention to these parameters.

The perspective view of an MML SWS presented in figure 5.1 provides a schematic representation of the structure along with its dimensional parameters. The values of the parameters are consistent with those in section 2.3, namely $\varepsilon_r = 6.5$, loss tangent $= 4 \times 10^{-4}$, $h = 0.15$ mm, $l = 0.8$ mm, $w = 0.08$ mm, $s = 0.1$ mm, and the thickness of the metal is 0.006 mm. The study of dispersion in this section is based on this illustrative structure.

First of all, we consider the width w of the metal strips. Figure 5.2(a) shows the dispersion of an MML SWS for different values of w. In general, a wider strip leads to increased dispersion in the microstrip line, according to equation (2.5) for dispersive microstrips. Although the effect is not particularly significant in this specific structure, a slight increase in dispersion can be observed when w changes from 0.04 to 0.08 mm. In addition to studying the dispersion properties, it is also

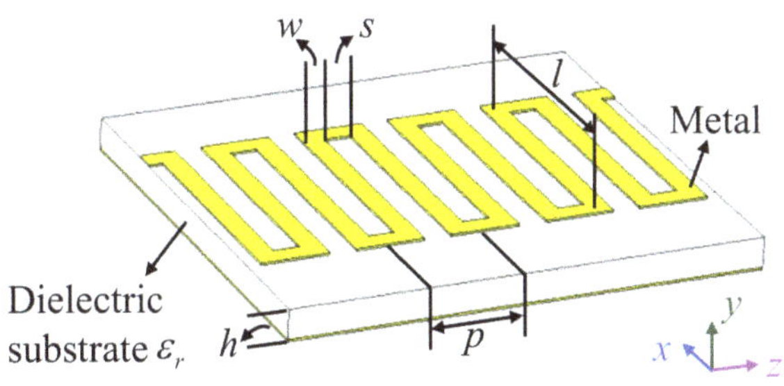

Figure 5.1. Configuration of an MML SWS.

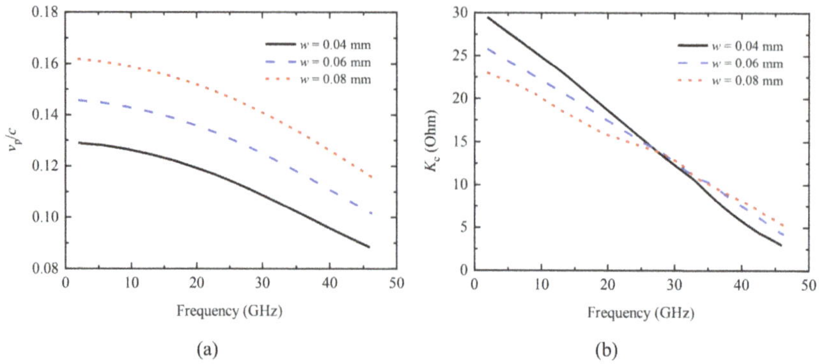

Figure 5.2. (a) Dispersion and (b) interaction impedance of the MML SWS for different values of *w*.

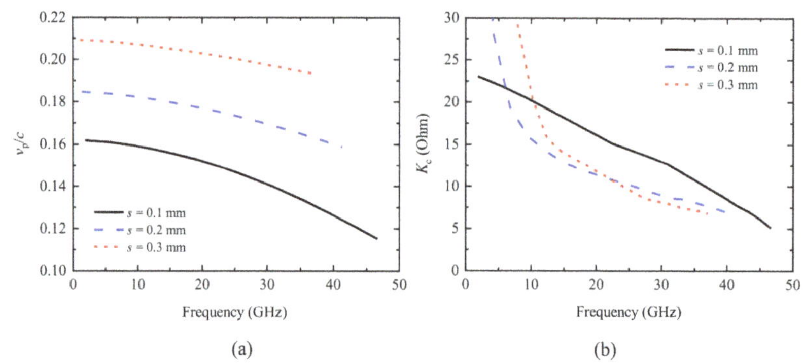

Figure 5.3. (a) Dispersion and (b) interaction impedance of the MML SWS for different values of *s*.

crucial to evaluate the interaction impedance of the MML SWS. The interaction impedance at the center, located 0.13 mm above the substrate, is also evaluated, and the results are presented in figure 5.2(b). The interaction impedance results for varying values of w are seen to be similar to each other.

Another factor that affects the dispersion of the MML SWS is the coupling between adjacent metal strips. The parameter s represents the distance between adjacent metal strips. The coupling is greater when the distance between the adjacent metal strips is smaller. Figure 5.3 shows how the dispersion properties change with varying s. As s changes from 0.1 to 0.3 mm, the phase velocity shown in figure 5.3(a) increases. This is because the increase of s appears as an increase in the period length p which equals $2w + 2s$. A longer period results in a reduced value of the propagation constant and higher phase velocity. In addition, the SWS is less dispersive when s is higher, i.e. when the coupling between the metal strips is lower. The interaction impedance values for varying values of s are presented in figure 5.3 (b). It is worth noting that the interaction impedance can increase as s gets smaller, i.e. when the SWS exhibits greater dispersion.

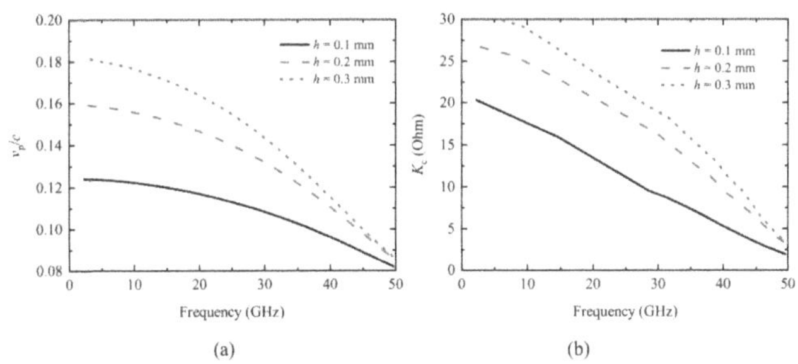

Figure 5.4. (a) Dispersion and (b) interaction impedance of the MML SWS for different values of h.

Furthermore, figure 5.4(a) depicts the changes in the phase velocity vs. frequency curve for varying substrate thickness h. It can be seen that the SWS becomes less dispersive when the substrate becomes thinner. This phenomenon can be explained by the electric field distribution in the SWS. First, when the thickness of the substrate is reduced, the ground plane is brought closer to the metal strip, resulting in a greater concentration of the fields within the dielectric [1]. Consequently, the phase velocity of the microstrip line decreases, with a more significant effect at lower frequencies than at higher frequencies. Thus, a thinner substrate leads to less dispersion. Second, when the substrate is thinner and the electric field is more concentrated in the substrate, the coupling between strips is reduced, further diminishing the dispersion of the SWS. The interaction impedance values for various values of h are presented in figure 5.4(b). While a thinner substrate yields a flatter dispersion curve, it is accompanied by a decrease in interaction impedance.

In summary, the dispersion of the MML SWS is significantly influenced by several parameters. By adjusting these parameters, the dispersion of the SWS can be effectively controlled. At the same time, it is important to consider the trade-off between the operating bandwidth and the interaction impedance for practical applications.

5.3 Dispersion control in PH-SEC SWSs

Since metal vanes have been found very effective in dispersion shaping for the circular helix SWS, this section first reports the use of metal vanes of different shapes for dispersion shaping of the PH-SEC SWS and compares the results with those for the circular helix. Furthermore, the PH-SEC offers an additional feature that can be used to modify the dispersion, namely the possibility of adding coplanar ground planes to the surfaces of the dielectric substrates which support the PH-SEC [16]. We also describe the effect of metal vanes combined with coplanar ground planes on the dispersion characteristics of the PH-SEC.

The abovementioned techniques of dispersion control are applied to the design of a Ka-band PH-SEC SWS that is based on a symmetric configuration of the PH-SEC. Both the cold-test and hot-test parameters of the designed structure are investigated. The content of this section is based on [17–19].

5.3.1 Circular helix and PH-SEC immersed in free space

This study examines how three types of metal vanes—solid, T-shaped, and thin—affect the dispersion characteristics and interaction impedance K_c of the PH-SEC. The corresponding structures for the circular helix and PH-SEC, both immersed in free space, are shown in figures 5.5 and 5.6, respectively. The dimensional parameters of the circular helix are as follows: the pitch length is 2.092 mm, the helix tape width is 1.45 mm, and the helix tape thickness is 0.2 mm. The metal shield's inner radius and the helix's inner radius are 6.24 and 3.32 mm, respectively, resulting in a shield–helix spacing of 2.72 mm. The vane–helix spacing is 1.26 mm. Simulation results have been obtained over the frequency range of 0.2–7 GHz and are presented in figure 5.7(a). As observed in the past, e.g. [7], the addition of vanes to the circular helix can produce a flatter phase velocity vs. frequency curve with only a moderate reduction of the interaction impedance, while T-shaped vanes provide flatter dispersion characteristics.

For the PH-SEC, a square cross section is considered, keeping the pitch and the perimeter of the helix as well as the inner perimeter of the shield the same as those of the circular helix. The shield–helix spacing and the helix–vane spacing are also kept

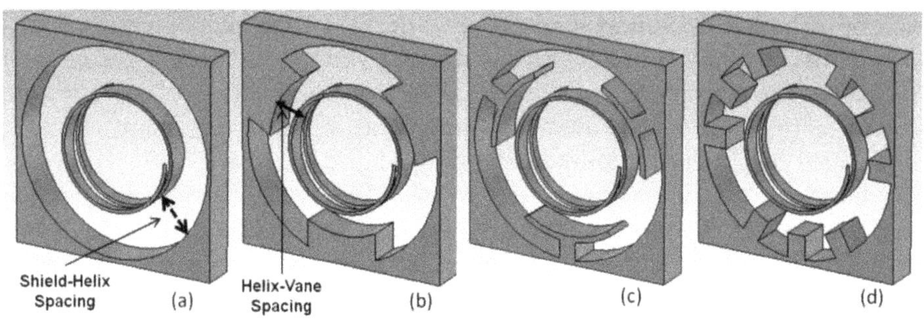

Figure 5.5. Model of a single-turn circular helix: (a) without vanes, (b) with solid vanes, (c) with T-shaped vanes, and (d) with thin vanes. © [2015] IEEE. Reprinted, with permission, from [17].

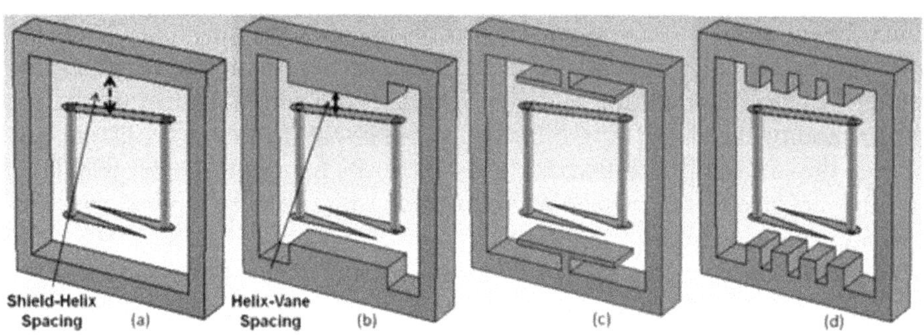

Figure 5.6. Model of a single-turn PH-SEC: (a) without vanes, (b) with solid vanes, (c) with T-shaped vanes, and (d) with thin vanes. © [2015] IEEE. Reprinted, with permission, from [17].

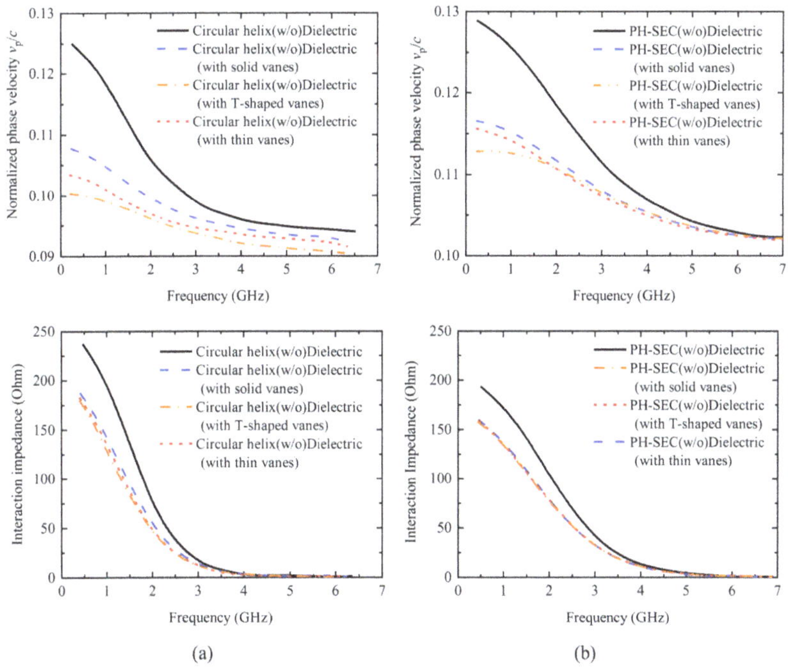

Figure 5.7. Dispersion and interaction impedance characteristics of three types of vanes incorporated in the metal shield: (a) circular helix, and (b) PH-SEC. © [2015] IEEE. Reprinted, with permission, from [17].

the same as those of the circular helix. The other dimensional parameters are as follows: the via (straight-edge connection) diameter is 0.5 mm, the helix strip width is 1 mm, and the helix strip thickness is 0.017 mm. The simulation results are presented in figures 5.7(a) and (b), respectively, for the circular helix and the PH-SEC. These results clearly show that the metal vanes have an influence on the phase velocity and interaction impedance values that is very similar to the case of the circular helix. For both SWSs, the T-shaped vanes produce the flattest phase-velocity curve, with only a moderate reduction in the interaction impedance. In the case of the PH-SEC, the reduction in the phase velocity is somewhat less, since the vanes are present only on the top and bottom of the SWS, not on the sides.

Comparing the PH-SEC with a square cross section to the circular helix reveals that when both structures have the same number of turns, along with comparable dispersion characteristics and interaction impedance, the cross section of the former is smaller by a factor of $\pi/4$. As a result, the square PH-SEC may have a relatively low output power. Of course, the main advantage of the PH-SEC is its compatibility with printed circuit and microfabrication techniques.

5.3.2 PH-SEC with dielectric substrates

5.3.2.1 Dispersion control techniques

When the PH-SEC is realized using printed circuit techniques, the effect of dielectric substrates on the dispersion characteristics can be significant. This is similar to the

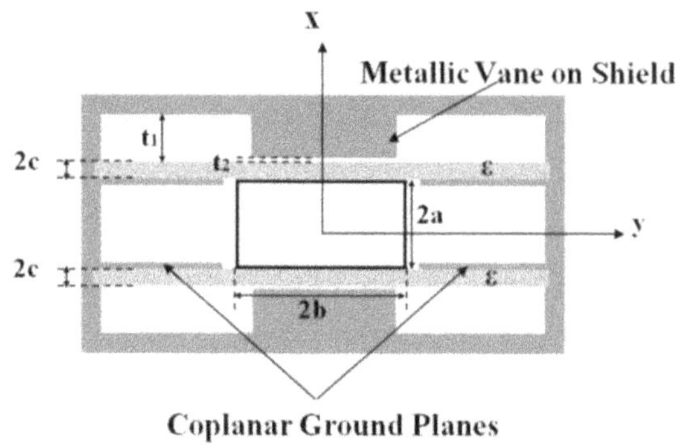

Figure 5.8. Cross-sectional view of the PH-SEC, including metallic vanes and coplanar ground planes on the dielectric substrates. The dielectric substrates are on the outer side of the PH-SEC. © [2015] IEEE. Reprinted, with permission, from [17].

case of the circular helix in the presence of dielectric support rods. While the vacuum-compatible substrate materials are alumina ($\varepsilon_r = 9.1$), beryllia ($\varepsilon_r = 6.5$), diamond ($\varepsilon_r = 5.7$), silicon ($\varepsilon_r = 11.9$), etc. we have chosen to use the non-vacuum-compatible Rogers RO4003 substrate ($\varepsilon_r = 3.55$) in this illustrative study due to the ease of subsequent proof-of-concept fabrication.

The cross section of the configuration chosen for this study, together with the dimensional parameters, is shown in figure 5.8. The following features of the configuration are noteworthy. Since the sheet beam offers many advantages for high-frequency TWTs [20], the aspect ratio b/a is kept greater than 3. The dielectric substrates are 'outside' the PH-SEC; this is similar to the configuration used in circular helix TWTs that have dielectric support rods outside the helix. Compared to the configuration in which the dielectric substrates are 'inside' the PH-SEC, the proposed configuration can reduce the dielectric charging problem when an electron beam flows through the SWS. In addition, previous studies of the dispersion characteristics of the PH-SEC have shown that the proposed configuration produces a flatter phase-velocity curve compared to the configuration in which the dielectric substrates are 'inside' the PH-SEC [21]. The metal vanes considered are the 'solid' type, once again due to their ease of fabrication. Figure 5.8 also shows coplanar ground planes on the 'inner' surfaces of the dielectric substrates. Such coplanar ground planes have been shown to provide a relatively flat phase-velocity curve [16] and are easy to integrate with a coplanar waveguide (CPW) feed.

The structural dimensions chosen are as follows: the height of the PH-SEC $2a = 2.44$ mm, the width of the PH-SEC $2b = 8.0$ mm (so that the aspect ratio $b/a = 3.28$), the substrate thickness $2c = 0.813$ mm, the helix period $= 2.092$ mm, the via diameter $= 0.5$ mm, the helix strip width $= 1$ mm, the thickness of the helix strips and coplanar ground planes $= 0.017$ mm, the shield–substrate spacing $t_1 = 1.95$ mm, the vane–substrate spacing $t_2 = 1.0$ mm, and the lateral gap between the coplanar ground planes and the PH-SEC is 0.2 mm. The internal width of the shielding enclosure is 22 mm.

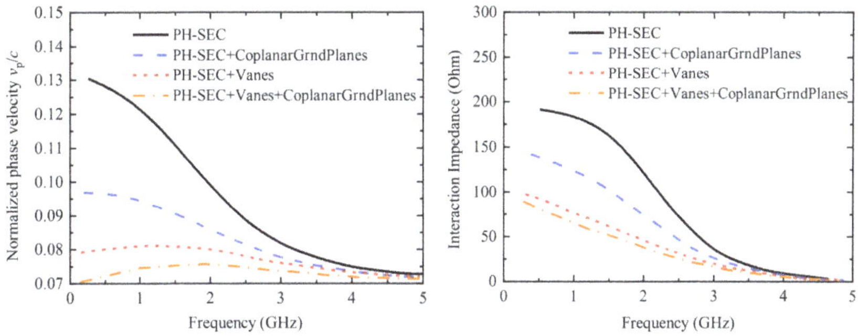

(a) (b)

Figure 5.9. (a) Dispersion and (b) interaction impedance characteristics of the PH-SEC illustrated in figure 5.8, showing the effect of coplanar ground planes, metal vanes, and coplanar ground planes together with metal vanes. © [2015] IEEE. Reprinted, with permission, from [17].

The simulation results for the phase velocity and interaction impedance of the PH-SEC printed on dielectric substrates are shown in figures 5.9(a) and (b), respectively, for the frequency range of 0.2–5 GHz. Four different possibilities are considered:

(i) the metal vanes and the coplanar ground planes are both absent;
(ii) only the coplanar ground planes are present;
(iii) only the metal vanes are present;
(iv) both the metal vanes and the coplanar ground planes are present.

For case (i), for the chosen dimensions, the low-frequency values of the phase velocity and interaction impedance are close to those of the PH-SEC in free space; however, as may be expected due to the presence of the dielectric substrates, there is more dispersion, and the values of both the phase velocity and interaction impedance decrease rather sharply at the high-frequency end. In general, compared to the PH-SEC in free space, the addition of vanes and/or coplanar ground planes in the presence of dielectric substrates leads to a greater reduction in the phase velocity and interaction impedance values. This effect can be attributed to the fact that metal vanes and coplanar ground planes concentrate the field in a region that contains the dielectric material.

The dispersion effects mentioned above can be countered to some extent by using thinner dielectric substrates and ensuring that the value of the dielectric constant is minimized. Moreover, dielectric substrates can be chosen to have a high value of thermal conductivity to facilitate the dissipation of heat from the SWS in a TWT. In this context, a promising choice for dielectric substrates would be diamond produced using chemical vapor deposition (CVD), which was proposed in [22]; it has a relatively low dielectric constant ($\varepsilon_r = 5.7$), can be made in very thin self-supporting membranes that have very good mechanical strength, and possesses high thermal conductivity.

As seen in figure 5.9(a), it is possible to achieve flatter dispersion characteristics in the presence of dielectric substrates compared to those of the PH-SEC in free space. Both coplanar ground planes (case ii) and metal vanes (case iii) are effective in doing this individually, the latter producing stronger flattening and even slightly negative dispersion at the low-frequency end. When both features are used together, it is possible to produce even more negative dispersion at the low-frequency end with only a small further reduction in the interaction impedance, as seen in figure 5.9(b). As mentioned earlier, negative dispersion can help to reduce the in-band harmonic content of a wideband TWT [2]. The degree of dispersion shaping achieved here can be improved further by optimizing the various dimensions and material parameters.

5.3.2.2 Experimental verification

Figures 5.10(a) and (b) show the configuration chosen for fabrication and testing to provide a proof of concept. The configuration with RO4003 substrates on the inner side of the planar helix and solid vanes is chosen due to its ease of fabrication. First, the model shown in figure 5.10(a) is simulated in the CST Microwave Studio (MWS) Eigenmode Solver to achieve reasonable dispersion and interaction impedance characteristics. Subsequently, the model shown in figure 5.10(b) is simulated in the CST MWS transient-mode solver; this model incorporates CPW input and output feeds and also considers material loss.

The structure is fabricated using three layers of substrates, each with a thickness of 0.813 mm. The inclined strips of the PH-SEC and the CPW feed are fabricated on the outer surfaces of the top and bottom layers, and the middle layer helps to achieve the overall height of the PH-SEC. The straight-edge connections (vias) of the PH-SEC are realized using silver-plated copper wire, which is inserted through 0.5 mm diameter holes and soldered to complete the PH-SEC. Figure 5.11 shows the CPW feed design and the associated dimensions. The CPW feed tapers from the helix end

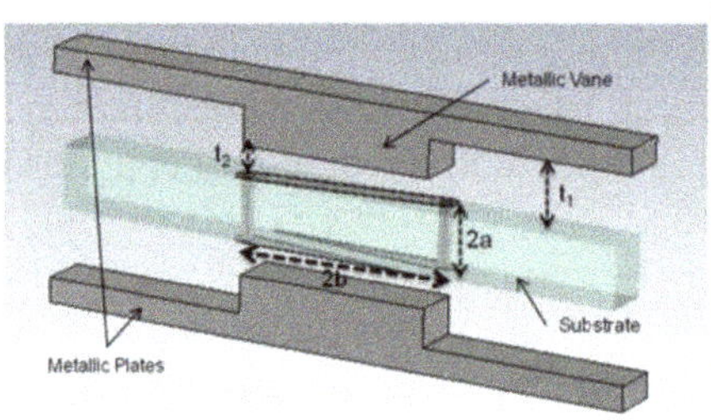

(a)

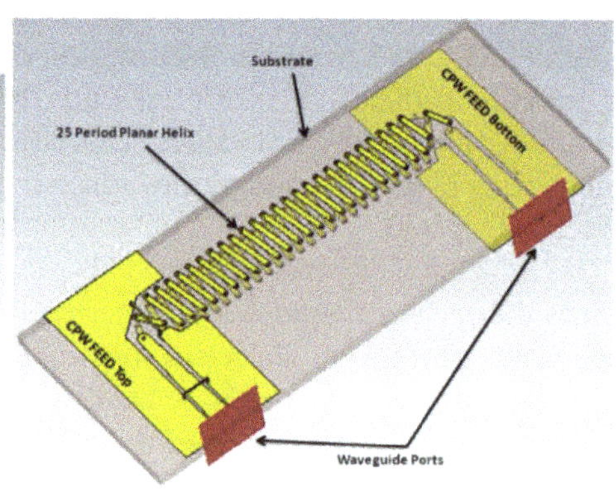

(b)

Figure 5.10. (a) Single period PH-SEC with solid metallic vanes; (b) PH-SEC on Rogers RO4003 substrates, including the CPW feed (shield with metal vanes is not shown). See table 5.1 for dimensions. The dielectric substrates are inside the PH-SEC. © [2015] IEEE. Reprinted, with permission, from [17].

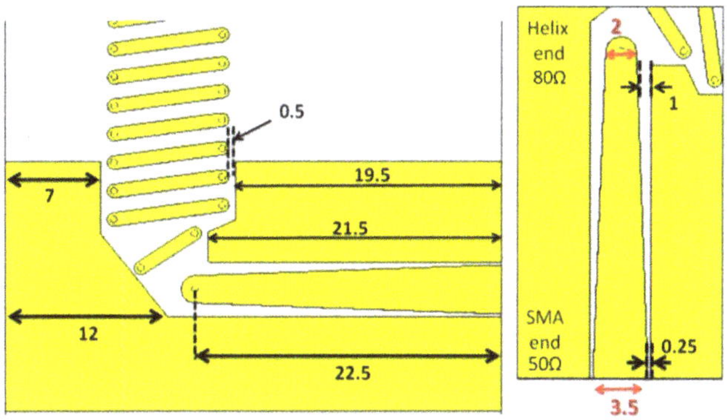

Figure 5.11. Schematic of the CPW feed design (all dimensions in millimeters). © [2015] IEEE. Reprinted, with permission, from [17].

Table 5.1. Dimensions of the fabricated structure [17].

Parameter	Value
$2a$	2.44 mm
$2b$	8 mm
Shield–substrate spacing (t_1)	3 mm
Helix period (S)	2.09 mm
Via diameter (VD)	0.5 mm
Helix strip width (SW)	1 mm
Helix strip thickness	0.017 mm
Substrate thickness	2.44 mm
ε_r (RO4003)	3.55
Vane width	8 mm
Vane–helix spacing (t_2)	1.263 mm
Substrate length	89 mm
Substrate width	36.5 mm

to a 50 Ω end connected to a SubMiniature version A (SMA) connector. The dimensions are listed in table 5.1.

Figure 5.12 depicts the dispersion and interaction impedance characteristics obtained by simulating the structure using the CST MWS Eigenmode Solver. The results show that a flatter dispersion curve is attained as the vane–helix spacing reduces. For smaller spacings of 0.763 and 0.263 mm, negative dispersion can also be obtained. However, this comes at the cost of significantly reduced interaction impedance. In addition, such close spacing values are difficult to achieve using the simple fabrication/assembly techniques adopted for this experiment. Therefore, the vane–substrate spacing used for fabrication is 1.263 mm.

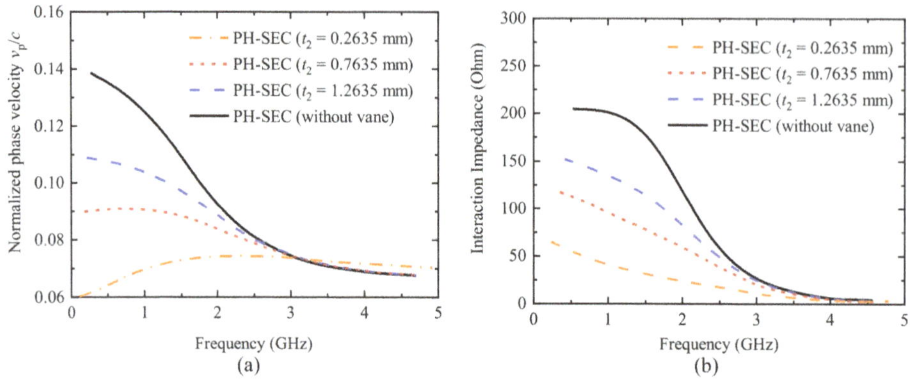

Figure 5.12. (a) Dispersion and (b) interaction impedance characteristics of the PH-SEC incorporating solid metal vanes (the structure shown in figure 5.10) for different values of the vane–substrate spacing. © [2015] IEEE. Reprinted, with permission, from[17].

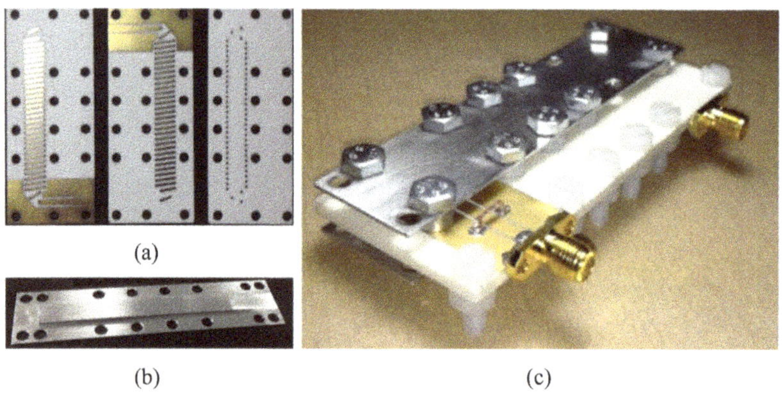

Figure 5.13. Photograph of (a) the three substrate layers, (b) one of the metal plates with a solid vane and (c) the fabricated structure with 25 periods. © [2015] IEEE. Reprinted, with permission, from [17].

A photograph of one of the fabricated structures is shown in figure 5.13. The three substrate layers are held together by nylon nuts and bolts. Air bridges are soldered to maintain the CPW ground planes at the same potential. Two aluminum plates with solid vanes extend along the length of the substrates. The vanes on the plates are tapered, as shown in figure 5.13, to avoid contact with the air bridges. Spacers 3 mm tall are used to support the metal plates and ensure the required vane–helix spacing. In order to short-circuit the metal plates and the ground planes, steel screws, nuts, and spacers are used to complete the assembly.

The simulation results for the S-parameters of the structure described above were obtained over a frequency range of 0–7 GHz using the CST MWS transient mode solver. Two otherwise identical structures, with 25 and 30 periods, respectively, are designed and fabricated to permit the calculation of the phase velocity from the measured phase of S_{21}. The measured and simulated S-parameter results are presented in figure 5.14. A simulated S_{11} value of less than -10 dB is observed

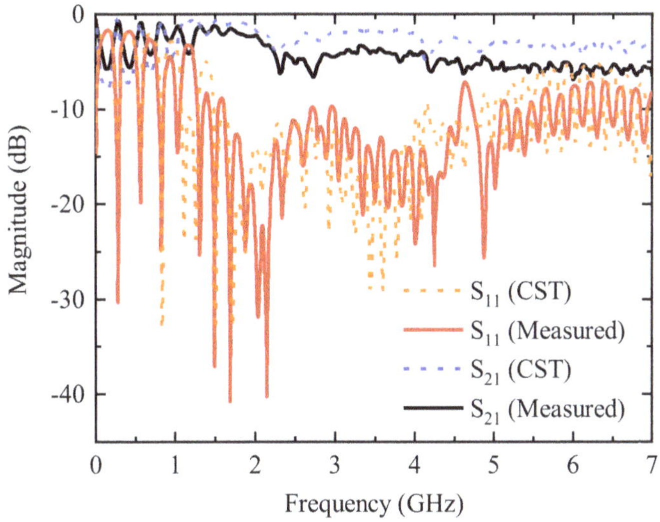

Figure 5.14. Measured and simulated S-parameters for the PH-SEC with 25 periods. © [2015] IEEE. Reprinted, with permission, from [17].

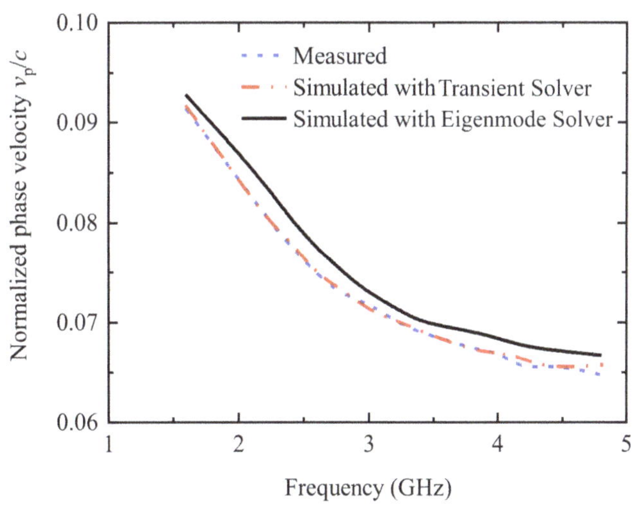

Figure 5.15. Comparison between the measured and simulated normalized phase velocities of the PH-SEC. © [2015] IEEE. Reprinted, with permission, from [17].

over the frequency range of 1.6–4.2 GHz; the measured S_{11} value also covers a similar frequency range. The shape of the measured S_{21} curve matches that of the simulation results quite well, but a higher attenuation is observed compared to the simulation values. This is attributed to the fact that the simulation values assume the bulk conductivity of copper and also ignore the additional loss that may be caused by the soldered joints. The measured phase-velocity characteristics, depicted in figure 5.15, closely match the simulation results obtained using the transient mode solver. The phase-velocity simulation results obtained using the CST MWS Eigenmode Solver, also included in figure 5.15, are slightly higher, since these do not consider loss.

5.3.3 Wideband PH-SEC suitable for microfabrication

The abovementioned techniques of dispersion control are applied to a novel Ka-band PH-SEC based on the symmetric PH-SEC [23]. Ka-band TWTs find applications in systems such as satellite communications and high-resolution military radars. In this section, both the cold-test parameters (propagation constant, phase velocity, interaction impedance, attenuation, feed design, etc.) and hot-test parameters including a sever (gain, output power, efficiency, etc.) are studied. Additional work on the same structure, such as its fabrication process, heat dissipation, as well as dielectric breakdown, is described in chapter 8.

5.3.3.1 Cold-test parameters and feed design

Parts of this section have been reprinted, with permission, from [18]. © [2016] IEEE.

As shown in figure 5.16, the proposed SWS consists of a PH-SEC sandwiched between two quartz substrates whose outer surfaces are coated with a good conductor and thus constitute the metal shield. Four coplanar ground planes are arranged beside the PH-SEC. Quartz is selected, since it has a relatively low dielectric constant of 4.43 and a very low loss tangent of 3×10^{-5} at 30 GHz. Moreover, quartz is vacuum compatible and amenable to microfabrication. A trench is made in the quartz substrates to reduce the dielectric loading effect that

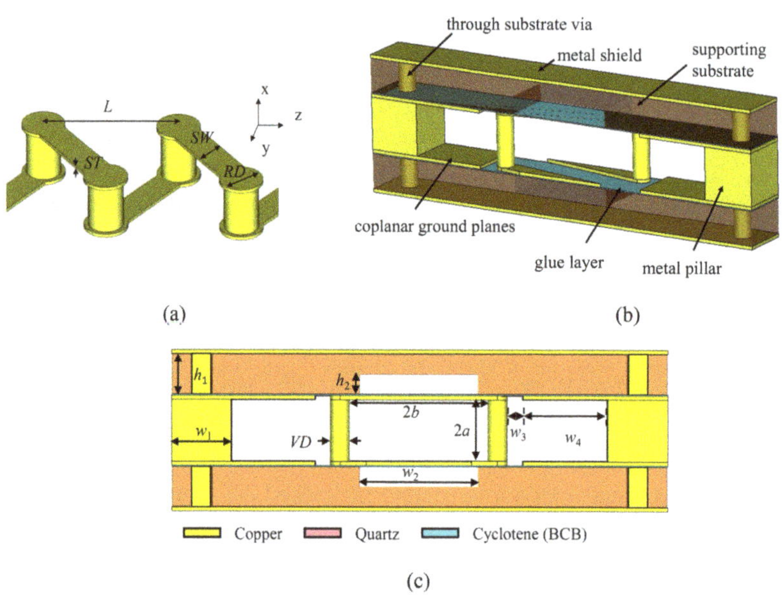

Figure 5.16. (a) Configuration of the PH-SEC SWS. (b) Perspective view of the proposed PH-SEC SWS. (c) Cross section of the proposed SWS. © [2016] IEEE. Reprinted, with permission, from [18].

Table 5.2. Dimensions of the proposed PH-SEC [23].

Parameter	Value in micrometers (μm)
$2a$	300
$2b$	700
w_1	300
w_2	600
w_3	100
w_4	390
L	300
ST	20
SW	120
VD	100
RD	120
h_1	200
h_2	100

would otherwise decrease the interaction impedance. The coplanar ground planes and the metal shield are maintained at the same potential with the help of copper sidewalls and through-substrate vias. The copper sidewalls also make the structure mechanically strong. For copper, as is common in the research literature, a conductivity of 2.9e7 S m^{-1} is used in the simulations. The dimensional parameters of the SWS are shown in figure 5.16. The dimensions used in the simulations are listed in table 5.2. As this structure is planned to be microfabricated, the feasibility of microfabrication guides the design.

Here, two measures are taken to reduce the dispersion of the SWS. First, the metal shield is placed directly on the outer surfaces of the quartz substrates. With such close spacing of the metal shield, the dispersion curve can be quite flat. Second, four coplanar ground planes are included to further reduce dispersion. The dispersion characteristics of the proposed SWS are simulated using the CST MWS Eigenmode Solver. Figure 5.17(a) shows the phase velocity of the proposed structure. The phase velocity shows only a small variation over a very wide frequency range. The interaction impedance is shown in figure 5.17(b) and has a relatively high value of about 25 Ω at 30 GHz.

The flat dispersion characteristics of the proposed structure are expected to provide velocity synchronism with an e-beam over a wide bandwidth. Furthermore, its flat dispersion characteristics imply that the electromagnetic field distribution does not change significantly with frequency. Therefore, it should be easy to obtain good values of the S-parameters without a complex impedance-matching network. In the following, we obtain the S-parameters for two different types of feed designs.

First, we use discrete ports, as shown in figure 5.18. Both the impedance and the length of the discrete port influence the match with the SWS. Optimum values of these parameters are found to be 80 Ω and 100 μm, respectively. The simulated

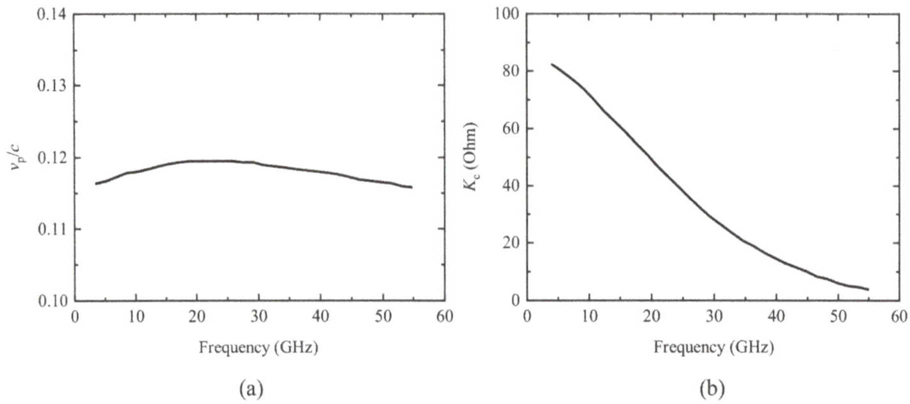

Figure 5.17. (a) Phase velocity and (b) interaction impedance of the proposed SWS. © [2016] IEEE. Reprinted, with permission, from [18].

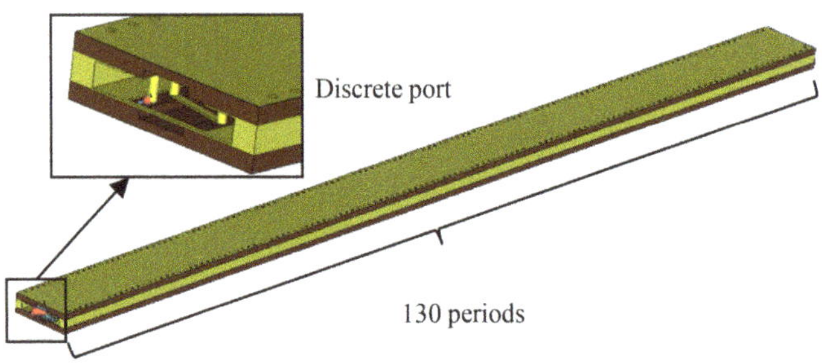

Figure 5.18. Perspective view of the SWS with discrete ports (130 periods). © [2016] IEEE. Reprinted, with permission, from [18].

S-parameters of a 130-period structure with discrete ports, including the conductor and dielectric losses, are shown in figure 5.19. The S_{11} values are below −20 dB over a frequency range from 23.9 to 36.9 GHz; this corresponds to a cold-test bandwidth of 42.8% centered at 30.4 GHz. The value of S_{21} at 30.4 GHz is −4.1 dB.

A design incorporating CPW ports was implemented. CPW ports permit cold tests to be performed on the wafer itself. The design details for a 130-period SWS are shown in figure 5.20. Both the input and output ports are on the top face of the upper quartz substrate. As can be seen in figure 5.20(b), a thin conducting via connects the last turn of the PH-SEC to the centre conductor of the CPW feed line. The impedance of the CPW is 76.8 Ω at the connection point with the PH-SEC, and it is gradually tapered to 50 Ω to match the standard CPW probes. The layout details and dimensions of the CPW feed line are shown in figure 5.20(c).

The SWS with CPW feed lines is simulated using the CST transient solver. The S-parameters of the SWS with 130 periods are shown in figure 5.21. It can be seen that S_{11} is below −20 dB from 25.4 to 33.1 GHz, corresponding to a 26.3% bandwidth centered at 29.25 GHz. The 4 dB insertion loss at 29.25 GHz continues to be quite low.

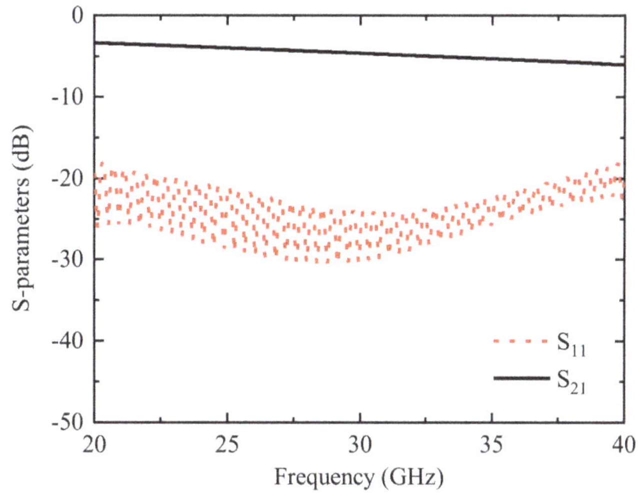

Figure 5.19. S-parameters of the 130-period SWS with with discrete ports of 80 Ω. © [2016] IEEE. Reprinted, with permission, from [18].

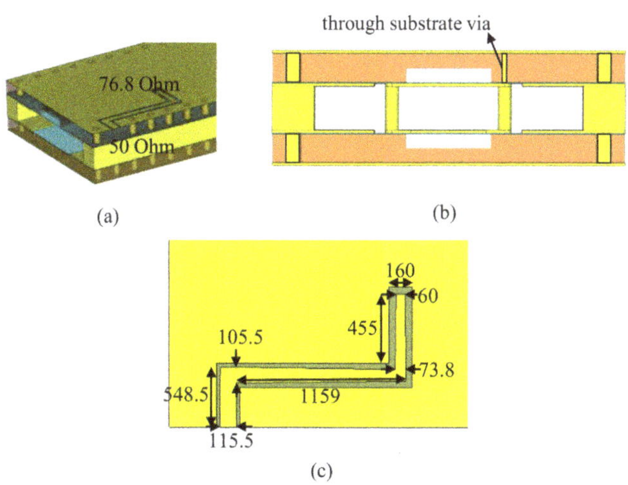

Figure 5.20. (a) Perspective view of the CPW port. (b) Cross-sectional view of the SWS with CPW port. (c) Dimensions of the tapered CPW port in μm. © [2016] IEEE. Reprinted, with permission, from [18].

5.3.3.2 Hot-test parameters

Parts of this section have been reprinted, with permission, from [18]. © [2016] IEEE.

The hot-test parameters of the proposed Ka-band PH-SEC SWS for TWT applications are estimated using the CST PIC Solver for the structure with discrete ports. An e-beam with a beam voltage of 3.72 kV and a current of 50 mA is applied centrally in the electron-beam tunnel. The elliptical cross section of the e-beam has a major axis of 350 μm and a minor axis of 150 μm. The focusing magnetic field is set to 0.8 T.

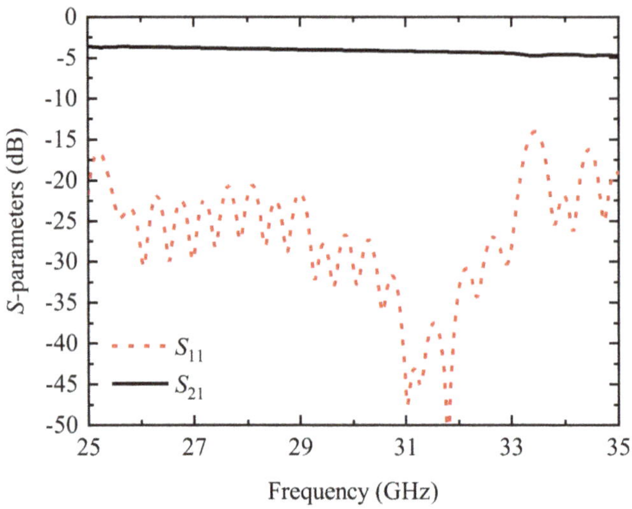

Figure 5.21. S-parameters of the SWS with 130 periods with CPW ports. © [2016] IEEE. Reprinted, with permission, from [18].

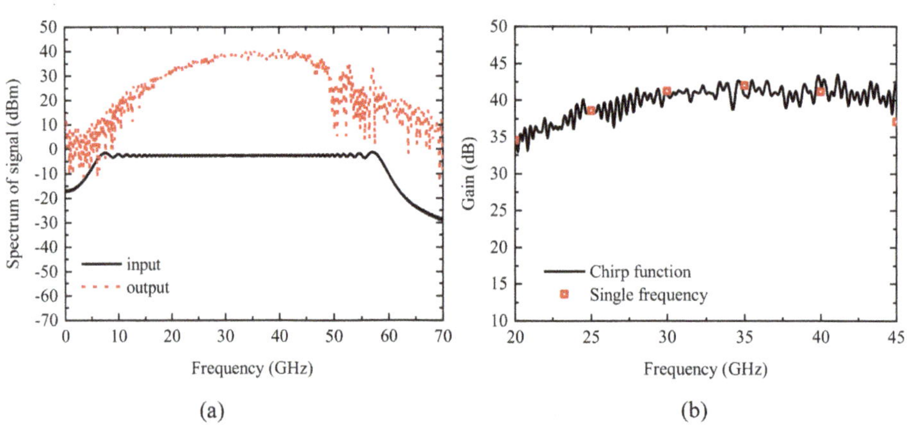

Figure 5.22. (a) Spectra of the input and output signals. (b) Gain vs. frequency curve. © [2016] IEEE. Reprinted, with permission, from [18].

The linear gain of the SWS with 130 periods is obtained at different frequencies for a −3 dBm average input power. A chirped input signal with gradually changing frequency is used to obtain wideband results in a single simulation run. The spectra of the input and output signals are shown in figure 5.22(a). The input signal has a uniform power level of −3 dBm from 10 to 57 GHz. The output signal is higher than the input signal over a wide frequency range. The gain values over the frequency range of 20–45 GHz are shown in figure 5.22(b). The red dots in the figure represent the gain obtained using single-frequency inputs in the simulation. The latter results show a very good agreement with those obtained using the chirped input. A maximum gain of 42 dB is achieved at 35 GHz, with a 3 dB bandwidth from

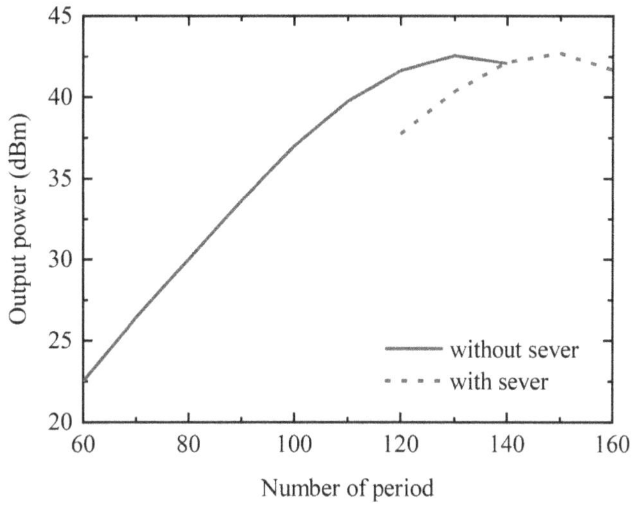

Figure 5.23. Output power vs. total number of periods for the SWS with and without a sever at 30 GHz. The input power is +7 dBm. © [2016] IEEE. Reprinted, with permission, from [18].

27 to 44.3 GHz. This corresponds to a hot-test bandwidth of 48.5%. This wide bandwidth is obtained because the flat phase velocity allows wideband interaction with the electron beam.

We also estimate the hot-test properties of the proposed SWS for high-gain applications using a sever. A sever is incorporated to reduce the possibility of oscillations due to the reflected wave as well as the backward wave. The sever consists of two discrete ports with a separation of 2000 μm and divides the TWT into two parts. The length of the first part is fixed at 40 periods, and the number of periods in the second part is varied. Figure 5.23 shows the output power with respect to the total number of periods both with and without the sever at 30 GHz for an input power of +7 dBm. We can see that the output power reaches saturation at 130 periods for the SWS without the sever. The saturation power is 18.1 W, corresponding to an RF efficiency of 9.7%. As expected, the severed SWS shows relatively lower output power for the same number of periods. However, this reduction can be offset by increasing the length of the SWS. A TWT with the severed SWS saturates at 150 periods with a very similar saturation power.

In order to further enhance the output power, a negative phase-velocity taper is applied. The pitch profile used for this is presented in figure 5.24. As shown, the pitch remains constant over the first 100 periods and then gradually decreases to 273.9 μm at the 143rd period. The chosen starting point of the taper is the position where the electron bunches start to disperse. The slope of the taper is optimized to achieve the maximum output power. The power in the z-direction for this design is presented in figure 5.25. The power increases in the z-direction and reaches a saturation value of 26.5 W at the end of the SWS. The corresponding RF efficiency is 14.2%.

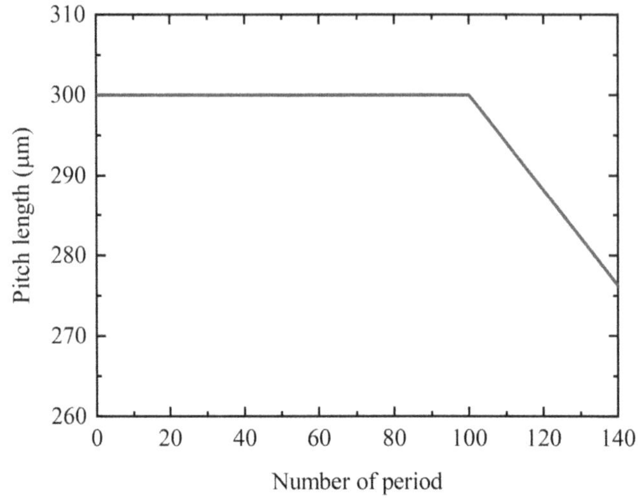

Figure 5.24. Pitch profile of the tapered SWS. © [2016] IEEE. Reprinted, with permission, from [18].

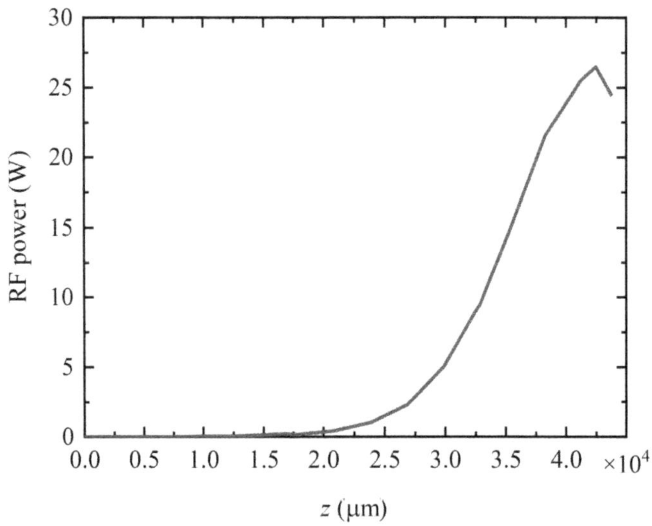

Figure 5.25. RF power vs. longitudinal position at 30 GHz for an SWS with 143 periods and pitch tapering. © [2016] IEEE. Reprinted, with permission, from [18].

5.4 Summary

In this chapter, we described dispersion control techniques for the MML SWS and the PH-SEC. For the MML SWS, the important factors that determine the dispersion characteristics were introduced, and a parametric study of some of the important parameters was carried out. For the PH-SEC, the use of metal vanes and coplanar ground planes for dispersion shaping was studied. Both coplanar ground planes and metal vanes were shown to be effective in dispersion shaping, both individually and in combination. A proof-of-concept structure operating over the frequency range of 1.6–4.2 GHz was designed and fabricated, and it was shown that the measured phase-velocity values closely match the simulated values. With appropriate design, these techniques can lead to flat dispersion characteristics without significantly reducing the interaction impedance values; such characteristics are important for broadband TWTs. These techniques can also provide negative dispersion at the low-frequency end; this characteristic can help to reduce the in-band harmonic content of a wideband TWT.

The abovementioned techniques of dispersion control were applied to a Ka-band PH-SEC consisting of a PH-SEC and quartz substrates. Quartz was chosen due to its low dielectric constant, very low loss, and compatibility with vacuum as well as microfabrication. A flat dispersion characteristic over a 42.8% bandwidth in the Ka-band was obtained using dispersion control techniques. Two types of feed ports, namely discrete ports and CPW ports, were used, both showing wideband transmission with low loss. We have shown that a TWT incorporating the proposed SWS exhibits a 3 dB gain bandwidth of 48.5%. PIC simulation results with and without a sever show that such a TWT can yield an output power of 26.5 W with the pitch-tapering technique.

References

[1] Gilmour A S 2011 *Klystrons, Traveling Wave Tubes, Magnetrons, Crossed-Field Amplifiers and Gyrotrons* (Boston, MA: Artech House)

[2] Lei W, Yang Z, Liao L and Liao P 2005 Analysis of a U-shaped vane-loaded helical slow-wave structure for wideband travelling-wave tubes *Int. J. Electron.* **92** 161–72

[3] Paik S F 1969 Design formulas for helix dispersion shaping *IEEE Trans. Electron Devices* **16** 1010–4

[4] Galuppi P and Lamesa C 1980 A new technique for ultra-broadband high power TWTs *Military microwaves '80; Proceedings of the Second Conference, London, England, October 22-24, 1980. (A82-18901 07-32)* **MM-80** (Richmond: Microwave Exhibitions and Publishers Ltd) 501–5 Military microwaves '80; Proceedings of the Second Conference, London, England, October 22-24, 1980. (A82-18901 07-32)

[5] Kumar L, Raju R S, Joshi S N and Basu B N 1989 Modeling of a vane-loaded helical slow-wave structure for broad-band traveling-wave tubes *IEEE Trans. Electron Devices* **36** 1991–9

[6] Ghosh S, Jain P K and Basu B N 1997 Analytical exploration of new tapered-geometry dielectric- supported helix slow-wave structures for broadband TWT's *Prog. Electromagn. Res.* **15** 63–85

[7] Yang J, Zhang Y, Cai X and Li L 2009 Study on effects of different metallic vane-loaded helix slow-wave structures in traveling-wave tubes *J. Infrared, Millimeter, Terahertz Waves* **30** 611–21

[8] Seshadri R *et al* 2010 P2-3: exploration of a broadband 'semi-vane' helical SWS *2010 IEEE Int. Vacuum Electronics Conf. (IVEC)* 227–8

[9] Lopes D T and Motta C C 2009 Design of vane loaded helical slow-wave structures for broad-band traveling-wave tubes *2009 IEEE Int. Vacuum Electronics Conf.* 459–60

[10] Parker R K, Abrams R H, Danly B G and Levush B 2002 Vacuum electronics *IEEE Trans. Microw. Theory Tech.* **50** 835–45
[11] Ives R L 2004 Microfabrication of high-frequency vacuum electron devices *IEEE Trans. Plasma Sci.* **32** 1277–91
[12] Krage M K and Haddad G I 1972 Frequency-dependent characteristics of microstriptransmission lines *IEEE Trans. Microw. Theory Tech.* **20** 678–88
[13] Getsinger W J 1973 Dispersion of parallel-coupled microstrip (short papers) *IEEE Trans. Microw. Theory Tech.* **21** 144–5
[14] Jansen R H 1978 High-speed computation of single and coupled microstrip parameters including dispersion, high-order modes, loss and finite strip thickness *IEEE Trans. Microw. Theory Tech.* **26** 75–82
[15] Garg R and Bahl I J 1979 Characteristics of coupled microstriplines *IEEE Trans. Microw. Theory Tech.* **27** 700–5
[16] Chua C, Aditya S, Shen Z and Tsai J M 2011 Planar helix slow-wave structure with straight-edge connections in the presence of coplanar ground planes *2011 8th Int. Conf. on Information, Communications and Signal Processing* 1–4
[17] Swaminathan K, Zhao C, Chua C and Aditya S 2015 Vane-loaded planar helix slow-wave structure for application in broadband traveling-wave tubes *IEEE Trans. Electron Devices* **62** 1017–23
[18] Zhao C, Aditya S, Wang S, Miao J and Xia X 2016 A wideband microfabricated Ka-band planar helix slow-wave structure *IEEE Trans. Electron Devices* **63** 2900–6
[19] Zhao C 2016 *Planar Helix-Based Slow-Wave Structures For Millimeter Wave Traveling-Wave Tubes* (Singapore: Nanyang Technological University)
[20] Srivastava V 2008 THz vacuum microelectronic devices *J. Phys. Conf. Ser.* **114** 012015
[21] Chua C, Aditya S and Shen Z 2010 Planar helix with straight-edge connections in the presence of multilayer dielectric substrates *IEEE Trans. Electron Devices* **57** 3451–9
[22] Lueck M R, Malta D M, Gilchrist K H, Kory C L, Mearini G T and Dayton J A 2011 Microfabrication of diamond-based slow-wave circuits for mm-wave and THz vacuum electronic sources *J. Micromech. Microeng.* **21** 065022
[23] Chen Z, Aditya S and Chua C 2014 Symmetric planar helix slow-wave structure with straight-edge connections for application in TWTs *IVEC* 291–2

Chapter 6

Variations of planar slow-wave structures

6.1 Introduction

Variations of the circular helix slow-wave structure (SWS) that offer particular advantages over the circular helix have been proposed in the research literature. For instance, the cross-wound helix [1] and its practical version, the ring–bar structure [2, 3], have been proposed to improve the operation of traveling-wave tube (TWT) amplifiers at high power levels. Coaxial contra-wound circular helices of different radii have been proposed to couple input and output power in a TWT [4, 5]. Moreover, a structure consisting of two circular helices connected by circular rings has been proposed to extend the operating frequency and improve the output power of TWTs [6].

To address fabrication challenges encountered in millimeter-wave and terahertz TWTs, SWSs with planar configurations such as the microstrip meander line (MML) and the planar helix with straight-edge connections (PH-SEC) have been proposed. Rapid advancements in fabrication techniques now enable the fabrication of planar SWSs with very small feature sizes through processes such as printed circuit or microfabrication techniques. However, for millimeter-wave applications, some issues and challenges still exist for planar SWSs, including MML and PH-SEC structures. Some of these challenges are mentioned in the following.

First, in contrast to solid-state devices operating at tens of volts, TWTs often need an electron gun with very high voltage power supplies which are costly and bulky. The need for an expensive and large power source hinders the application of millimeter-wave TWTs in comparison to solid-state devices, especially for civilian applications. This necessitates the development of TWTs with reduced operating beam voltages and smaller sizes.

Second, both MML and PH-SEC SWSs involve the utilization of dielectric substrates to provide structural support. While a dielectric material with high

doi:10.1088/978-0-7503-5764-7ch6

thermal conductivity aids in effective thermal dissipation, the dielectric also introduces the dielectric loading effect. When the electromagnetic wave travels along such an SWS, more electromagnetic power concentrates in the substrate, resulting in less power in the vacuum where the electron beam is located. This weakens the interaction between the electromagnetic wave and the electron beam. Additionally, for the MML SWS, the electric field normally concentrates on the surface of the substrate and diminishes rapidly above the substrate. This effect further reduces the interaction impedance. Consequently, it is important to enhance the electric field and beam–wave interaction in planar SWSs.

Furthermore, similar to the circular helix, the PH-SEC SWS is likely to have undesirable backward-wave oscillation and lower interaction impedance in the fundamental mode under high-power operation, which produces stronger space harmonics.

In addition, at millimeter-wave frequencies, the size of the SWS and the electron-beam tunnel become very small. Errors are very likely to occur in fabrication, alignment, and beam focusing. With the presence of dielectric substrates, there is a higher probability of electron collisions with the substrate, causing dielectric charging that may adversely affect the performance of the TWT.

Analogous to the variations of the circular helix, variations of planar SWSs that offer some advantages over the original planar SWSs are also possible. This chapter presents a few variations of the MML and the PH-SEC SWSs that address some of the abovementioned problems.

6.2 Meander-line (ML) SWS with low operating voltage and compact size

6.2.1 Logarithmic spiral

The radial-beam TWT was proposed by Soviet Union scientists in the 1960s. Such TWTs could operate under very low beam voltages. Unlike traditional TWTs, where the electron beam travels in a straight line, the radial electron beam expands radially from a ring-shaped cathode. The operation of a radial-beam TWT requires the electron beam to have radial synchronization with the electromagnetic wave. One typical SWS is the logarithmic spiral, which originates from a frequency-independent spiral antenna. The configuration of this structure is presented in figure 6.1. Thc shape of the structure is defined by $r = ae^{b\varphi}$, where a is the initial radius and b is a constant value that controls the distance between successive laps.

Due to the long path of the transmission line, the SWS exhibits very low phase velocity, making it suitable for low-voltage applications. This structure has been used in a TWT operating in the frequency range from 200 to 600 MHz. The beam voltage was only 35–40 V. The output power ranged from 50 to 100 mW, with a corresponding gain of 10 to 22 dB [7]. Similar work was done by Varian Associates in 1974 [8]. Theoretical analysis has shown that a TWT based on the logarithmic helix can work with a relatively flat gain over an octave bandwidth. In addition, it could generate a maximum gain of 10 dB with a beam voltage of 300 V and a beam

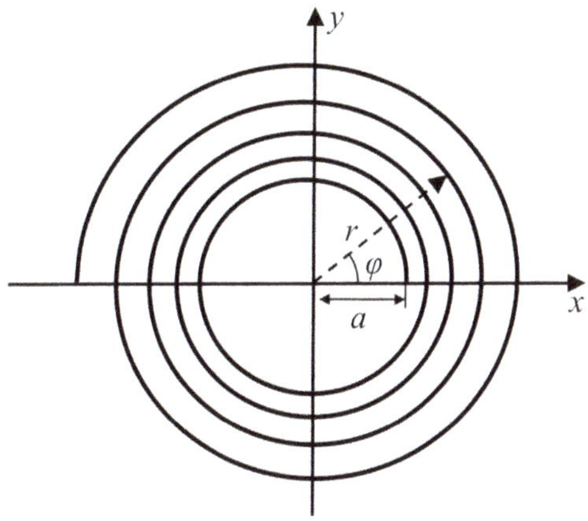

Figure 6.1. Configuration of the logarithmic spiral.

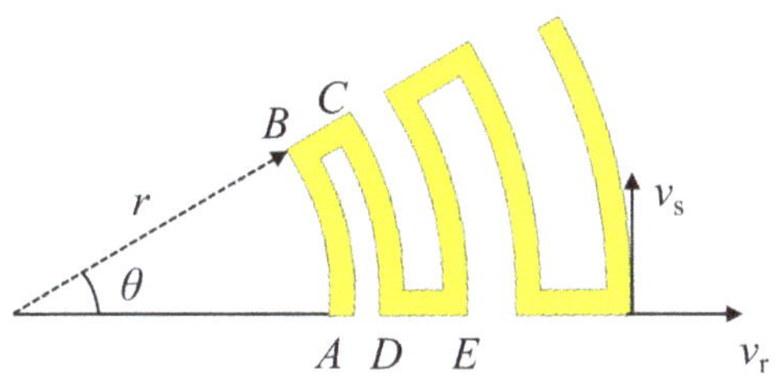

Figure 6.2. Configuration of the angular log-periodic meander-line SWS.

current of 4 A at 1.04 GHz. An experiment was also conducted; however, because of issues with oscillation, mode competition, and dielectric charging, the experimental results were not as good as expected.

6.2.2 Angular log-periodic meander-line SWS

6.2.2.1 Structural configuration

As the operating voltage of the logarithmic spiral is low, the gain and output power of the corresponding TWT are also very limited. This prevents the logarithmic spiral TWT from working at millimeter-wave frequencies. To address this issue, the angular log-periodic meander-line SWS was proposed [9, 10].

The configuration of the angular log-periodic meander-line SWS is illustrated in figure 6.2. This structure can be seen as a derivative of the logarithmic spiral SWS obtained by slicing the logarithmic spiral in the radial direction and alternately connecting its edges with segments. The length of the microstrip line is less than that of the logarithmic spiral, so its phase velocity is higher. This, in turn, leads to a higher beam voltage compared to the voltage used with the logarithmic spiral, which, however, is still lower than that of the traditional TWT.

6.2.2.2 Dispersion characteristics

The dispersion characteristics of the angular log-periodic meander line can be derived through a circuit analysis method, which is similar to the method employed for analyzing the logarithmic spiral [11]. Assuming the phase velocity of the electromagnetic wave moving along the microstrip line is v_s and the phase velocity of the electromagnetic wave moving in the radial direction is v_r, these two velocities satisfy the condition that the travel time of the radial wave moving between adjacent laps equals that of the wave moving along the microstrip line:

$$\frac{l_{AB} + l_{BC} + l_{CD} + l_{DE}}{v_S} = \frac{d_{AE}}{v_r} + nT, \quad n = 0, \pm 1, \pm 2 \tag{6.1}$$

where l_{AB} and l_{CD} are the lengths of the arcs AB and CD; l_{BC}, l_{DE}, and d_{AE} are the lengths of the segments BC, DE, and AE; and n represents the spatial harmonic number of the radial wave. The lengths of the various segments can be expressed as:

$$l_{AB} = ae^{b\phi}\frac{\sqrt{b^2+1}}{b}(e^{b\theta} - 1) \tag{6.2}$$

$$l_{BC} = ae^{b\phi}e^{b\theta}(e^{2b\pi} - 1) \tag{6.3}$$

$$l_{CD} = ae^{b\phi}\frac{\sqrt{b^2+1}}{b}e^{2b\pi}(e^{b\theta} - 1) \tag{6.4}$$

$$l_{DE} = ae^{b\phi}e^{2b\pi}(e^{2b\pi} - 1) \tag{6.5}$$

$$d_{AE} = ae^{b\phi}(e^{4b\pi} - 1) \tag{6.6}$$

where ϕ is the angle of point A in the polar coordinate system (r, θ), while a and b are the constants of the equation $r = ae^{b\varphi}$. The relationship between v_s and v_r can be obtained by substituting equations (6.2)–(6.6) into (6.1):

$$\frac{ae^{b\phi}\left[\frac{\sqrt{b^2+1}}{b}(e^{b\theta} - 1)(1 + e^{2b\pi}) + (e^{2b\pi} - 1)(e^{b\theta} + e^{2b\pi})\right]}{v_S} = \frac{ae^{b\phi}(e^{4b\pi} - 1)}{v_r} + nT. \tag{6.7}$$

It is evident from this equation that the radial velocity of the higher-order space harmonics ($n \neq 0$) depends on the values of a and ϕ. Consequently, the radial phase velocity changes as the radius and angle increase. Therefore, the higher-order space harmonics are not suitable for TWT applications, since the electron beam does not remain synchronized with the electromagnetic wave as its phase velocity changes. However, for the fundamental mode ($n = 0$), the equation can be simplified to:

$$\frac{v_r}{v_s} = \frac{e^{4b\pi} - 1}{\left[\frac{\sqrt{b^2+1}}{b}(e^{b\theta} - 1)(1 + e^{2b\pi}) + (e^{2b\pi} - 1)(e^{b\theta} + e^{2b\pi})\right]}. \tag{6.8}$$

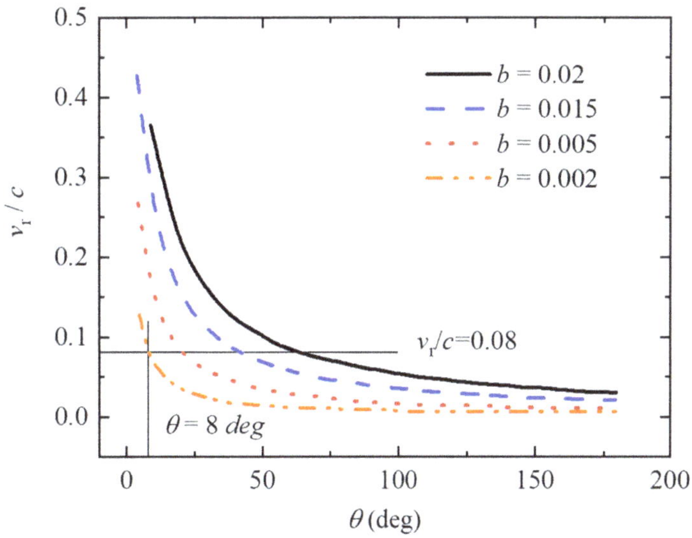

Figure 6.3. Normalized phase velocity for different values of b and θ. © [2013] IEEE. Reprinted, with permission, from [9].

In this case, it is apparent that v_r is independent of the values of a and ϕ. Once the values of b and θ are established, the ratio of v_r and v_s remains constant. Since the phase velocity of the microstrip line does not change much with frequency, the phase velocity of the fundamental wave in the radial direction is similar over a wide range of frequencies, enabling a wide operational frequency range for the TWT.

The normalized radial phase velocity of the angular log-periodic meander line without a dielectric substrate is calculated using equation (6.8). In this case, v_s equals the speed of light c. The normalized phase velocity versus θ is depicted for different values of b in figure 6.3. Once the operating voltage is determined, the values of b and θ can be derived from the figure. For example, if the beam voltage is 1642 V, the beam line with a normalized velocity of 0.08 intersects with the four phase velocity curves corresponding to different values of b. The value of θ can then be obtained at these intersection points. It should be noted that there is more than one solution of (b, θ) for a given operating voltage. When θ increases, the sheet beam widens, supporting a higher electron-beam current but posing more challenges for focusing. In practical applications, the solution should be selected based on the trade-off between these two factors.

6.2.2.3 Design and RF characterization

A model of the angular log-periodic meander line with a dielectric substrate was constructed based on a theoretical analysis using values of $b = 0.002$ and $\theta = 8°$. The dielectric substrate utilized was boron nitride with a permittivity of 4 and a loss tangent of 2.5×10^{-4}. The configuration of a 30-period SWS with input/output ports is presented in figure 6.4. The S-parameters were simulated by the CST Microwave Studio (MWS) Transient solver, and the results are presented in figure 6.5. S_{11} is less

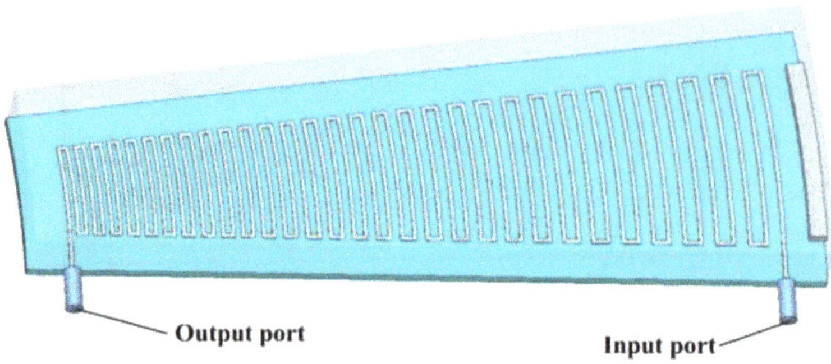

Figure 6.4. Model of the microstrip angular log-periodic meander-line SWS. © [2013] IEEE. Reprinted, with permission, from [9].

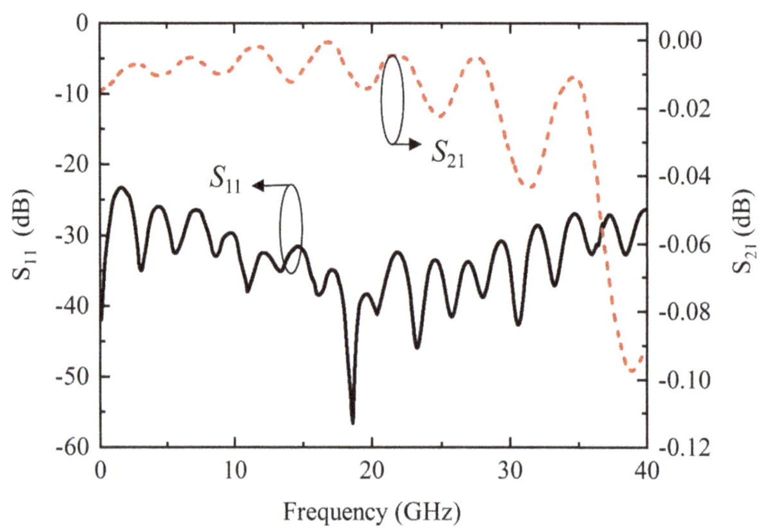

Figure 6.5. Transmission characteristics of the SWS. © [2013] IEEE. Reprinted, with permission, from [9].

than −22.5 dB, and S_{21} is higher than −0.1 dB over the frequency range from 0 to 40 GHz.

For this nonperiodic SWS, the dispersion properties cannot be obtained from the eigenmode method, which relies on periodic boundaries. In this case, the normalized phase velocity can be calculated from the simulation results using:

$$v_r = \frac{\Delta r}{\Delta t c} \tag{6.9}$$

where Δr is the radial distance between the input port and the output port and Δt is the delay between the input and output signals. Δr can be calculated using the geometrical equation:

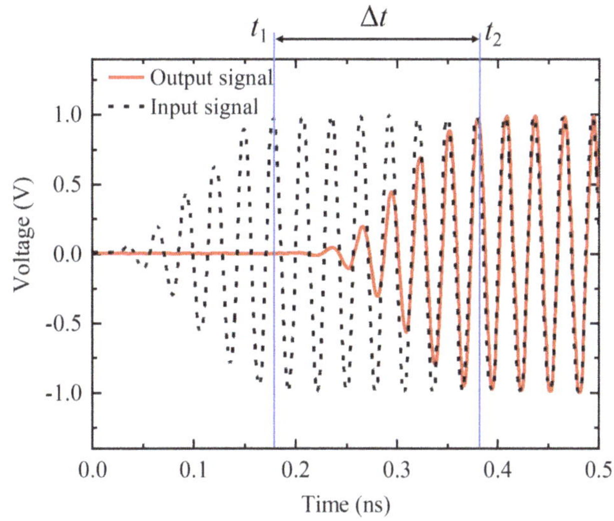

Figure 6.6. Input and output signals at 35 GHz. © [2013] IEEE. Reprinted, with permission, from [9].

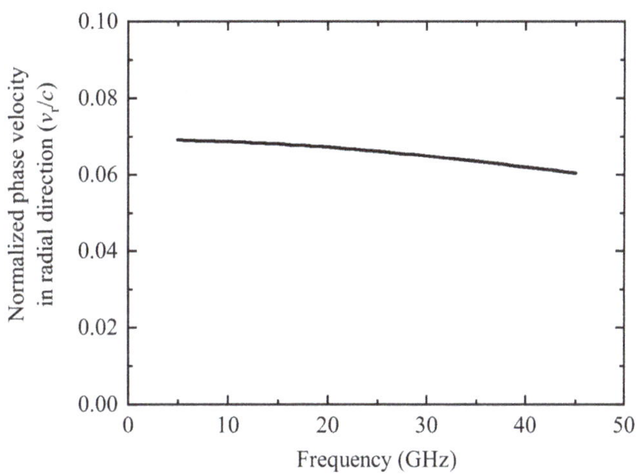

Figure 6.7. Dispersion characteristics of the microstrip angular log-periodic meander-line SWS. © [2013] IEEE. Reprinted, with permission, from [9].

$$\Delta r = a(e^{2bn\pi} - 1). \tag{6.10}$$

The simulated input and output signals of a sine wave can be obtained from time-domain simulations. Δt can be derived from the time difference between the input and output signals. For example, the input and output signals of a sine wave at 35 GHz are illustrated in figure 6.6, where Δt is approximately 0.2 ns and Δr is about 3.3 mm. Subsequently, the normalized phase velocity is calculated to be 0.063. This calculation process can be applied to obtain the phase velocities at other frequencies. A plot of phase velocity with varying frequency is shown in figure 6.7.

As can be seen, the phase velocity remains relatively constant over a wide range of frequencies.

6.2.2.4 Simulation results for hot-test parameters

The hot-test parameters of this structure are simulated using CST Particle Studio. A radial sheet beam is emitted by an arc-shaped cathode at the outer side of the SWS, while a cylindrical collector is located at the inner side of the SWS. The beam exhibits a cross-sectional area of 0.88×0.66 mm^2 at the cathode, with a voltage of 1624 V and a current of 0.5 A. The electron beam is located approximately 0.02 mm above the surface of the SWS. The radial magnetic field is 0.65 T, roughly 1.25 times greater than the Brillouin magnetic field for the radial sheet beam [12]. Particle-in-cell (PIC) simulations are carried out for the SWS at an input power of 1 W. The single microstrip angular ML SWS is able to produce an output power of more than 100 W over the frequency range from 20 to 40 GHz with a very flat gain of around 20 dB.

6.3 Meander-line SWSs with increased interaction impedance

As has been mentioned, due to the dielectric loading and the predominant concentration of the electric field on the surface of the MML SWS, the interaction impedance of the MML SWS is quite low. This is one of the main challenges in applying this SWS in practical TWTs. Consequently, the pursuit of MML SWS designs with improved interaction impedance is of great importance. This section introduces several variations of the MML SWS that aim to achieve higher interaction impedance.

Parts of this section have been reprinted, with permission, from [16]. © [2013] IEEE.

6.3.1 V-shaped meander-line SWS

6.3.1.1 Structural configuration and electromagnetic properties

The V-shaped MML SWS was proposed due to its advantage of higher interaction impedance [13]. A perspective view of this SWS is depicted in figure 6.8.

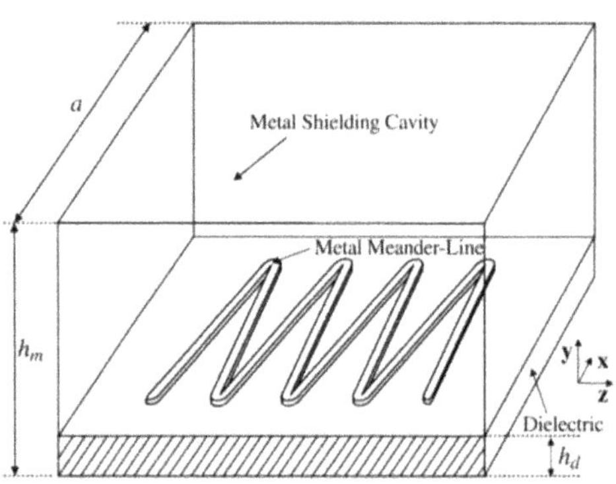

Figure 6.8. Configuration of the V-shaped MML. © [2011] IEEE. Reprinted, with permission, from [13].

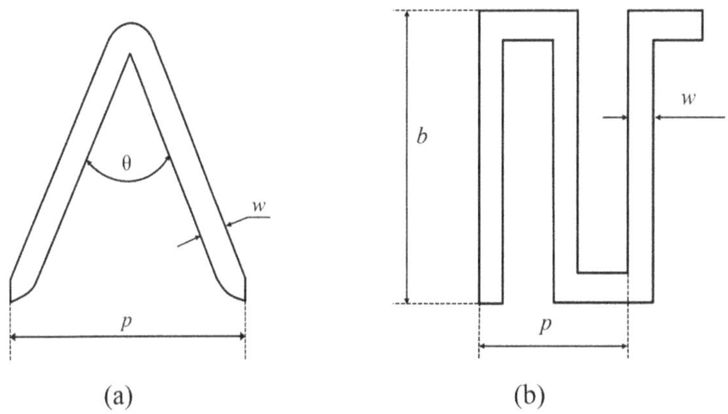

Figure 6.9. Top view of one period of: (a) the V-shaped MML; (b) the U-shaped MML. © [2011] IEEE. Reprinted, with permission, from [13].

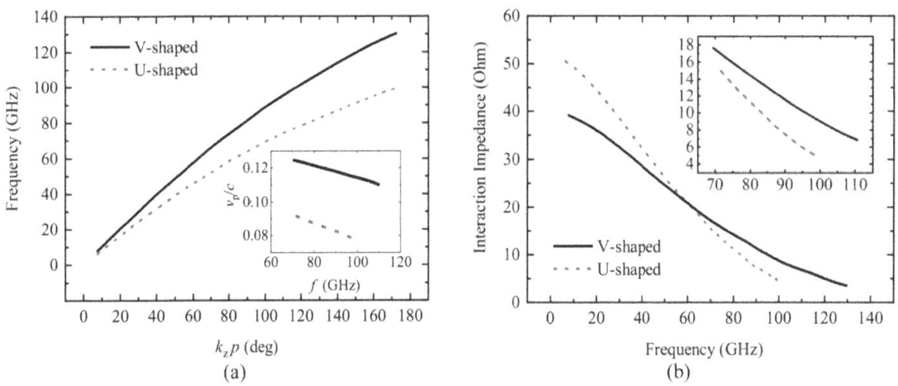

Figure 6.10. Comparisons of the (a) dispersion characteristics and (b) interaction impedances of two MML SWSs. © [2011] IEEE. Reprinted, with permission, from [13].

Furthermore, top views of one period of the V-shaped MML SWS and the traditional U-shaped MML SWS are shown in figure 6.9. As can be seen, the V-shaped MML SWS is quite similar to the traditional meander line except that the microstrip follows a V-shaped path.

The dimensional parameters of a V-shaped MML SWS are shown in figures 6.8 and 6.9. The dimensions of a W-band SWS are as follows: $b = 0.46$ mm, $p = 0.112$ mm, $\theta = 9.8$, $w = 0.02$ mm, $t = 0.005$ mm, $a = 0.92$ mm, $h_m = 0.75$ mm, and $h_d = 0.05$ mm. The supporting dielectric substrate is boron nitride with a permittivity of 4. The conductivity of copper is set to 2.25×10^7 S m^{-1}, considering the surface roughness of the metallization.

The structure is simulated using the CST MWS Eigenmode Solver and compared with a U-shaped MML SWS that has the same values of b, p, w, and t. The dispersion diagram and interaction impedance are presented in figure 6.10. It can be noted that the V-shaped MML SWS has a wider operating bandwidth than the

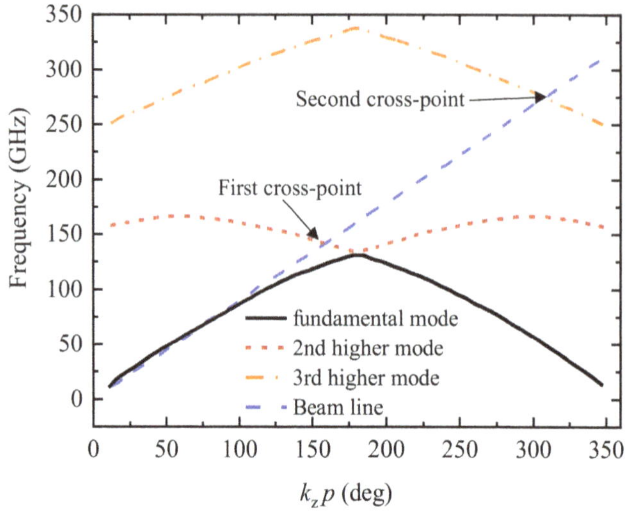

Figure 6.11. Dispersion characteristics of the SWS, showing higher-order modes. © [2011] IEEE. Reprinted, with permission, from [13].

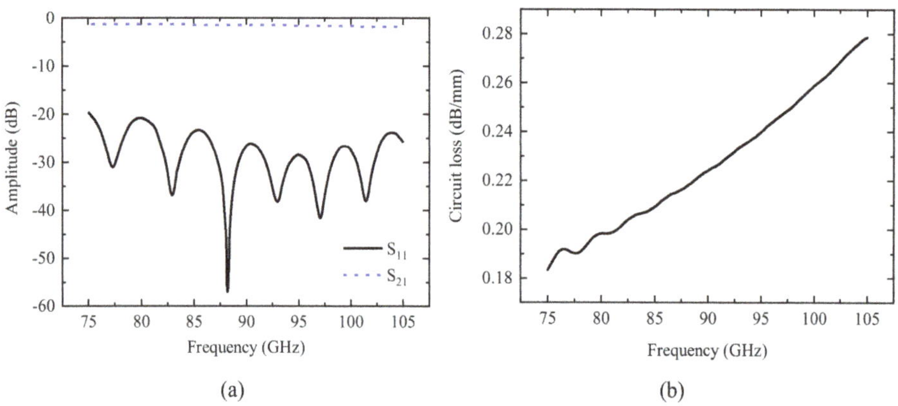

Figure 6.12. Transmission characteristics: (a) S-parameters, (b) circuit loss. © [2011] IEEE. Reprinted, with permission, from [13].

U-shaped MML SWS. In addition, the interaction impedance of the V-shaped MML SWS is higher than that of the U-shaped MML SWS at higher frequencies.

A dispersion diagram of the SWS, including higher-order modes, is presented in figure 6.11. The line representing the 3.7 kV beam intersects the fundamental mode, confirming that forward-wave interaction is possible between the electron beam and the electromagnetic wave. Furthermore, the line representing the beam also intersects the second and third higher-order modes in the backward-wave region at around 140 and 275 GHz, respectively. As a result, backward-wave oscillation may occur when the SWS is used for high-gain applications.

The structure is fed via a microstrip line that has the same strip width as the MML SWS. The S-parameters and circuit loss are presented in figure 6.12.

A good impedance match is achieved. S_{11} is below −20 dB, and S_{21} is higher than −1.9 dB over the frequency range from 75 to 105 GHz. The circuit loss is less than 0.28 dB mm^{-1}.

6.3.1.2 Hot-test parameters

An SWS with 136 periods is used to investigate beam–wave interaction. To obtain high gain and low backward-wave oscillation, the whole structure is divided into four sections with 25, 25, 25, and 61 periods, respectively. Three severs are involved, which consist of attenuators with a length of 15 periods. The overall length of the SWS is 21 mm.

The structure is simulated with CST Particle Studio. A rectangular sheet beam with a cross-sectional size of 0.46 × 0.06 mm^2 is used at a height of 0.06 mm above the surface of the SWS. The beam voltage is 3.7 kV, and the beam current is 100 mA. The focusing magnetic field is 1.4 T.

The simulation results indicate that an input power of 40 mW yields an output power exceeding 30 W over the frequency range from 75 to 100 GHz. A peak power of 90 W is achieved at 97 GHz, corresponding to a maximum gain of 33.5 dB and a maximum efficiency of 24.3%. When the input power is lower than 20 mW, the output power increases linearly with the input power. When the input power reaches 40 mW, a saturation power of around 90 W is obtained. When the severs are not incorporated, backward-wave oscillations are observed at around 140 and 265 GHz.

6.3.2 Double V-shaped meander-line SWS

It has been shown that the V-shaped MML SWS has an increased interaction impedance compared with the traditional U-shaped MML SWS at higher frequencies. However, both structures suffer from the problems of asymmetric field distribution and rapid decay of the axial electric field above the surface of the SWS. Such a field distribution requires a high-current-density electron beam, or the electron beam needs to be located in close proximity to the SWS to maintain high efficiency. This results in additional challenges in beam transportation and alignment. In order to further improve the interaction impedance of the V-shaped MML SWS, a double V-shaped MML SWS has been studied and proposed [14].

6.3.2.1 Structural configuration and electromagnetic properties

Figure 6.13 illustrates the configuration of the double V-shaped meander-line SWS. Two V-shaped MML SWSs are symmetrically positioned on the top and bottom surfaces of the metal shield. The dimensional parameters are presented in figure 6.14, and the dimensions for a W-band design are listed in table 6.1. The dielectric substrate consists of boron nitride with a permittivity of 4 and a loss tangent of 3 × 10^{-4}. The conductivity of copper is taken to be 2.25 × 10^7 S m^{-1}.

The SWS is simulated with CST MWS Eigenmode Solver. The longitudinal electric field at the center along the y-direction is presented in figure 6.15 and compared with that of the single V-shaped MML SWS. The electric field is distributed symmetrically with respect to the xz-plane. The average longitudinal

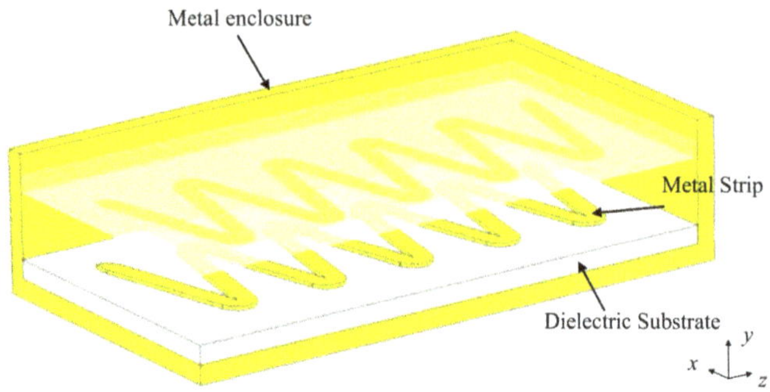

Figure 6.13. Perspective view of the double V-shaped MML SWS [14].

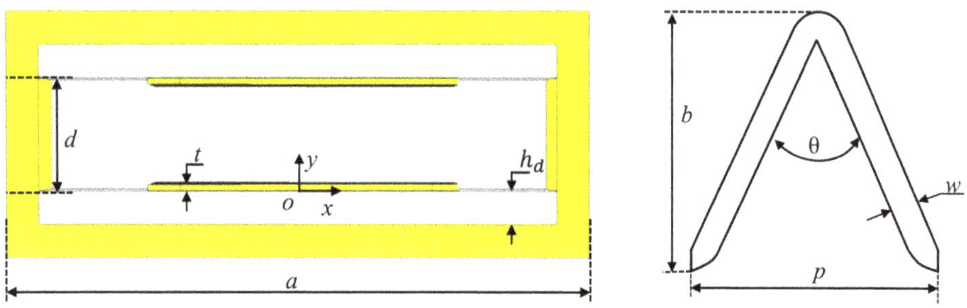

Figure 6.14. Side view and dimensional parameters of the double V-shaped MML SWS. © [2012] IEEE. Reprinted, with permission, from [14].

Table 6.1. Optimized parameters of the V-shaped MML SWS [14].

Parameter	Value (mm)
a	0.9
b	0.44
p	0.124
w	0.02
t	0.01
d	0.17
h_d	0.05
θ	12 (deg)

electric field is higher than that of the single MML SWS, which can lead to higher interaction impedance.

The simulated normalized phase velocity and interaction impedance of both the double and single V-shaped MML SWSs are presented in figure 6.16. The interaction impedance of the double V-shaped MML SWS is higher than that of

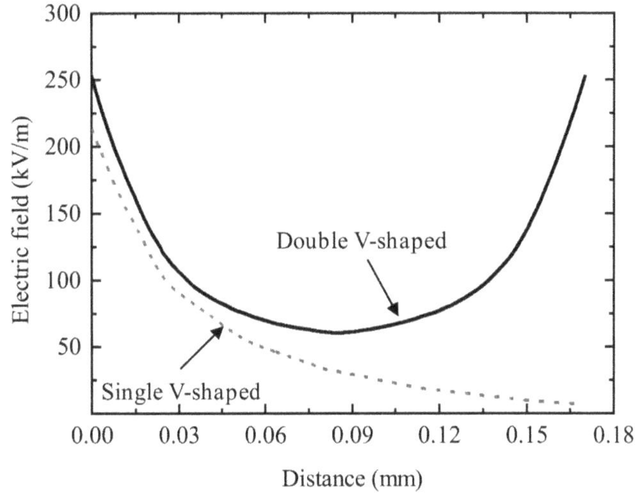

Figure 6.15. Longitudinal electric field along the *y*-axis. © [2012] IEEE. Reprinted, with permission, from [14].

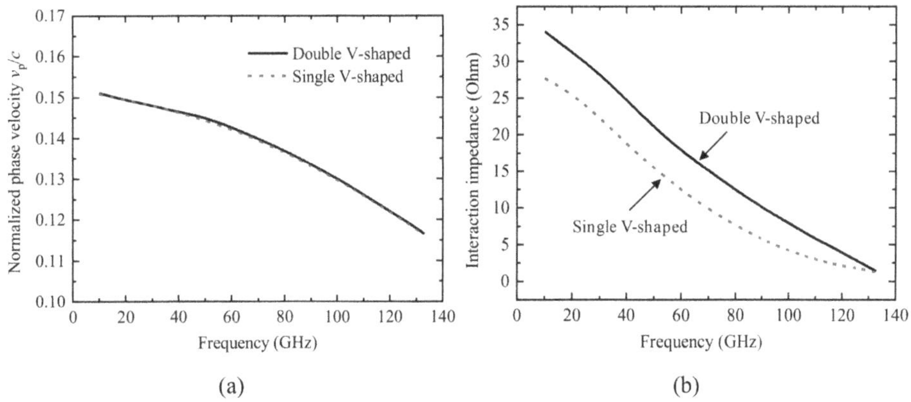

Figure 6.16. (a) Normalized phase velocity and (b) interaction impedance of the double V-shaped MML SWS and the single V-shaped MML SWS. © [2012] IEEE. Reprinted, with permission, from [14].

the single V-shaped MML SWS, whereas the phase velocities of both structures are almost the same.

The individual V-shaped MML SWSs are fed by a microstrip line and then connected to SubMiniature version A (SMA) connectors. The two V-shaped MML SWSs are fed separately by two input ports and two output ports. The simulated reflection coefficient and attenuation are presented in figure 6.17. S_{11} is below −20 dB from 80 to 125 GHz. The circuit loss is less than 0.41 dB mm^{-1}.

6.3.2.2 Hot-test parameters

Similar to the single V-shaped MML SWS in section 6.3.1, the SWS is divided into four sections, with 25, 25, 25, and 60 periods, respectively. Three severs consisting of attenuators with a length of 15 periods are applied between the sections. The model

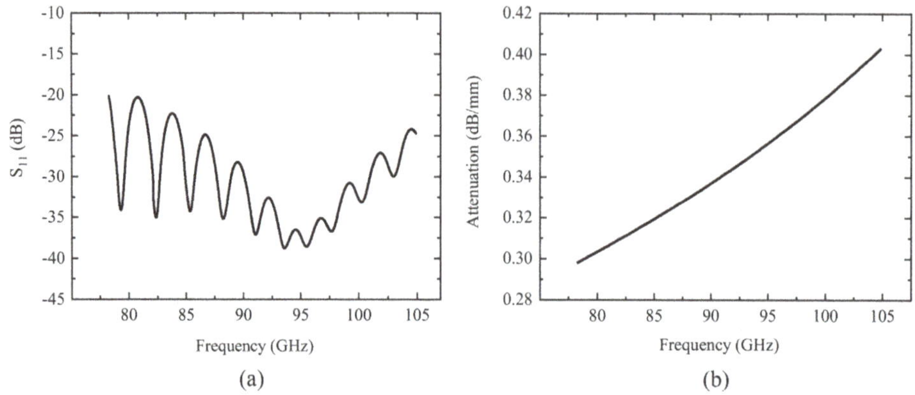

Figure 6.17. (a) S_{11} of the double V-shaped MML SWS. (b) Attenuation of the double V-shaped MML SWS. © [2012] IEEE. Reprinted, with permission, from [14].

Figure 6.18. Model of the double V-shaped MML SWS used in PIC simulations. © [2012] IEEE. Reprinted, with permission, from [14].

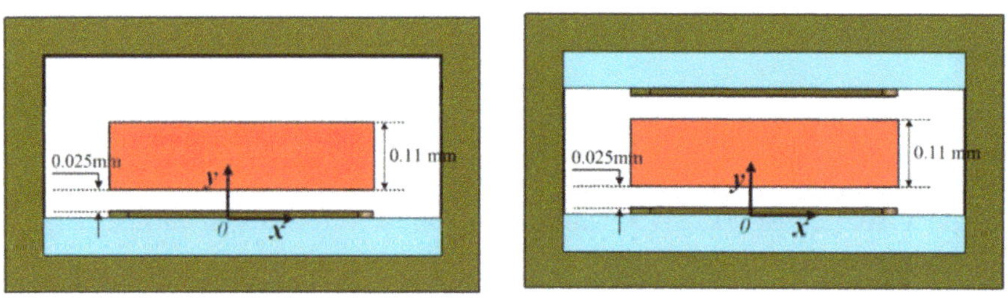

Figure 6.19. Cross-sectional views of the single and double V-shaped MML SWSs with a sheet electron beam. © [2012] IEEE. Reprinted, with permission, from [14].

used for PIC simulation is presented in figure 6.18. The total input power is the sum of the input powers provided at the two input ports, while the total output power is the sum of the output powers leaving the two output ports. In the simulations, the signals at the two input ports have the same amplitude and phase.

A sheet beam with a rectangular cross section of $0.44 \times 0.11\ \text{mm}^2$ is employed. The beam is located at the center of the SWS, as shown in figure 6.19. The beam

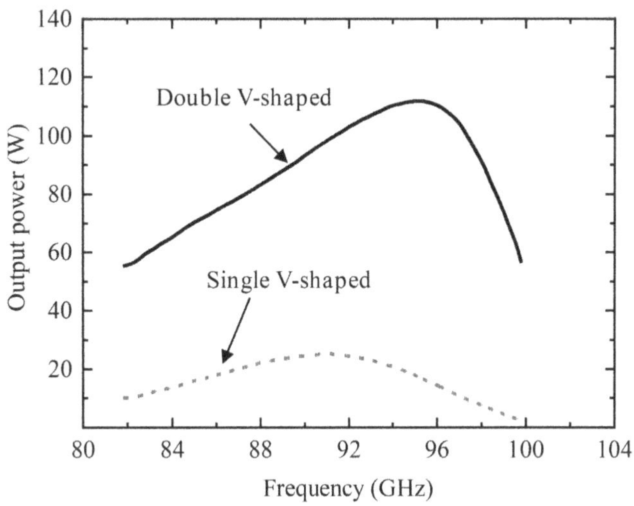

Figure 6.20. Output power versus frequency for the single and double V-shaped MML SWS. © [2012] IEEE. Reprinted, with permission, from [14].

voltage is 4570 V and the beam current is 0.1 A. Operating at a frequency of 94 GHz, the input power at each input port is 40 mW, corresponding to a total input power of 80 mW. The focusing magnetic field is 0.7 T. The current density in this case is around 206 A cm^{-2}, which contrasts with the higher current density of 362 A cm^{-2} used for the single V-shaped MML SWS in section 6.3.1.

The hot-test parameters are demonstrated with CST Particle Studio. The output power versus frequency curve is presented in figure 6.20. The results are compared with those of the single V-shaped MML SWS with the same simulation specifications. For the double V-shaped MML SWS, the 3 dB gain bandwidth ranges from 82 to 100 GHz. The peak power is 110 W, corresponding to a gain of 31.4 dB. The average efficiency is 12%. On the other hand, for the single V-shaped MML SWS, the 3 dB gain bandwidth is 15 GHz and the peak power is only 25 W. The output power increases further with a higher beam current. When the beam current increases to 200 mA, a saturation power of 190 W is achieved.

6.3.3 Three-dimensional U-shaped meander-line SWS

Despite the advantages of the double V-shaped MML SWS, it has some shortcomings. These include the difficulty of aligning the two V-shaped structures and the difficulty of designing an RF feed which matches the phase and amplitude of the input power coupled into the two V-shaped circuits. Another problem mentioned is the high RF loss due to significant dielectric loading. To reduce dielectric loading, an approach similar to that in [15] has been suggested, in which a raised serpentine (or meandering) ridge of the dielectric is selectively metallized. However, such an approach makes the fabrication relatively more complicated and essentially modifies the original 2D MML SWS into a 3D structure. To address these issues, a 3D U-shaped MML SWS has been proposed and studied [16]. The content of this subsection closely follows [16, 17], which reported the work done by the research group of the authors of this book.

Parts of this section have been reprinted, with permission, from [16]. © [2013] IEEE.

6.3.3.1 *Structural configuration*

Perspective views of the 3D U-shaped and the symmetric double V-shaped MML SWSs are shown in figures 6.21(a) and (b), respectively. As depicted in figure 6.21(a), the 3D U-shaped MML SWS consists of top and bottom arrays of horizontal conductors; the conductors in the top array, oriented along the *y*-direction, resemble the shape '|' and the conductors in the bottom array, oriented in the *y*- and *z*-directions, resemble the shape '*S*.' All metal strips have a width of *SW* and a thickness of *ST*. The conductors in the top and bottom arrays are connected at their sides by vertical conductors (or cylindrical vias) oriented along the *x*-direction. Each cylindrical via has a diameter of *VD* and connects the top and bottom ring pads, which have a diameter of *RD* (>*VD*). The top and bottom arrays are separated by 2*a* in the *x*-direction. The via-to-via distances are 2*b* and *L* in the *y*- and *z*-directions, respectively. The cylindrical vias are similar to the PH-SEC, which can easily be fabricated using printed circuit or microfabrication techniques.

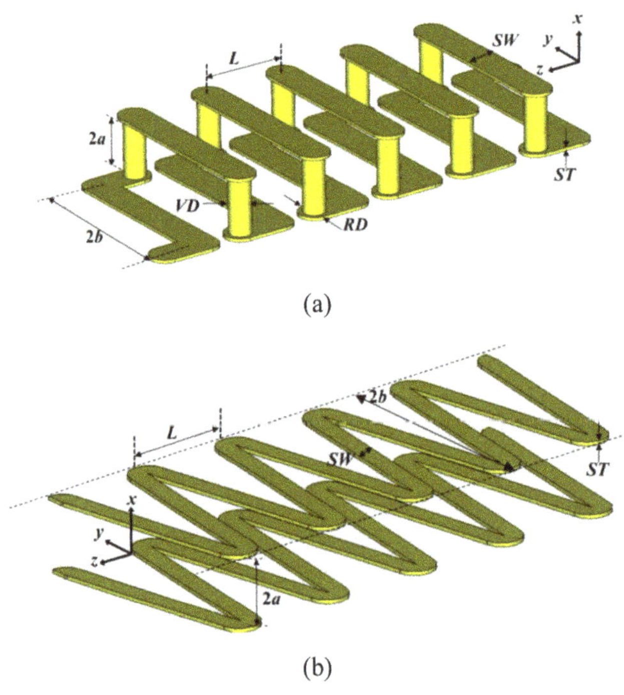

Figure 6.21. Perspective views of the (a) 3D U-shaped MML and (b) symmetric double V-shaped MML SWSs. The dielectric substrates are not shown in the figure. © [2013] IEEE. Reprinted, with permission, from [16].

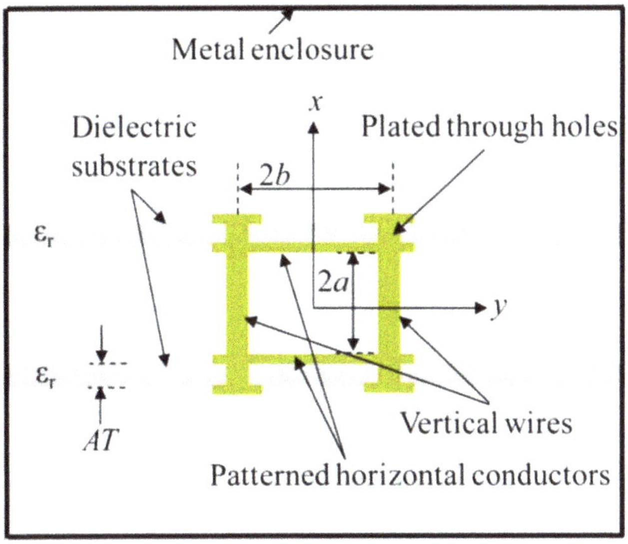

Figure 6.22. Cross-sectional view of the proposed configuration for the printed circuit fabrication of the 3D U-shaped MML SWS. © [2013] IEEE. Reprinted, with permission, from [16].

For a simple proof-of-concept practical implementation of the 3D U-shaped MML SWS, a configuration suitable for printed circuit fabrication is proposed here. A cross-sectional view of the proposed printed circuit configuration is shown in figure 6.22. It consists of two dielectric substrates, a metal enclosure, and the 3D U-shaped MML SWS. The dielectric substrates have a dielectric constant ε_r and a thickness AT. The 3D U-shaped MML SWS can be formed by patterning the top and bottom arrays of horizontal conductors on the metallized dielectric substrates. The patterned surfaces of the dielectric substrates face each other. The conductors in the top and bottom arrays are connected by vertical wires. For easier fabrication, it is proposed to use plated-through holes (PTHs) to extend the contact points of the inner ring pads to the outer surfaces of the dielectric substrates in the x-direction. This allows vertical wires to be inserted through the hollow PTHs and to be soldered or welded easily. The 3D U-shaped MML SWS is enclosed in a metal housing. An electron beam can pass through the center of the 3D U-shaped MML SWS. By applying microfabrication techniques, the frequency of operation of the proposed 3D U-shaped MML SWS can be extended to millimeter and submillimeter frequencies.

6.3.3.2 Electromagnetic properties

A period of the 3D U-shaped MML SWS with the printed circuit configuration shown in figure 6.22 is simulated using the CST MWS Eigenmode Solver. The 3D U-shaped MML SWS designed for S-band frequencies has dimensions of $2a = 2b = 3.048$ mm, $SW = 0.5$ mm, $ST = 17$ μm, $AT = 0.813$ mm, $VD = 0.3$ mm, $RD = 0.5$ mm, and $L = 1.616$ mm. A Rogers RO4003 printed circuit board (PCB) with $\varepsilon_r = 3.55$ and a loss tangent of 0.0027 is chosen for the simulations. The symmetric double V-shaped MML

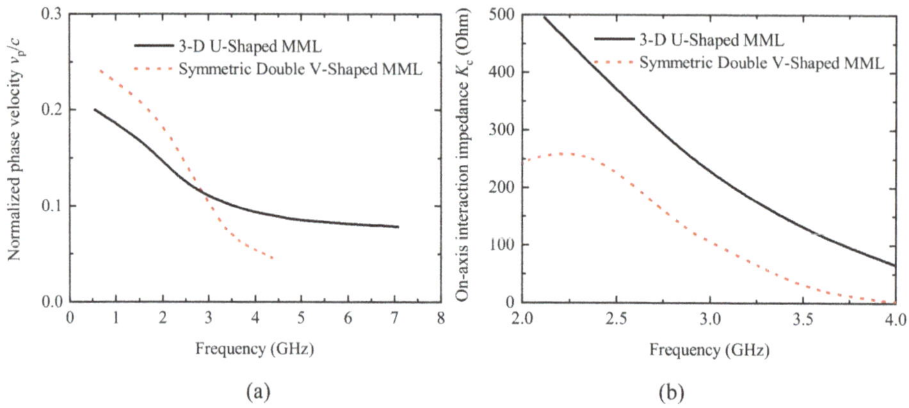

Figure 6.23. Comparison between the (a) phase velocities and (b) on-axis interaction impedances of the 3D U-shaped and symmetric double V-shaped MML SWSs. © [2013] IEEE. Reprinted, with permission, from [16].

SWS is assumed to be printed on two dielectric substrates with the dielectric constant, loss tangent, thickness, and separation identical to those of the 3D U-shaped MML SWS. A period of the symmetric double V-shaped MML SWS, with $SW = 0.5$ mm and $ST = 17$ μm, is also simulated and compared with a period of the 3D U-shaped MML SWS. The dimensions $2b$ and L, which are 14.59 and 1.616 mm, respectively, for the symmetric double V-shaped MML SWS, are chosen such that the phase velocity at 3 GHz is similar to that of the 3D U-shaped MML SWS. A metal enclosure with dimensions of 20×20 mm^2 is included in the structures used for simulations.

Figure 6.23(a) shows the normalized phase velocities of the fundamental forward-wave mode of the 3D U-shaped and the symmetric double V-shaped MML SWS. As seen in figure 6.23(a), the bandwidth for ±5% phase velocity variation for the 3D U-shaped MML SWS is around 20% centered at 3 GHz, while that for the symmetric double V-shaped MML SWS is around 5%. The greater amount of dispersion in the latter structure indicates that the field is more strongly concentrated in the dielectric material as the frequency increases. The larger bandwidth of the 3D U-shaped MML SWS implies that this structure can sustain the velocity synchronism between the electromagnetic wave and the electron beam over a wider bandwidth. As shown in figure 6.23(b), while both SWSs show significant values of on-axis interaction impedance at 3 GHz, the values for the 3D structure are twice those for the double V-shaped MML SWS.

Both MML SWSs were also simulated using the CST MWS transient solver. A simplified coplanar waveguide (CPW) feed was designed to couple the RF signal to and from the 3D U-shaped MML SWS. For the symmetric double V-shaped MML SWS, the RF signal is fed simultaneously to the top and bottom halves using two waveguide ports. All metal parts are assumed to be copper with a bulk conductivity of 5.9×10^7 S m^{-1}. The simulated circuit attenuations for the SWSs are shown in figure 6.24. The circuit attenuation in each case is obtained using the difference in the magnitude of S_{21} for two structures that are identical except for having different numbers of periods. In general, the attenuation increases as the frequency increases

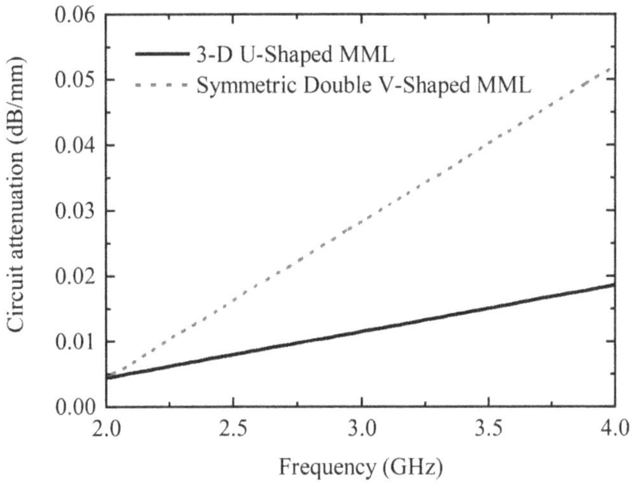

Figure 6.24. Comparison of simulated circuit attenuations for the 3D U-shaped and symmetric double V-shaped MML SWSs. © [2013] IEEE. Reprinted, with permission, from [16].

for both structures. However, at higher frequencies, the symmetric double V-shaped MML SWS has a higher circuit attenuation than the 3D U-shaped MML SWS. The likely reason for the higher circuit attenuation in the case of the symmetric double V-shaped MML SWS is that the conduction current flows very close to the dielectric substrates at higher frequencies, thus increasing the effect of dielectric loading. On the other hand, in the case of the 3D U-shaped MML SWS, a significant part of the current path is situated away from the dielectric substrates, even at higher frequencies. The dispersion characteristics in figure 6.23(a) support this reasoning.

6.3.3.3 Hot-test parameters

PIC simulations were carried out for both MML SWSs using CST Particle Studio. The simulation models for both MML SWSs are the same as those in the MWS transient solver. The interaction length is 72.72 mm (45 periods). A cylindrical e-beam with a beam voltage of 3.7 kV and a current of 250 mA is assumed to be centered on the z-axis. The beam diameter is 1.524 mm, corresponding to a filling factor of 50%, and the current density is 13.7 A cm^{-2}. The beam is assumed to propagate in the $+z$-direction under a homogeneous axial magnetic field of 0.25 T. The beam interacts with the fundamental forward-wave mode of both structures, and the linear and nonlinear gain values are examined. The beam voltage for the symmetric double V-shaped MML SWS has been adjusted so that the frequency for peak gain is similar to that of the 3D U-shaped MML SWS. Figure 6.25(a) shows the simulated saturated gains for both MML SWSs. It can be seen that the saturated gain for the symmetric double V-shaped MML SWS is much lower than that of the 3D U-shaped MML SWS. This is attributed to the higher circuit attenuation and lower interaction impedance values for the symmetric double V-shaped MML SWS. In addition, the 3 dB gain bandwidth for the 3D U-shaped MML SWS is much

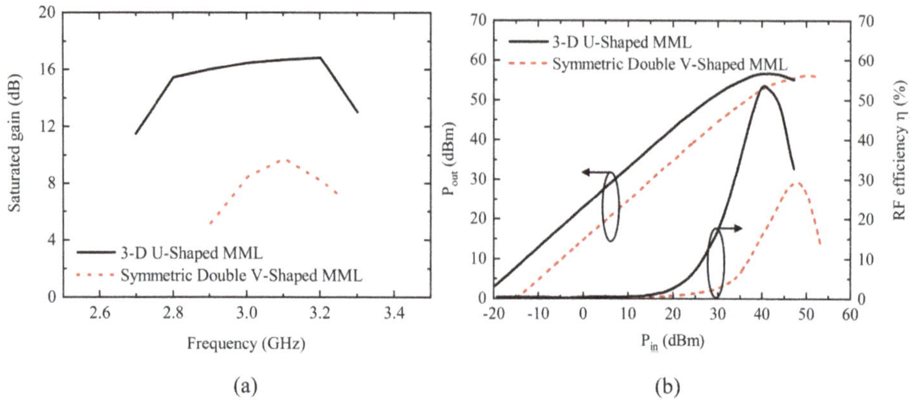

Figure 6.25. Simulated (a) saturated gain versus frequency and (b) output power (peak) and RF efficiency versus input power at 3 GHz for the 3D U-shaped and symmetric double V-shaped MML SWSs. © [2013] IEEE. Reprinted, with permission, from [16].

wider than that for the symmetric double V-shaped MML SWS. The output power (peak) and RF efficiency for both MML SWSs at 3 GHz are shown in figure 6.25(b). The saturated output peak power is around 56 dBm (400 W) for both MML SWSs. The saturated RF efficiency for the 3D U-shaped MML SWS is 53%, which is significantly higher than that of the symmetric double V-shaped MML SWS. More studies need to be carried out to assess the thermal issues that are bound to arise at high power levels.

6.3.3.4 Fabrication and measurement

The 3D U-shaped MML SWS with the proposed printed circuit configuration was designed and fabricated by incorporating microstrip feed lines. The dimensions of the fabricated 3D U-shaped MML SWS follow the dimensions described in the previous section. A Rogers RO4003 PCB was selected for this proof-of-concept fabrication. Two pieces of double-sided copper-clad RO4003 PCB were processed to form the horizontal metal strips, microstrip-line feeds, and PTHs. As depicted in figures 6.26(a) and (b), respectively, two right-angle-bent 50 Ω microstrip lines, tapers, and ring pads are printed on the top face of the first PCB, while the ground plane (for the microstrip lines), ring pads, and 'S'-shaped metal strips are printed on the bottom face of the first PCB. As depicted in figures 6.26(c) and (d), respectively, the top face of the second PCB is printed with '|'-shaped metal strips, and the ring pads are printed on the bottom face of the second PCB. Both PCBs are assembled such that the faces containing the metal strips with the shapes 'S' and '|' face each other. As shown in figure 6.26(e), the PCBs are fixed using screws, spacers, and nuts. Silver-plated copper wires, with a diameter of 0.29 mm, are inserted through the PTHs of both PCBs and soldered to the ring pads on the outer faces of the PCBs. SMA connectors are then connected to the 50 Ω microstrip-line input/output feeds.

Figure 6.27 shows the measured and simulated return losses (S_{11}) of the fabricated 3D U-shaped MML SWS over the frequency range of 2–4 GHz. The simulation

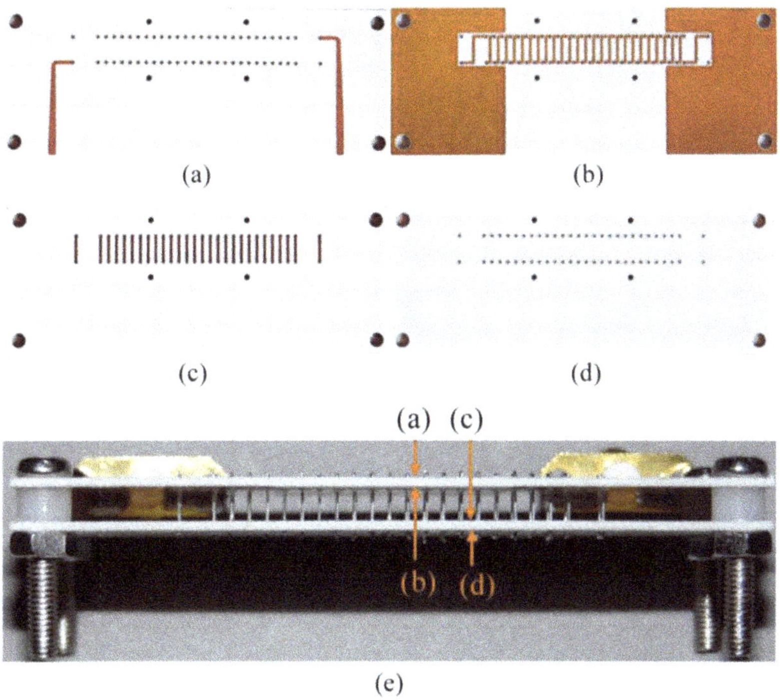

Figure 6.26. Photos of the fabricated 3D U-shaped MML SWS showing the top and bottom faces of the PCBs (a), (b), (c), and (d) with 27 periods and $b/a = 1$. (e) Side view of the assembled 3D U-shaped MML SWS.

results were obtained using the CST MWS transient solver and include dielectric and conductor losses. The number of mesh cells was selected to ensure that the simulation results converge. The measured S_{11} values match the simulated ones very well. The measured -19 dB S_{11} bandwidth covers the frequency range from 2.7 to 3.3 GHz. The measured phase velocity and circuit attenuation are shown in figure 6.28, together with the simulated ones. The measured phase velocity and circuit attenuation are obtained using the differences in the phase and magnitude of S_{21}, respectively, for two fabricated 3D U-shaped MML structures that are identical except for differing numbers of periods (22 and 27). Half the bulk conductivity value (2.95×10^7 S m^{-1}) for copper is used in this simulation for comparison with the measured results. The nature of the frequency variation of the measured and simulated values is quite similar over the frequency range considered. In these results, the measured phase velocity values are generally lower than the simulated values, whereas the measured circuit attenuation values are higher. These differences are attributed to fabrication tolerances, such as the spacing between the two dielectric substrates, the extra via lengths at the solder joints, and the poorer conductivity of the solder material with respect to copper. Also, the conductivity of copper can be quite different from the bulk conductivity, depending on the method of deposition. Many of the aforementioned problems can be resolved by using precision techniques for fabrication and assembly.

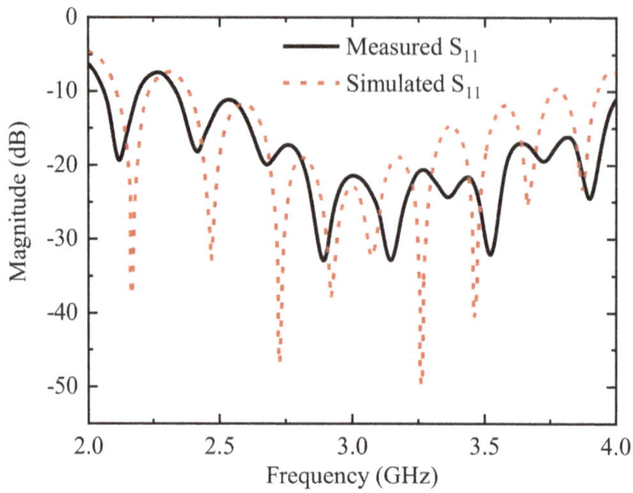

Figure 6.27. Measured and simulated return losses (S_{11}) of the fabricated structure shown in figure 6.26. © [2013] IEEE. Reprinted, with permission, from [16].

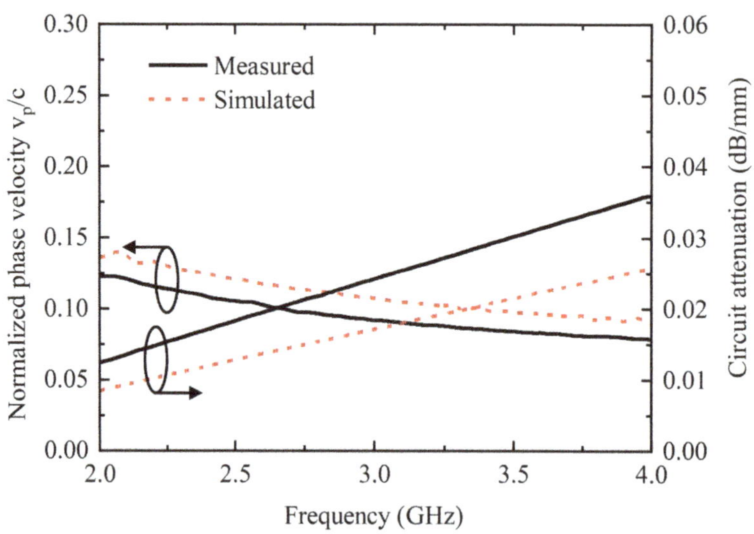

Figure 6.28. Measured and simulated phase velocity and circuit attenuation of the fabricated structure shown in figure 6.26. © [2013] IEEE. Reprinted, with permission, from [16].

6.3.4 Meander-line SWS with shaped substrates

The use of a dielectric substrate inevitably leads to the dielectric loading effect. A thick dielectric substrate with high permittivity leads to low phase velocity, low interaction impedance, high dispersion, and high circuit loss. This limits the performance of the MML for TWT applications. To reduce the dielectric loading effect, it is often preferable to use a thin dielectric substrate with low permittivity.

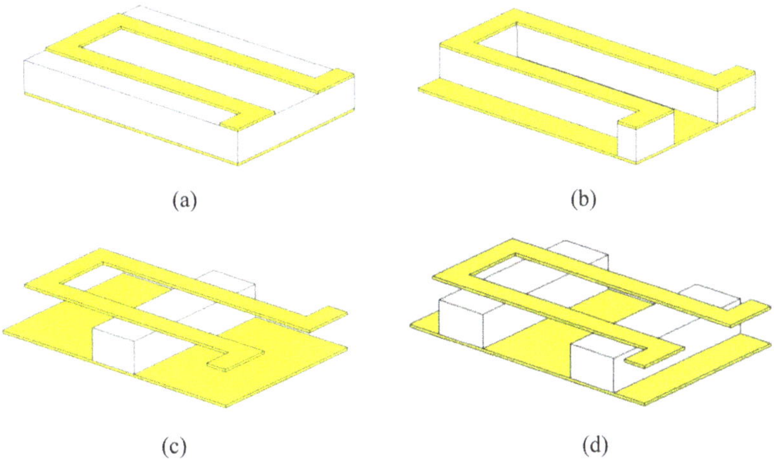

Figure 6.29. Configuration of one period of three types of MML SWS with (a) continuous substrate; (b)–(d) shaped substrates.

Apart from that, it is conceptually simple to tackle the dielectric loading problem by removing some parts of the dielectric from the substrate. As a decrease in dielectric loading leads to a decrease in the stored energy in the dielectric substrate, the MML SWS correspondingly has higher phase velocity, higher interaction impedance, higher characteristic impedance, and lower dielectric loss. The main issue lies in how to implement this concept in practice.

Figure 6.29 shows three distinct configurations of MML SWSs with shaped substrates. Figure 6.29(a) depicts the traditional U-shaped MML SWS with a continuous substrate, which was introduced in detail in chapter 2. Figure 6.29(b) shows the MML SWS with metallization only on top of a raised serpentine ridge of dielectric, which can also be called a raised meander line. This structure has been studied and realized for use in the L and S bands with precision laser machining [18] and for use in the W band with microfabrication techniques [15]. The work reported in [15] is discussed in detail in chapter 9. Figures 6.29(c) and (d) show MML SWSs with single and double supporting dielectric rods, respectively [19–21]. These two structures exhibit similar electromagnetic properties. Unlike the raised MML SWS, the meander line and the supporting rods must be fabricated separately and subsequently assembled together. PIC simulation results for an X-band SWS with double supporting ceramic rods indicate that its output power exceeds 500 W and its gain is 16 dB [19].

6.4 PH-SEC with reduced backward-wave oscillation

Although the helical SWS has the advantage of wide bandwidth, its application has been limited to relatively low-power TWTs. For high-power applications, the beam voltage is required to be high (>5 kV). To achieve velocity synchronism, this corresponds to a helix with a large pitch angle. Under such conditions, (i) the interaction impedance of the fundamental forward-wave mode is reduced due to non-interacting space harmonics, and (ii) the interaction impedance of the backward-wave mode is increased. These effects may result in oscillation [1]. In 1955, Chodorow and Chu [1] proposed the cross-wound twin helix to overcome the difficulties of the single tape helix. In 1956, Birdsall and Everhart [2] proposed the circular ring–bar structure as a practical form of the cross-wound twin helix. In high-power applications, the PH-SEC also suffers from similar problems to those of the circular helix [22]. In this section, two variations of PH-SEC SWS with reduced backward-wave oscillation are introduced.

6.4.1 Planar ring–bar SWS

To address the abovementioned problems, the authors' research group has proposed a new structure, a rectangular ring–bar with straight-edge connections (RRB-SEC); the square shape is considered a special case of the rectangular shape. Analogous to the circular ring–bar, the new structure has the potential to enable high-power operation of planar TWTs operating at millimeter-wave and higher frequencies.

The basic configuration of the RRB-SEC is shown in figure 6.30. It is inspired by the circular ring–bar structure [2], which is shown in figure 6.31 and is a practical form of the cross-wound twin helices. In the symmetric waveguide mode of the cross-wound twin helices, the axial electric fields of the fundamental forward-wave modes in the two helices combine and result in a high value of the axial interaction impedance; the space harmonics carry mainly the magnetic energy and result in low interaction impedance for the backward wave [1]. The RRB-SEC is a planar version of the circular ring–bar structure that may be suitable for sheet beam applications.

The RRB-SEC, as shown in figure 6.30, consists of three metal layers—the bottom horizontal layer, the top horizontal layer, and the vertical straight-edge connections (or vias) that connect the two horizontal layers. The top and bottom horizontal layers consist of conducting strips, oriented along the y-direction, with a thickness of ST and a width of SW. Adjacent conducting strips are joined by a conducting bar oriented along the z-direction, with a thickness of ST and a width of BW. The conducting bars are analogous to the crossovers (overlaps) of the cross-

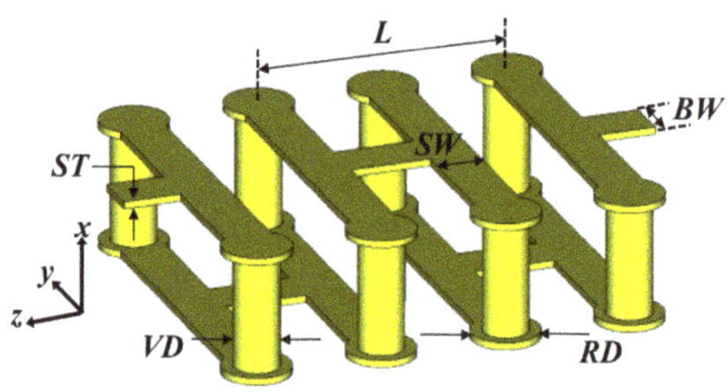

Figure 6.30. Perspective view of the basic configuration of an RRB-SEC with cylindrical vias. © [2015] IEEE. Reprinted, with permission, from [22].

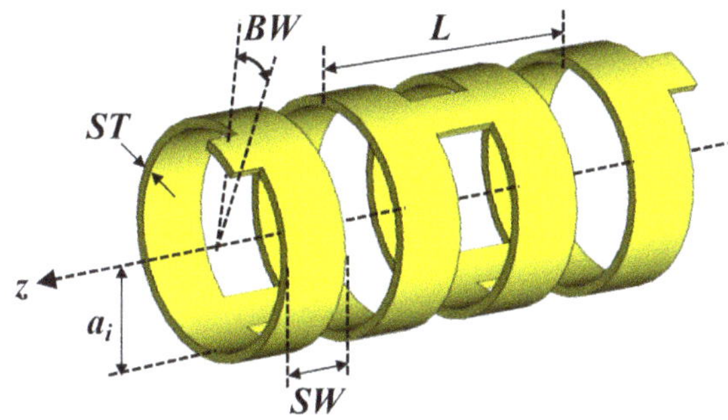

Figure 6.31. Perspective view of the circular ring–bar structure. © [2015] IEEE. Reprinted, with permission, from [22].

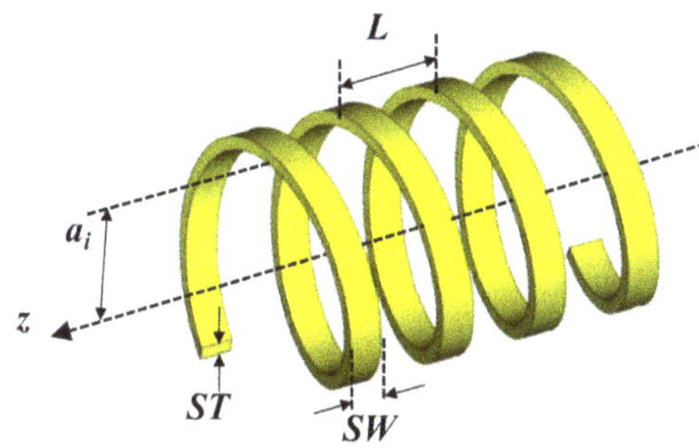

Figure 6.32. Perspective view of a circular tape helix. © [2015] IEEE. Reprinted, with permission, from [22].

wound twin helices. Each cylindrical via has a diameter of *VD* and connects the top and bottom ring pads, which have a diameter of *RD* (>*VD*). A rectangular ring is formed by the top and bottom conducting strips and the two vias at the edges of the metal strips. *L* is the period of the structure. The center-to-center distance between two adjacent rectangular rings is *L*/2. The cross-sectional dimensions of the RRB-SEC along the x- and y-directions are $2a$ and $2b$, respectively. The centers of the rectangular rings coincide with (0, 0). For the circular ring–bar structure shown in figure 6.31, the inner radius and bar width (arc length) are denoted by a_i and *BW*, respectively. The circular tape helix is shown in figure 6.32.

Figures 6.33(a) and (b) respectively show the phase velocity and interaction impedance of the circular helix, circular ring–bar, and RRB-SEC with $b/a = 1$, 23, and 1/23, respectively. The RRB-SEC is considered with both cylindrical vias and tape vias. We assume that the metal shield used in simulations is far away from the structures so that the structures can effectively be considered to be immersed in free space. In addition, *P*, *L*, *SW*, *ST*, and *BW* are the same for all the structures. The detailed dimensions of each structure are listed in table 6.2. The frequency is normalized using an equivalent radius a_{eq}. For the square and rectangular geometries, a_{eq} is $P/2\pi$; while for the circular geometry, a_{eq} is the inner radius a_i. For the circular and square ($b/a = 1$) ring–bar structures, the phase velocity and frequency range of operation are close to each other, since the period and cross-sectional

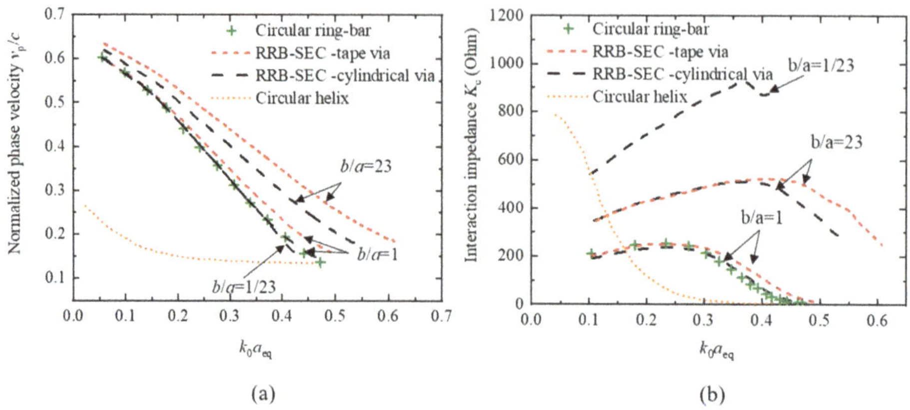

Figure 6.33. Simulated (a) normalized phase velocity and (b) interaction impedance for the circular helix, circular ring–bar, and RRB-SEC for $b/a = 1$, 23, and 1/23, respectively, with fixed P and L. © [2015] IEEE. Reprinted, with permission, from [22].

Table 6.2. Dimensions of various SWSs (μm). © [2015] IEEE. Reprinted, with permission, from [22].

Description	P	L	ST	SW	RD	VD
Circular helix	1382.4	200	20	50	—	—
Rectangular helix	1382.4	200	20	50	—	—
PH-SEC (tape via)	1382.4	200	20	50	—	—
PH-SEC (cylindrical via)	1382.4	200	20	50	70	50

perimeter are the same for both structures. The RRB-SEC with a square cross section tends to have a slightly faster phase velocity and higher interaction impedance than that of the circular cross section due to strong field coupling at the four right-angle corners. For the RRB-SEC with cylindrical vias, the dimensions of the vias and ring pads are greater than those of the tape vias; hence, there is a slight difference in the phase velocity and the interaction impedance for the two cases. In addition, the phase velocity and interaction impedance for the circular and square ring–bar structures are significantly greater than those of the circular helix.

Unlike the circular ring–bar structure, the RRB-SEC allows an additional degree of freedom with respect to the aspect ratio b/a. For $b/a = 23$, the phase velocity, interaction impedance, and bandwidth are greater than those of the square RRB-SEC. For $b/a = 1/23$, the phase velocity does not change much compared to that for the square case; however, interestingly, the interaction impedance is significantly greater than that for $b/a = 23$. The reason for this is that for $b/a = 1/23$, the straight-edge connections (along the x-direction) in a particular ring are quite close to each other, with the current flowing in the same direction.

A microfabricated RRB-SEC SWS operating in the W-band frequencies is described in chapter 8, together with measured results for cold-test S-parameters and simulation results for hot-test parameters.

6.4.2 Coupled PH-SECs

Two further variations of the PH-SEC are shown in figures 6.34(b) and (c). We refer to the structure in figure 6.34(b) as an unconnected pair of PH-SECs. This structure is similar to the pair of square helices proposed in [23], which can provide a larger space for the electron beam and thus has the potential to increase the frequency of operation. The unconnected pair of PH-SECs is expected to offer the same advantages as the structure in [23], together with easier fabrication due to the straight-edge connections. The structure in figure 6.34(c) is referred to as a coaxial pair of PH-SECs; similarly to its circular counterpart, it is expected to have applications in coupling input and output power in a TWT. This structure also offers easier fabrication using printed circuit or microfabrication techniques.

An analysis of the PH-SEC was reviewed in chapter 4. The PH-SEC can be considered to be derived from a pair of unidirectionally conducting (UC) screens by incorporating truncation in the transverse direction in the form of straight-edge connections. In a similar manner, the coupled PH-SEC structures shown in figures 6.34(b) and (c) can be considered to be derived from the generic structure shown in figures 6.34(d) and (e), which consists of four parallel infinite UC screens. This generic structure can lead to the structures in figures 6.34(b) and (c) by the incorporation of straight-edge connections that connect the appropriate screens with suitable directions of conduction.

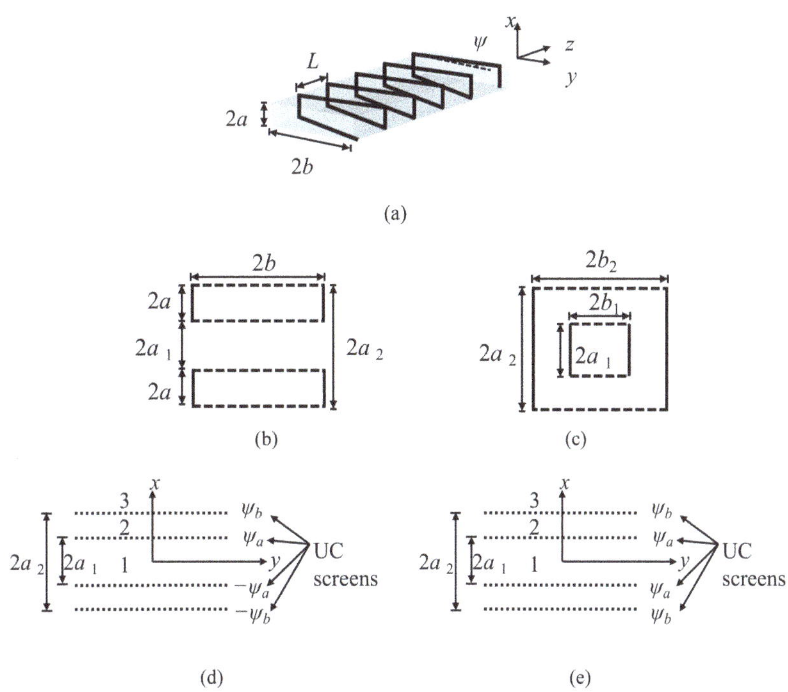

Figure 6.34. (a) Perspective view of the PH-SEC. (b) Cross-sectional view of the unconnected pair of PH-SECs. (c) Cross-sectional view of the coaxial pair of PH-SECs. Cross-sectional view of four parallel infinite unidirectionally conducting (UC) screens: (d) case I, (e) case II. Reprinted with permission from [24].

In the following, we describe a novel and simple technique to obtain the cold-test parameters of the unconnected pair of PH-SECs (figure 6.34(b)) and the coaxial pair of PH-SECs (figure 6.34(c)). In section 6.4.2.1, two cases of the generic structure shown in figures 6.34(d) and (e) are considered, and the dispersion characteristics, including the modes of propagation and field distribution, are calculated. In section 6.4.2.2, we obtain the dispersion characteristics and interaction impedance of the unconnected pair of PH-SECs and the coaxial pair of PH-SECs from the results of the generic structure. A comparison of the results based on simple analytical approximations with those obtained from simulations for the two structures is presented. The simulation results are obtained using the CST MWS Eigenmode Solver and show a good match with the analytical results. In addition, a comparison between the unconnected pair of PH-SECs and a corresponding single PH-SEC is presented, and it is shown that the former has reduced interaction with backward waves. The content of this subsection is based on the authors' work in [24–26].

6.4.2.1 Analysis of four UC screens

The concept of UC screens was described in chapter 3. The generic structure consists of four UC screens. Each of the screens in the structure shown in figures 6.34(d) and (e) is considered to be parallel to the yz-plane and infinite in the y-direction. The separation between the inner two screens along the x-direction is $2a_1$, while it is $2a_2$ for the outer two screens. The screens are symmetrically located with respect to the $x = 0$ plane.

We consider two cases of UC screens with different arrangements of pitch angles (i.e. directions of conduction). In the first case (figure 6.34(d)), the inner two UC screens are assumed to be perfectly conductive in different but symmetric directions $\pm\psi_a$ with respect to the y-axis; similarly, the outer two UC screens conduct in different but symmetric directions $\pm\psi_b$ with respect to the y-axis. In the second case (figure 6.34(e)), the inner two UC screens are assumed to be perfectly conductive in the same direction ψ_a with respect to the y-axis, and the outer two UC screens conduct in the same direction ψ_b with respect to the y-axis. Both structures admit modes that are symmetric or antisymmetric with respect to the $x = 0$ plane. Just as a single pair of UC screens supports two modes (transverse symmetric and transverse antisymmetric) [27–29], the structure with two pairs of UC screens is expected to support four distinct types of propagation mode.

Case I

An analysis of the structure described for case I has been reported in the literature using the equivalent circuit approach [30]. However, only two of the propagating modes were mentioned and discussed. The field distributions were not discussed; neither was there a discussion of how the results for the infinite structure could be applied to practical structures that are finite in the transverse direction. Here, using the field theory approach, we derive the characteristic equations and present the dispersion characteristics for all four modes, together with details of the field distribution for all the modes. These results show clearly how each mode may be excited. This information is very helpful for obtaining approximate analytical results for the coupled PH-SEC structures shown in figures 6.34(b) and (c).

We assume that the wave is propagating with a phase constant β and the fields have a variation of $e^{j\omega t - j\beta z}$. The structure is assumed to be immersed in air. From the point of view of TWT applications, the modes of interest are those that have symmetric variation of E_z with respect to the $x = 0$ plane, i.e. transverse antisymmetric (or longitudinal symmetric) modes. The field expressions for these modes in different regions can be written as follows [27]:

$$E_z = \begin{cases} A\cosh(k_x x) & (0 \leqslant x \leqslant a_1) \\ B_1 \cosh k_x(x - a_1) + B_2 \sinh k_x(x - a_1) & (a_1 \leqslant x \leqslant a_2) \\ Ce^{-k_x(x-a_2)} & (x \geqslant a_2) \end{cases} \tag{6.11}$$

$$H_z = \begin{cases} D\cosh(k_x x) & (0 \leqslant x \leqslant a_1) \\ E_1 \cosh k_x(x - a_1) + E_2 \sinh k_x(x - a_1) & (a_1 \leqslant x \leqslant a_2) \\ Fe^{-k_x(x-a_2)} & (x \geqslant a_2) \end{cases} \tag{6.12}$$

where $k_0^2 = \omega^2 \varepsilon_0 \mu_0$, $k_x^2 = \beta^2 - k_0^2$, and A, B_1, B_2, C, D, E_1, E_2, and F are unknown amplitude constants. For slow-wave solutions, k_x is considered to be a positive real number. Expressions for E_y and H_y can be obtained from equation (3.8). Only fields in the half-space $x \geqslant 0$ are mentioned here; fields in the other half-space can easily be obtained through the use of symmetry. Expressions for the transverse symmetric modes can be written in a similar manner.

Using the properties of the UC screens [27], which are similar to the sheath-helix approximation, the following boundary conditions must be satisfied for the structure:

(1) E_y is continuous across $x = \pm a_1$ and $x = \pm a_2$;
(2) E_z is continuous across $x = \pm a_1$ and $x = \pm a_2$;
(3) The electric field in the direction of conduction is zero at $x = \pm a_1$ and $x = \pm a_2$;
(4) The magnetic field in the direction of conduction is continuous across $x = \pm a_1$ and $x = \pm a_2$.

The characteristic equations for different modes of the structure can be obtained by looking for a nontrivial solution of the set of equations that arise by substituting the field expressions into the boundary conditions. Equations (6.13) and (6.14) are the characteristic equations for the transverse antisymmetric and transverse symmetric modes, respectively.

$$\begin{aligned} &\left[\cosh(k_x a_1) + \left(1 - \frac{k^2}{k_0^2}\tan^2\psi_b\right)X\right]e^{k_x(a_2 - a_1)} + \\ &\frac{k_x^2}{k_0^2}\tan^2\psi_b[\sinh(k_x a_2 - k_x a_1)\sinh(k_x a_1) - \cosh(k_x a_2)] - \cosh(k_x a_1)U = 0 \end{aligned} \tag{6.13}$$

$$\left[\sinh(k_x a_1) + (1 - \frac{k_x^2}{k_0^2}\tan^2\psi_b)Y\right]e^{k_x(a_2 - a_1)} + \frac{k_x^2}{k_0^2}\tan^2\psi_b[\sinh(k_x a_2 - k_x a_1)\cosh(k_x a_1) - \sinh(k_x a_2)] - \sinh(k_x a_1)U = 0 \tag{6.14}$$

where

$$X = \frac{-k_0^2\sinh(k_x a_1) + k_x^2\coth(k_x a_1)\cosh(k_x a_1)\tanh^2\psi_a + k_x\cosh(k_x a_1)V}{-k_0^2 + k_x^2\tan\psi_a\tan\psi_b} \tag{6.15}$$

$$Y = \frac{-k_0^2\cosh(k_x a_1) + k_x^2\tanh(k_x a_1)\sinh(k_x a_1)\tanh^2\psi_a + k_x\sinh(k_x a_1)V}{-k_0^2 + k_x^2\tan\psi_a\tan\psi_b} \tag{6.16}$$

$$U = \frac{-k_0^2}{k_x}\sinh(k_x a_2 - k_x a_1)\frac{\cosh^2(k_x a_2 - k_x a_1)\tan^2\psi_b - \tan\psi_a\tan\psi_b}{\sinh^2(k_x a_2 - k_x a_1)} \tag{6.17}$$

$$V = k_x\coth(k_x a_2 - k_x a_1)(\tan^2\psi_a - \tan\psi_a\tan\psi_b) \tag{6.18}$$

The constants A to F can be expressed in terms of a single constant using field expressions and boundary conditions; for example, the constants B to F can be expressed in terms of A. The expressions are shown as follows:

$$B_1 = A\cosh(k_x a_1) \tag{6.19}$$

$$B_2 = AX \tag{6.20}$$

$$C = A\cosh(k_x a_1)\cosh(k_x a_2 - k_x a_1) \tag{6.21}$$

$$D = A\frac{k_x}{j\omega\mu}\coth(\gamma a_1)\tan\psi_a \tag{6.22}$$

$$E_1 = -A\frac{k_x}{j\omega\mu}\cosh(k_x a_1)\coth(k_x a_2 - k_x a_1)(\tan\psi_a - \tan\psi_b) + AX\frac{k_x}{j\omega\mu}\tan\psi_b \tag{6.23}$$

$$E_2 = A\frac{k_x}{j\omega\mu}\cosh(k_x a_1)\tan\psi_a \tag{6.24}$$

$$F = -A\cosh(k_x a_1)\cosh(k_x a_2 - k_x a_1)\frac{k_x}{j\omega\mu}\tan\psi_b \tag{6.25}$$

Substituting (6.19)–(6.25) into the field expressions in (6.13) and (6.14), one can obtain the field distributions for various modes of the structure.

The dispersion characteristics for the various modes are calculated using (6.13) and (6.14) and plotted for a structure with $a_2/a_1 = 2.2$, $\psi_a = -\psi_b = 2.29°$. As shown in figure 6.35, four modes can be supported by this structure, with different phase

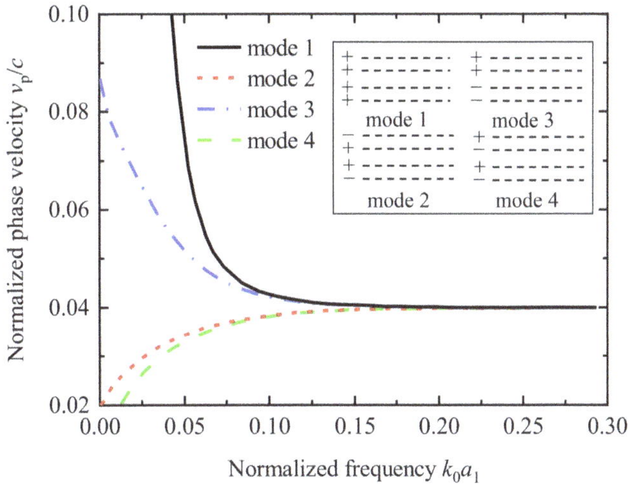

Figure 6.35. Normalized phase velocity for the modes supported by the generic structure in figure 6.34(d) (case I). © [2013] IEEE. Reprinted, with permission, from [25].

velocities. Two of the modes are transverse antisymmetric (modes 1 and 2) and the other two are transverse symmetric (modes 3 and 4). Mode 1 has the highest phase velocity, while mode 4 is the slowest. The phase velocities of the four modes are similar when $k_0a_1 > 0.2$, since the coupling between each screen is weak at higher frequencies. The phase velocity approaches $\sin\psi$ asymptotically. Dispersion curves for the structure with the same dimensions as those used in [30] were also obtained for comparison, and the curves for modes 1 and 2 turned out to be identical to those in [30].

The x-variation of the normalized E_z for $k_0a_1 = 0.13$ and $a_2/a_1 = 2.2$ for all the modes is shown in figure 6.36. From the field distribution, as well as the symmetry of the structure, the different modes can be categorized by the sign of the voltage on each screen. Following the convention adopted in [4], the symmetric variation in E_z between two adjacent screens corresponds to like signs of the voltages on the screens (+ +). On the other hand, the antisymmetric variation in E_z between two adjacent screens corresponds to unlike signs of the voltages on the screens (+ −). Thus, as shown in figure 6.36(a), mode 1 can be described as the '+ + + +' mode; this is represented in the inset in figure 6.35. In the same manner, modes 2, 3, and 4 can be described as the '− + + −,' '+ + − −,' and '+ − + −' modes, respectively. Mode 4 involves a strong coupling between all adjacent screens and correspondingly has the lowest phase velocity (see figure 6.35). In addition, mode 1 is of the most interest here, since it has a relatively large longitudinal electric field in the center, where an electron beam can interact with this field component.

Case II

Case II has different pitch angles compared to case I, which leads to a difference in the field expressions in region 1 ($-a_1 \leqslant x \leqslant a_1$) of the four UC screens. Assuming

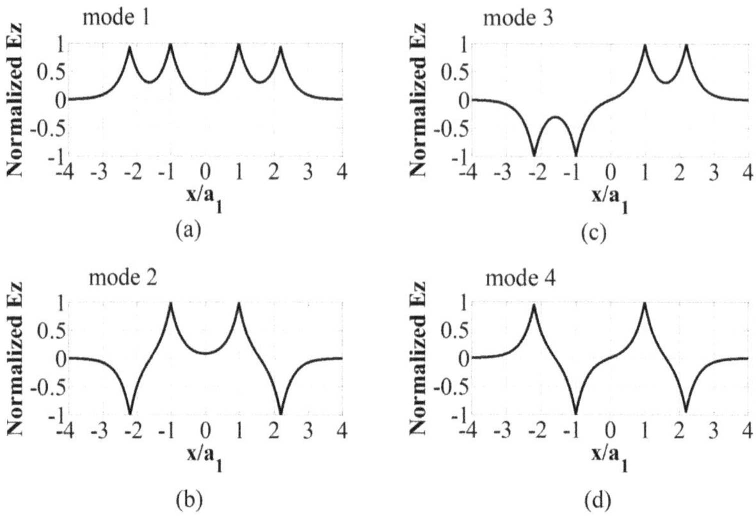

Figure 6.36. x-variation of the normalized E_z for $k_0 a_1 = 0.13$ for all the modes shown in figure 6.35. © [2013] IEEE. Reprinted, with permission, from [25].

that E_z is symmetric with respect to $x = 0$, we have $E_{z1}(a_1) = E_{z1}(-a_1)$. For case I, according to the boundary conditions, at $x = a_1$,

$$E_{y1} \cos \psi_a + E_{z1} \sin \psi_a = 0; \tag{6.26}$$

at $x = -a_1$,

$$E_{y1} \cos \psi_a - E_{z1} \sin \psi_a = 0. \tag{6.27}$$

As a result, E_{y1} is antisymmetric with respect to $x = 0$, i.e. $E_{y1}(a_1) = -E_{y1}(-a_1)$. According to the relationship between E_y and H_z:

$$E_y = -\frac{j\beta}{k_c^2}\frac{\partial E_z}{\partial y} + \frac{j\omega\mu}{k_c^2}\frac{\partial H_z}{\partial x} = \frac{j\omega\mu}{k_c^2}\frac{\partial H_z}{\partial x}. \tag{6.28}$$

H_{z1} is symmetric with respect to $x = 0$, i.e. $H_{z1}(a_1) = H_{z1}(-a_1)$, which is consistent with the field expressions presented in (6.11) and (6.12). But for case II at $x = a_1$,

$$E_{y1} \cos \psi_a + E_{z1} \sin \psi_a = 0; \tag{6.29}$$

at $x = -a_1$,

$$E_{y1} \cos \psi_a + E_{z1} \sin \psi_a = 0. \tag{6.30}$$

Contrary to case I, here E_{y1} is symmetric and H_{z1} is antisymmetric with respect to $x = 0$. Considering this, the field expressions for case II should be written as:

$$E_z = \begin{cases} A \cosh(k_x x) & (0 \leqslant x \leqslant a_1) \\ B_1 \cosh k_x(x - a_1) + B_2 \sinh k_x(x - a_1) & (a_1 \leqslant x \leqslant a_2) \\ Ce^{-k_x(x-a_2)} & (x \geqslant a_2) \end{cases} \tag{6.31}$$

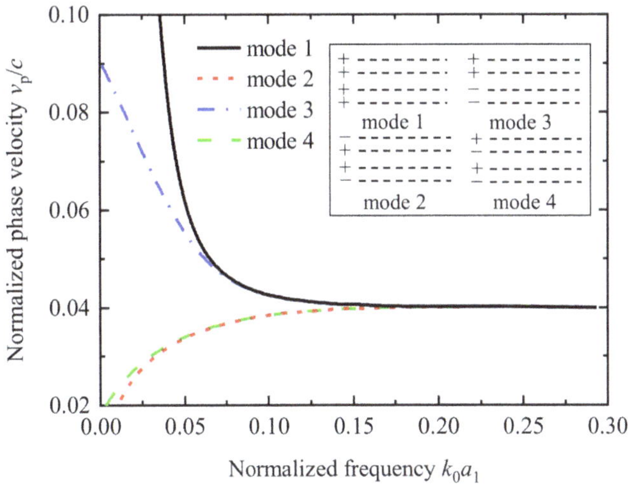

Figure 6.37. Normalized phase velocity for the modes supported by the generic structure in figure 6.34(e) (case II). Reprinted with permission from [24].

$$H_z = \begin{cases} D\sinh(k_x x) & (0 \leqslant x \leqslant a_1) \\ E_1 \cosh k_x(x - a_1) + E_2 \sinh k_x(x - a_1) & (a_1 \leqslant x \leqslant a_2)\ . \\ Fe^{-k_x(x-a_2)} & (x \geqslant a_2) \end{cases} \tag{6.32}$$

The remaining steps in the analysis for case II are exactly the same as for case I. The dispersion diagram (figure 6.37) is plotted with the same dimensions as for case I: $a_2/a_1 = 2.2$, $\psi_a = -\psi_b = 2.29°$. Similar to case I, there are four modes, including two symmetric modes and two antisymmetric modes. These modes can also be described as the '+ + + +,' '− + + −,' '+ + − −,' and '+ − + −' modes, respectively. The dispersion curves for the various modes are quite close to the curves in case I—at least for this set of dimensions. One difference should be noted: in case I, mode 4 is the slowest among all modes, but for case II, mode 4 is a little faster than mode 2. The x-variation of the normalized E_z for $k_0a_1 = 0.13$ and $a_2/a_1 = 2.2$ for all the modes is shown in figure 6.38. We can see that although cases I and II involve different natures of H_z, the distribution for E_z is exactly the same; this leads to comparable interaction impedance values for both cases. As a result, both cases of the generic structure have similar potential for application in TWTs.

6.4.2.2 Two coupled PH-SECs

For practical application, the screens need to be confined in the transverse direction. This can be achieved by truncating the screens in the y-direction and connecting the edges of the truncated screens with straight-edge connections in different ways. In this section, we apply the results of the previous section to obtain approximate analytical results for two practical structures that can be derived from the generic structure consisting of four parallel infinite UC screens. Two of these possibilities, shown in figures 6.34(b) and (c), are considered in the following.

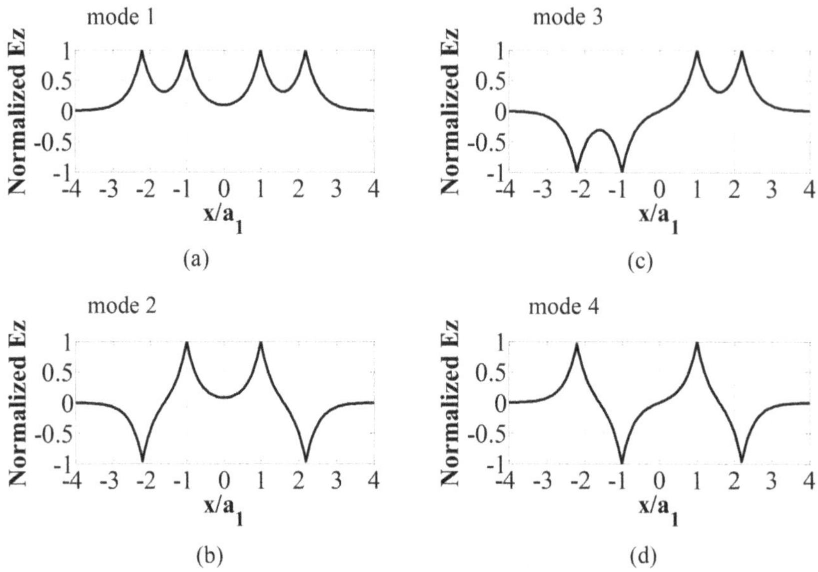

Figure 6.38. x-variation of the normalized E_z for $k_0a_1 = 0.13$ for all the modes shown in figure 6.37. Reprinted with permission from [24].

Unconnected pair of PH-SECs

The unconnected pair of PH-SECs is obtained from the generic structure when $\psi_a = -\psi_b$, the top two screens are connected together, and the bottom two screens are connected together. As presented in figure 6.34(b), this results in an unconnected pair of identical PH-SECs. In general, we expect to observe only two modes in this structure, since the signs of the voltages on the screens connected with straight-edge connections cannot be different. As a result, the two modes of this structure should correspond to mode 1 (+ + + +) and mode 3 (+ + − −). To apply the analysis of the previous section to this structure, we calculate an effective pitch angle which, by analogy with a circular helix, is defined as the arctangent of the ratio of the period L and the perimeter $4a + 4b$ of the cross section of each PH-SEC (see figure 6.34(a)). Thus:

$$\psi_{a\_\text{eff}} = -\psi_{b\_\text{eff}} = \tan^{-1}\left(\frac{L}{4(a+b)}\right). \tag{6.33}$$

As this structure involves two identical PH-SECs, we just need one value of the effective pitch angle. The above value of the effective pitch angle is used to replace the pitch angles ψ_a and ψ_b in the expressions in (6.13) and (6.14) for case I. Furthermore, the value of the interaction impedance K_c at the center can also be calculated using equation (1.2). For a structure with infinite width along the y-direction, the power flow per unit width, P_{uw}, is often used [31]:

$$P_{uw} = \frac{1}{2}\frac{\beta}{\omega\mu}\int\left(\frac{\mu_0}{\varepsilon_0}\left|H_y\right|^2 + \left|E_y\right|^2\right)dx. \tag{6.34}$$

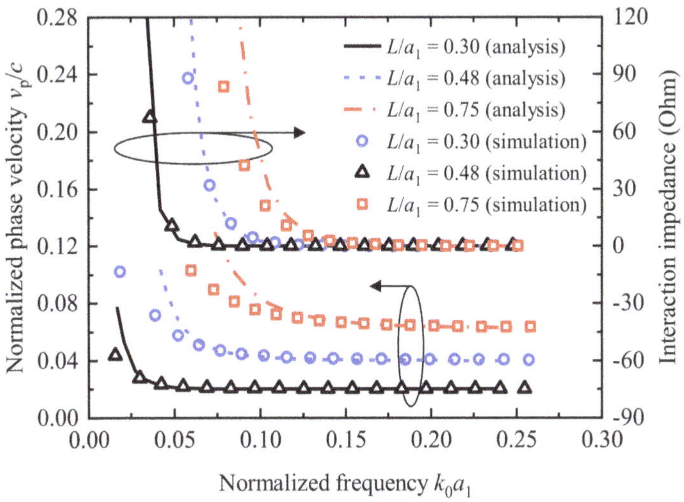

Figure 6.39. Normalized phase velocity and interaction impedance for mode 1 (+ + + +) of the unconnected pair of PH-SECs for various values of the normalized period L/a_1. © [2013] IEEE. Reprinted, with permission, from[25].

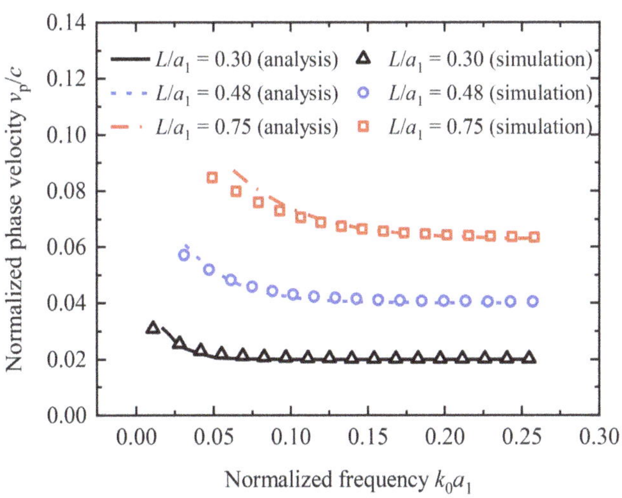

Figure 6.40. Normalized phase velocity for mode 3 (+ + − −) of the unconnected pair of PH-SECs for various values of the normalized period L/a_1. © [2013] IEEE. Reprinted, with permission, from [25].

The total power P can be obtained by multiplying P_{uw} by the width in the y-direction:

$$P = P_{uw} \times \text{width}. \tag{6.35}$$

Figures 6.39 and 6.40 show the results calculated for case I from the approximate analysis mentioned above as well as the simulation results obtained from the Eigenmode Solver of CST MWS. The geometrical parameters of the structure are:

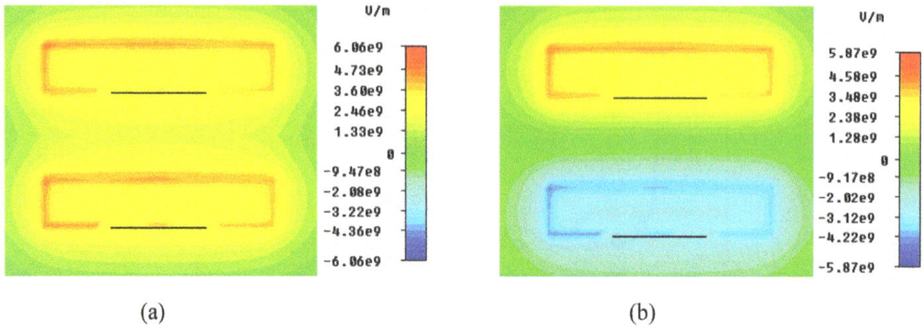

Figure 6.41. Distribution of E_z on the xy-plane for the unconnected pair of PH-SECs: (a) mode 1 (+ + + +) and (b) mode 3 (+ + − −). © [2013] IEEE. Reprinted, with permission, from [25].

$a_2/a_1 = 2.2$, $b/a_1 = 2.4$, and $L/a_1 = 0.24$, 0.48, and 0.75. The results based on the approximate analysis match very well with the simulation results, except for low frequencies, where the effect of the transverse truncation is more severe and is not captured in the approximate analysis, which is based on infinite screens. As expected, the phase velocity increases as the period increases. Further, as the period gets larger, the actual structure deviates more and more from the sheath-helix model. As a result, the accuracy of the results based on the analysis reduces when the period is large. The interaction impedance of mode 3 (+ + − −) is not shown in figure 6.40, since E_z is zero at $x = 0$ in this mode.

For mode 1 (+ + + +) and mode 3 (+ + − −), which are supported by the unconnected pair of PH-SECs, the simulated field distribution of E_z over the xy-plane is also obtained. As shown in figures 6.41(a) and (b), for $L/a_1 = 0.48$ and a phase shift per period of 35°, the other dimensions remaining the same as those for figures 6.39 and 6.40, the simulated field distribution is consistent with the analytical results presented for these two modes in figures 6.36(a) and (c), respectively.

We only present the results for case I here; the results for case II can be obtained in a similar manner.

Study of backward-wave oscillation

It is known that in a TWT based on the single circular helix or the single PH-SEC, there is a possibility of oscillations due to interaction with the backward waves, especially at high power levels. We compare the cold-test parameters of an unconnected pair of PH-SECs with those of a single PH-SEC; both have the same electron-beam tunnel cross section of 4.8 × 4.8 mm with respect to the interaction with backward waves. We use $2a = 0.6$ mm and a period of $L = 1.5$ mm for the unconnected pair of PH-SECs and $L = 3.8$ mm for the single PH-SEC, so that both structures have the same phase velocity at 6 GHz for the forward wave. Figure 6.42 (a) presents the complete simulated dispersion characteristics for the first three transverse antisymmetric modes. To study the beam–circuit interaction, we consider two beam lines with different velocities, beam 1 and beam 2. As can be seen, the beam lines intersect the first modes of both structures at 6 GHz. They also intersect

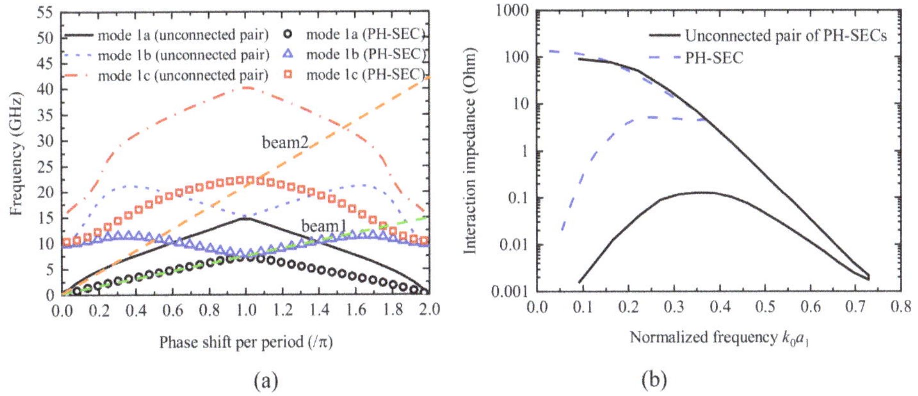

Figure 6.42. (a) Dispersion characteristics for the first three transverse antisymmetric modes, and (b) interaction impedance of a pair of unconnected PH-SECs and a single PH-SEC with the same beam tunnel cross section. © [2013] IEEE. Reprinted, with permission, from [25].

the backward waves of the second modes at around 8 and 17 GHz, respectively. Thus, the working frequency of the single PH-SEC is much closer to the frequency of backward-wave oscillation. Figure 6.42(b) shows that the interaction impedance for the backward wave for the single PH-SEC is much larger than that of the unconnected pair of PH-SECs throughout the frequency range. Thus, compared to the single PH-SEC, the unconnected pair of PH-SECs has better potential for avoiding backward-wave oscillations in a TWT at high power levels.

Coaxial pair of PH-SECs

For this structure, $\psi_a = -\psi_b$; however, here the two inner UC screens are connected together and the two outer UC screens are connected together. As shown in figure 6.34 (c), this results in a coaxial pair of PH-SECs. Similar to a coaxial pair of circular helices, this structure can be used to couple input and output power in a TWT. Using the same reasoning as for the previous structure, this structure also admits only two modes, which can be labeled as mode 1 $(++++)$ and mode 2 $(-++-)$. In this case, we calculate two different values of the effective pitch angles $-\psi_{a\_\mathrm{eff}}$ for the inner PH-SEC and ψ_{b_eff} for the outer PH-SEC. With reference to figure 6.34(c), these effective angles can be written as:

$$\psi_{a\_\mathrm{eff}} = \tan^{-1}\left(\frac{L}{4(a_1 + b_1)}\right) \tag{6.36}$$

$$\psi_{b\_\mathrm{eff}} = \tan^{-1}\left(\frac{L}{4(a_2 + b_2)}\right). \tag{6.37}$$

To determine the interaction impedance, the total power P in this structure is calculated by multiplying the power flow per unit width P_{uw} by an average width $2b = b_1 + b_2$. Figures 6.43 and 6.44 show the results obtained from the above considerations as well as simulations. The interaction impedance of mode 2 $(-++-)$

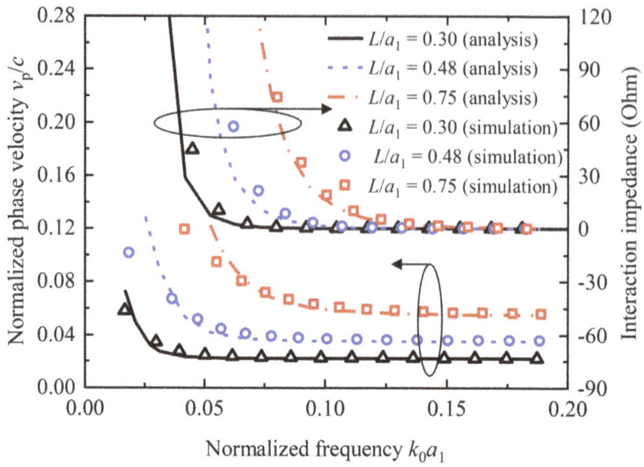

Figure 6.43. Normalized phase velocity and interaction impedance for mode 1 (+ + + +) in a coaxial pair of PH-SECs for different values of the normalized period L/a_1. © [2013] IEEE. Reprinted, with permission, from [25].

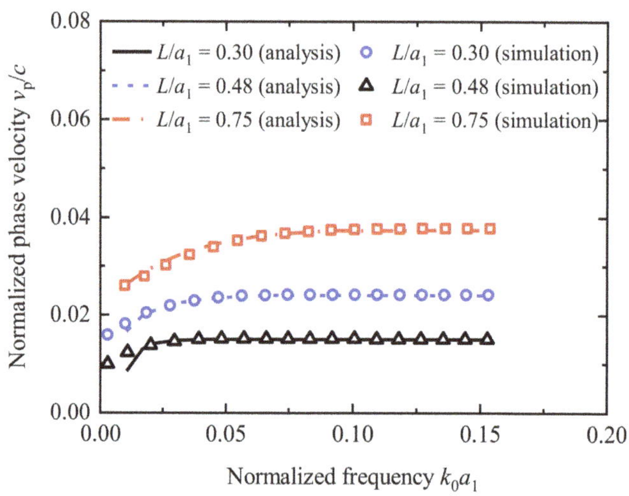

Figure 6.44. Normalized phase velocity for mode 2 (− + + −) in a coaxial pair of PH-SECs for different values of the normalized period L/a_1. © [2013] IEEE. Reprinted, with permission, from [25].

is not included in figure 6.44, since the E_z field components of the inner and outer helices roughly cancel out, resulting in interaction impedance values close to zero. The dimensions of the structure used for these two figures are: $a_2/a_1 = 2.2$, $b_1/a_1 = 2.4$, $b_2/a_1 = 2.8$, and $L/a_1 = 0.3$, 0.48, and 0.75. Once again, the results of the approximate analysis and simulations match very well except at low frequencies. As expected, a larger period results in a higher phase velocity. It can also be seen that, unlike the previous case, this case has different values of the phase velocities for the two modes at high frequencies. A similar observation was made in [29], and the difference is caused

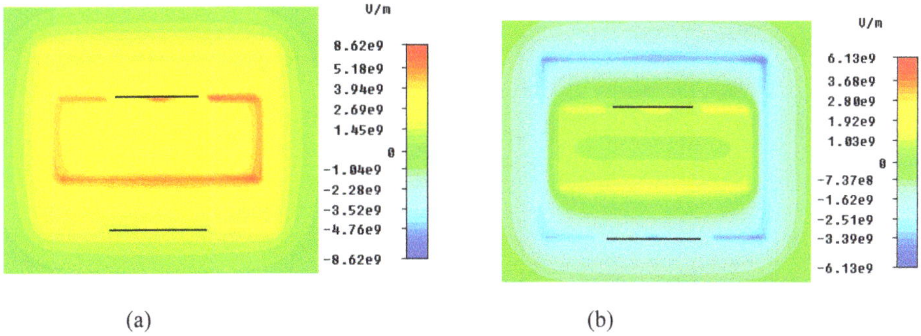

Figure 6.45. Distribution of E_z in the xy-plane for the coaxial pair of PH-SECs: (a) mode 1 (+ + + +), and (b) mode 2 (− + + −). © [2013] IEEE. Reprinted, with permission, from [25].

by the different values of the effective pitch angles for the inner and the outer PH-SECs. This is explained more clearly with the help of the field distribution, which is considered next.

The simulated distribution of E_z over the xy-plane for mode 1 (+ + + +) and mode 2 (− + + −), supported by the coaxial pair of PH-SECs, is shown in figures 6.45(a) and (b) for $L/a_1 = 0.48$ and a phase shift per period of 35°, the other dimensions remaining the same as for figures 6.43 and 6.44. The simulated field distribution is qualitatively consistent with the analytical results presented for these two modes in figures 6.36(a) and (b), respectively. Quantitatively, there are differences in the field values, since the results in figure 6.36 are for the case of equal-magnitude pitch angles for the four screens. Furthermore, it can be seen in figure 6.45 (a) that for mode 1 (+ + + +), the E_z values are stronger close to the inner PH-SEC, showing that at high frequencies, the inner PH-SEC dominates for this mode. On the other hand, in figure 6.45(b), for mode 2 (− + + −), the E_z values are stronger close to the outer PH-SEC, implying that the outer PH-SEC dominates for this mode. Equations (6.36) and (6.37) show that the value of the effective pitch angle, ψ_{a_eff}, for the inner PH-SEC is larger compared to the value of the effective pitch angle, ψ_{b_eff}, for the outer PH-SEC. Accordingly, the phase velocity values at high frequencies for mode 1 (+ + + +) in figure 6.43 are higher than those for mode 2 (− + + −) in figure 6.44. The coaxial pair of PH-SECs does not have the advantage of reducing the risk of backward-wave oscillation compared to the single PH-SEC. However, it is useful for coupling energy from one helix to the other, just like the coaxial circular helices [4, 5].

6.4.2.3 Realization of a connected pair of PH-SECs

The dispersion characteristics of the unconnected pair of PH-SECs were obtained using field analysis, and it has been shown that this structure has a larger electron-beam tunnel and a better potential for avoiding backward-wave oscillations compared to a single PH-SEC. However, similar to other coupled SWSs, such as the multi-helix [23] and the double V-shaped MML [14], the unconnected pair of PH-SECs also supports both a symmetric mode and an antisymmetric mode. In

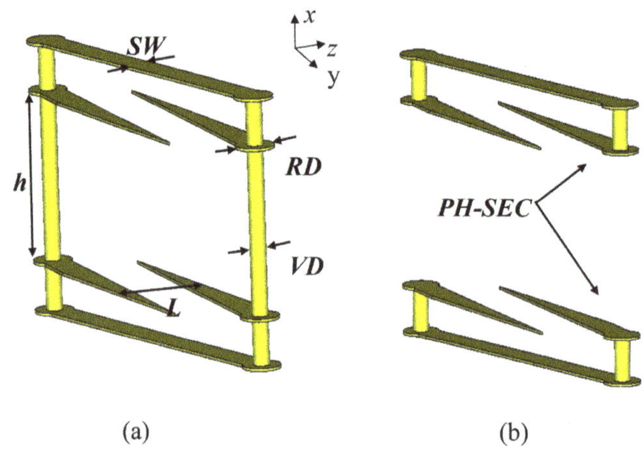

Figure 6.46. Perspective view of one period of (a) a connected pair of PH-SECs; (b) an unconnected pair of PH-SECs. © [2014] IEEE. Reprinted, with permission, from [26].

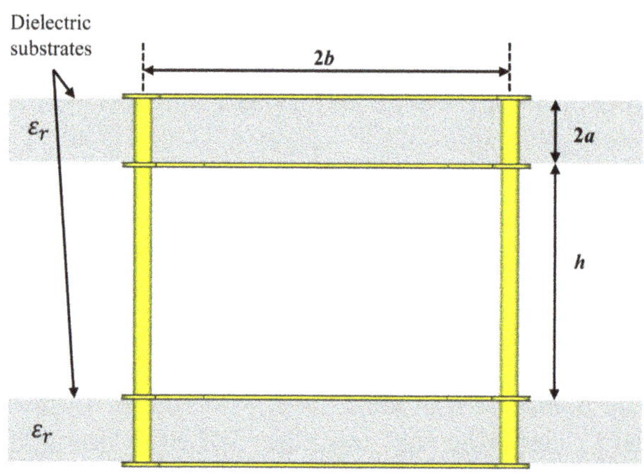

Figure 6.47. Cross-sectional view of the printed circuit configuration of the connected pair of PH-SECs. © [2014] IEEE. Reprinted, with permission, from [26].

order to excite the symmetric mode in these structures, which is the mode useful for beam–wave interaction, one needs to provide equi-amplitude and equi-phase inputs and outputs for the coupled SWSs. This often requires a power divider and power combiner, making the entire structure large and complex. Moreover, any errors in the performance of the power dividers adversely affect the performance of the SWS. This problem may become more serious for higher frequencies of operation.

As shown in figure 6.46(a), a connected pair of PH-SECs can be obtained by connecting the two PH-SECs in the unconnected pair (figure 6.46(b)) by extending the straight-edge connections at the edges. In this section, we present a proof-of-concept design and the fabrication of the connected pair of PH-SECs using printed circuit techniques; the target frequency range is S band. Figure 6.47 shows the cross section of the connected pair of PH-SECs; it incorporates two dielectric substrates, with one PH-SEC on each substrate. We use Rogers RO4003 ($\varepsilon_r = 3.55$, loss tangent = 0.0027, thickness = 0.813 mm) as the dielectric substrates.

Electromagnetic properties

Simulations of both connected and unconnected pairs of PH-SECs with the same dimensions were carried out using the CST MWS Eigenmode Solver. The dimensions of both structures are listed in table 6.3. The current distributions for the symmetric mode on the two structures are presented in figure 6.48 for a frequency of 4 GHz. We can see that in the connected pair, the current mainly flows along the upper and lower PH-SECs. Only a small amount of current flows from the upper PH-SEC to the lower PH-SEC, and this can be neglected compared with the current on each PH-SEC. Thus, the current distribution for the connected pair of PH-SECs is very similar to that of the symmetric mode of the unconnected pair of PH-SECs.

The phase velocity and interaction impedance of the two structures are presented in figures 6.49(a) and (b), respectively. It can be seen that the phase velocity and interaction impedance for the connected pair of PH-SECs are almost the same as

Table 6.3. Dimensions of the connected and unconnected pairs of PH-SECs [24].

Parameter	Value (mm)
h	2
$2a$	0.813
$2b$	6
L	1.5
SW	0.5
VD	0.1
RD	0.35

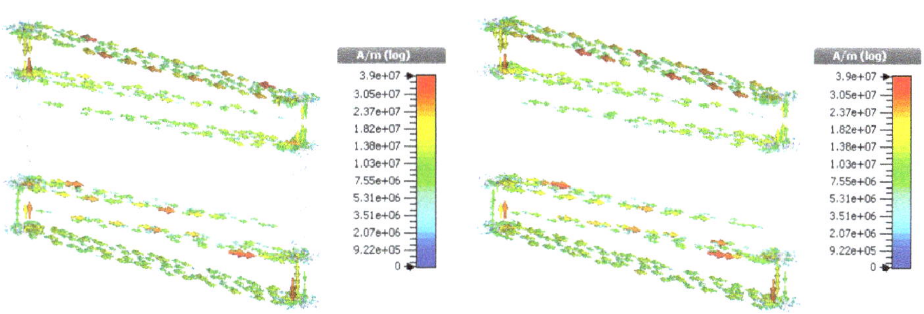

(a) (b)

Figure 6.48. Current distribution at 4 GHz (a) on the connected pair of PH-SECs and (b) on the unconnected pair of PH-SECs for the symmetric mode. © [2014] IEEE. Reprinted, with permission, from [26].

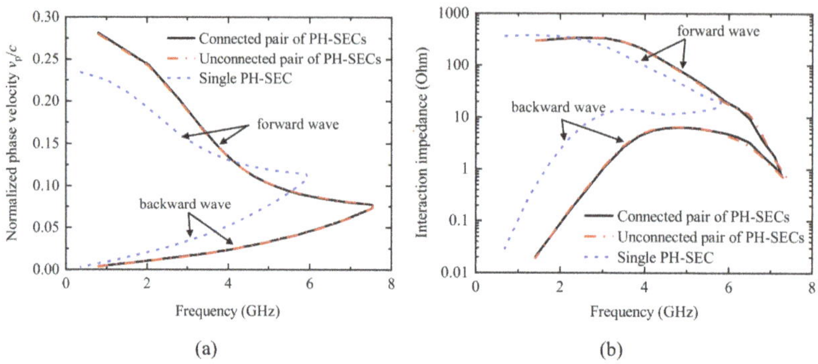

Figure 6.49. (a) Normalized phase velocity and (b) interaction impedance of the single PH-SEC and connected and unconnected pairs of PH-SECs. © [2014] IEEE. Reprinted, with permission, from [26].

those for the symmetric mode of the unconnected pair of PH-SECs. These results are sufficient to establish that extended conductors connecting the upper and lower PH-SECs do not have much influence on the characteristics of the connected pair for the lowest-order mode of propagation. At the same time, since the two PH-SECs are connected in the connected pair, this structure should be able to support the symmetric mode without a complex feed.

In figure 6.49, the dispersion and interaction impedance characteristics of the connected pair of PH-SECs are also compared with those of a single PH-SEC that has the same size of electron-beam tunnel as the connected pair. The period of the single PH-SEC is set to 2.8 mm in order to obtain a comparable phase velocity to that of the connected pair. The substrate and other parameters are the same for both structures. The dispersion characteristics results in figure 6.49(a) show that the forward wave for the connected pair of PH-SECs and the single PH-SEC have the same phase velocity at around 4 GHz. Figure 6.49(b) shows that the interaction impedance of the connected pair for the forward wave is generally higher than that of the single PH-SEC; this should result in a higher gain growth rate in a TWT using the connected pair. In addition, the interaction impedance values for backward waves in the connected pair are lower than those for the single PH-SEC; this should result in a lower risk of backward-wave oscillation in TWTs.

Fabrication

The connected pair of PH-SECs was fabricated on RO4003 substrates using printed circuit techniques. A perspective view of the printed structure is shown in figure 6.50 (a). A photograph of the side view of the fabricated connected pair of PH-SECs is shown in figure 6.50(b). Instead of two bulky power dividers that would be required in the input/output feed for the unconnected pair, only simple microstrip-line feeds are needed to excite the longitudinal symmetric mode in the connected pair.

Both the upper and the lower PH-SECs are first connected to 67 Ω microstrip lines, which are then tapered to 50 Ω microstrip lines in order to connect to standard SMA connectors. The widths w_1 and w_2 of the microstrip lines are 1.1 and 1.81 mm for the two impedance levels, respectively. The taper length L_g is 18.95 mm. The size

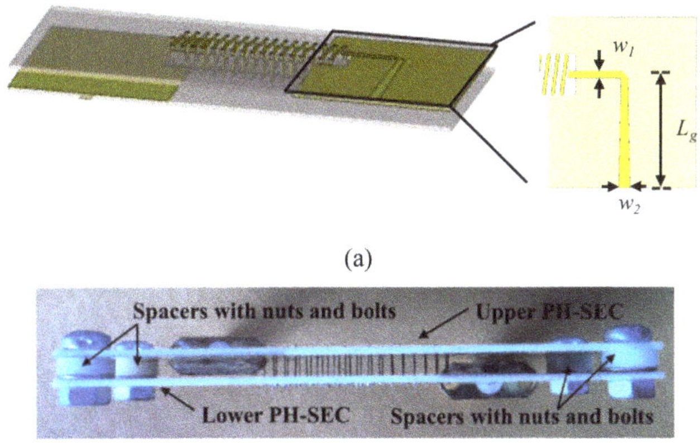

(b)

Figure 6.50. Connected pair of PH-SECs fabricated using printed circuit techniques. (a) Perspective view showing two dielectric substrates with printed PH-SECs and microstrip feed; (b) side view showing the wire connections between the two PH-SECs and input/output SMA connectors. © [2014] IEEE. Reprinted, with permission, from [26].

and position of the ground planes for the microstrip lines have a great influence on the return loss and therefore need to be designed carefully. After optimization using the CST MWS transient solver, a design with low return loss is achieved. The inclined horizontal conductors of the two PH-SECs, together with PTHs, are printed on dielectric substrates which are then aligned and fixed together with nuts and bolts. Thin metal wires are inserted through the PTHs and soldered to connect the two PH-SECs and also to act as the vertical straight-edge connections.

To measure the phase velocity of the connected pair, two structures with 16 and 21 periods are designed and fabricated. These two structures are identical except for the number of periods.

Figure 6.51(a) shows the simulated and measured S_{11} and S_{21} values for the connected pair of PH-SECs with 16 periods over the frequency range from 1 to 7 GHz. The simulation results are obtained using the Transient solver in CST MWS. The measurement results match well with the simulations. The measured S_{11} indicates a wide -15 dB bandwidth ranging from 1.81 to 4.98 GHz. Simulated as well as measured phase velocity results over the frequency range of 2–5 GHz are shown in figure 6.51(b). The phase velocity is obtained using equation (1.22). The measured phase velocity values are slightly lower than the simulated values. This difference may be attributed to fabrication tolerances, extra lengths of connecting wires at the solder joints, and the difference in the conductivity values used in simulation and those achieved in fabrication.

Hot-test parameters

Hot-test parameters for both the connected pair of PH-SECs and the single PH-SEC were obtained by PIC simulations using CST Particle Studio. The substrate parameters and the dimensions follow the values mentioned in the previous section.

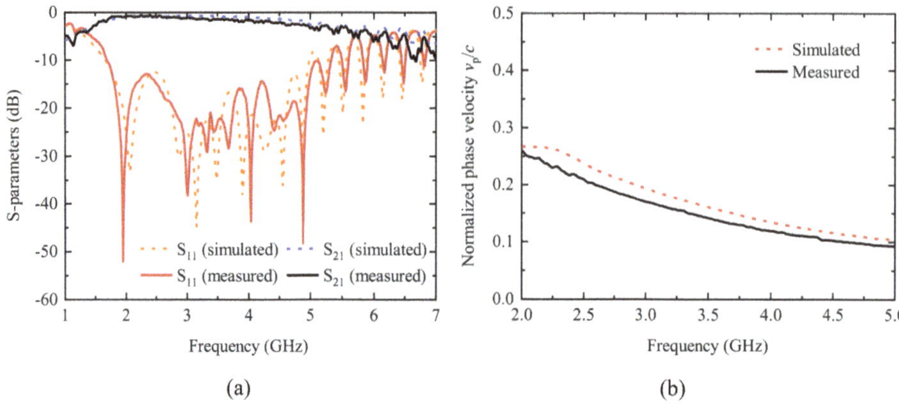

Figure 6.51. Simulated and measured results: (a) S-parameters for the connected pair of PH-SECs with 16 periods; (b) normalized phase velocity for the connected pair of PH-SECs. © [2014] IEEE. Reprinted, with permission, from [26].

In order to make a fair comparison between the hot performances of the two structures, the interaction lengths of the two structures are set to the same value. Thus, we use 50 periods with a total interaction length of 75 mm for the connected pair of PH-SECs and 27 periods with a total interaction length of 75.6 mm for the single PH-SEC. For simplicity, discrete ports are used to achieve good values of S_{11}. After optimization using the CST MWS Transient solver, the S_{11} values are lower than −20 dB for both structures over the frequency range from 3.6 to 4.2 GHz. An e-beam with a beam voltage of 5.35 kV and a current of 200 mA is applied at the center of the electron-beam tunnel. The elliptical cross section of the e-beam has a semi-major axis of 1 mm and a semi-minor axis of 0.5 mm. The focusing magnetic field is set to 0.5 T. Figure 6.52 shows the variation in linear gain with frequency for the connected pair of PH-SECs and the single PH-SEC with an input signal power of 100 mW. The linear gain is the small-signal gain, where the output power increases linearly with the input power. At the peak gain frequency, the gain values for the connected pair and the single PH-SEC are 18.5 and 12.5 dB, respectively; the corresponding gain growth rate values for the two structures are 0.25 and 0.17 dB mm^{-1}, respectively. Thus, as expected from the interaction impedance values, the connected pair has a higher gain growth rate than the single PH-SECs. However, since the design of the connected pair presented here is more dispersive than the single PH-SEC, it also has a smaller 3 dB bandwidth. The dispersion can be reduced, e.g. by incorporating a suitably designed metal shield [32].

In addition, the linear gain of the connected pair of PH-SEC-based TWTs is also investigated using Pierce theory. Some of the important parameters, such as the propagation constant and the interaction impedance, can be captured from figure 6.49. The detuning parameter b, which represents the level of electron synchronism, can easily be calculated using the beam voltage. The attenuation parameter d can be calculated from the S-parameters in figure 6.51. A value of 0.1 is used for the space charge parameter QC. Details of the calculation process can be

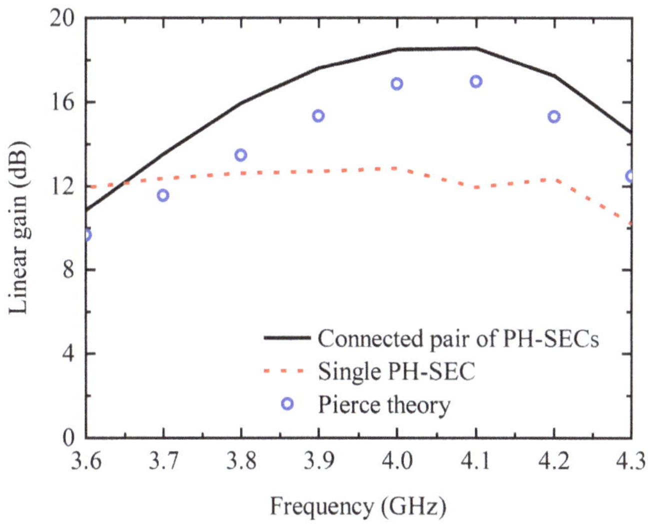

Figure 6.52. Linear gain vs. frequency for the connected pair of PH-SECs and the single PH-SEC for an input signal power of 100 mW. © [2014] IEEE. Reprinted, with permission, from [26].

found in section 1.7.1.2. The calculated gain is also presented in figure 6.52. The maximum gain is achieved at 4.1 GHz, which is very close to the PIC simulation results. It is noted that the gain calculated using Pierce theory is less than the simulated gain values. This may be caused by growing electron bunching and space charge effects along the circuit [33].

The hot-test performance of the connected pair of PH-SECs for application in a high-gain TWT is also investigated. In order to reduce the reflected signal in the TWT, the structure is divided into two sections that are 30 periods and 80 periods long, and an idealized sever is applied between the two sections. The sever consists of two ports with a separation of 7 periods. The two ports are used to absorb the forward and reflected signals. The e-beam parameters and the feed design are the same as those mentioned in the previous paragraph. Figure 6.53 shows the output power and gain of the connected pair as a function of the input power at 4 GHz. The maximum gain of this structure is around 42.5 dB, and the maximum peak output power is 56.7 dBm (468 W) at the saturation point for an input power of 17.8 dBm (60 mW). The corresponding maximum RF efficiency is 21.9%. Figure 6.54 shows the simulated phase space parameter values as a function of the longitudinal position at saturation at 4 GHz after 20 ns from the start. We can see that more electrons are decelerated than accelerated, clearly indicating that the e-beam loses energy to the electromagnetic wave.

The design and simulation results presented here serve as a proof of concept. In practical applications, the issues of outgassing, dielectric charging, and thermal dissipation need to be addressed. Outgassing may cause cathode poisoning. The electron beam can charge up the dielectric and may eventually cause breakdown.

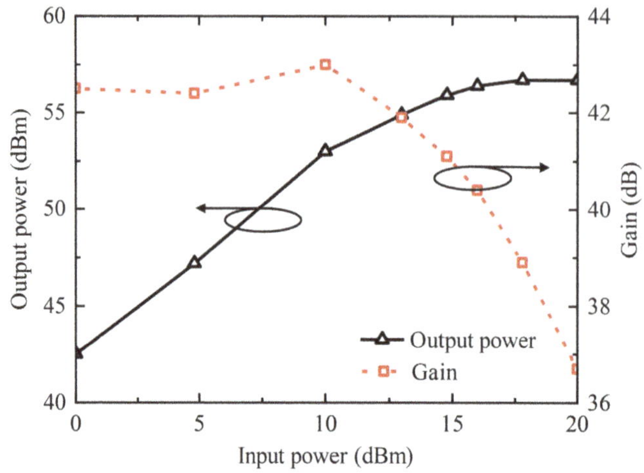

Figure 6.53. Output power and gain vs. input power at 4 GHz for the connected pair of PH-SECs. © [2014] IEEE. Reprinted, with permission, from [26].

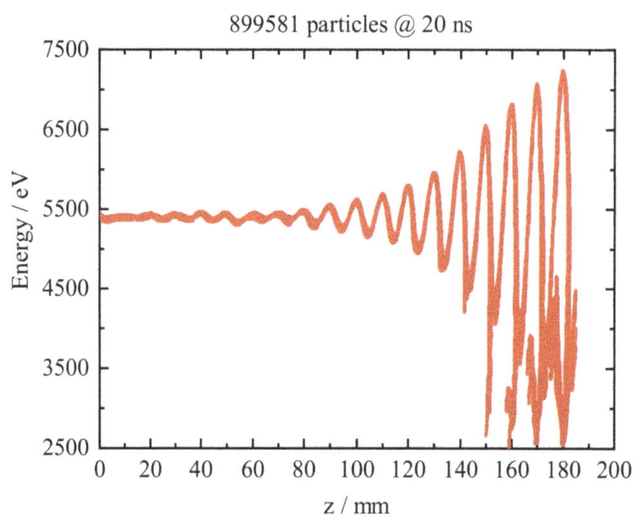

Figure 6.54. Phase space energy vs. the longitudinal position. © [2014] IEEE. Reprinted, with permission, from [26].

Thermal issues are also very important in determining the power handling capability of the TWT. These issues can be addressed with suitable dielectric material selection and electrical and thermal design.

6.5 PH-SEC with reduced dielectric charging effect

The dielectric charging problem is very important for millimeter-wave or terahertz TWTs, since it is difficult to control the alignment of different parts and achieve an adequately high magnetic field at such high frequencies. If the electrons hit the SWS and the supporting dielectric, they create a voltage difference between the SWS and the supporting dielectric. This voltage may defocus the electron beam and even cause dielectric breakdown when the voltage is too high.

Some methods have been proposed in the past to address the problem of dielectric charging. One proposed method is to coat the dielectric with a thin layer of conductive material [34]. However, the thickness of the coating is difficult to control and may cause excessive RF loss in the circuit. Another method is to coat the dielectric with BeO, which has been proven to have a relatively lower dielectric charging effect [35, 36], but this also cannot fully avoid the problem in some cases. Yet another method proposed is to replace the material of the rods with a lossy dielectric material that exhibits a relatively high electrical conductivity at low frequencies [37]; but this may again cause high loss at millimeter-wave frequencies.

In this section, we first propose a Ka-band symmetric PH-SEC SWS that can be fabricated using microfabrication techniques [39]. Next, two modifications to the symmetric PH-SEC are proposed to avoid the dielectric charging problem. As a first step in solving this problem, a simple single-dielectric model is proposed to study the dielectric charging phenomenon; this model is analyzed using an equivalent circuit and also simulated using CST Particle Studio. This exercise demonstrates that the phenomenon of dielectric charging can be simulated accurately. Then, a more realistic model consisting of two dielectric layers, Si and silicon dioxide (SiO_2), is studied; this model shows that even a thin layer of SiO_2 plays a dominant role in dielectric charging. This provides guidelines for modifications to the symmetric PH-SEC SWS to avoid the dielectric charging problem. First, the SiO_2 layer is removed between the metal strips to prevent the electrons from landing on the SiO_2 layer directly. Second, the Si conductivity is appropriately enhanced to avoid the buildup of charge and voltage. The effects of these modifications are presented, and it is shown that the modified SWS exhibits substantially reduced dielectric charging while maintaining a low insertion loss. The modifications proposed here are compatible with microfabrication. The content of this section closely follows [24, 38, 39], which reported the work done by the research group of the authors of this book.

6.5.1 Symmetric PH-SEC

The microfabrication of a PH-SEC intended for use in the W band has been demonstrated on a thick Si substrate (figure 6.55(b)) [40]; because of the asymmetry of the structure, the interaction impedance turns out to be asymmetric, which may cause high circuit loss and mode competition problems in a TWT [40]. One way to avoid this problem is to make a trench in the Si substrate beneath the PH-SEC [41]. The symmetric configuration of the PH-SEC shown in figure 6.55(a) can solve the problem of asymmetry in an alternative manner. Unlike a structure with a thick substrate at the bottom, the PH-SEC is symmetrically placed between two thin Si substrates. The entire structure is placed in a metal enclosure. The size of the metal enclosure can affect the dispersion, interaction impedance, etc.

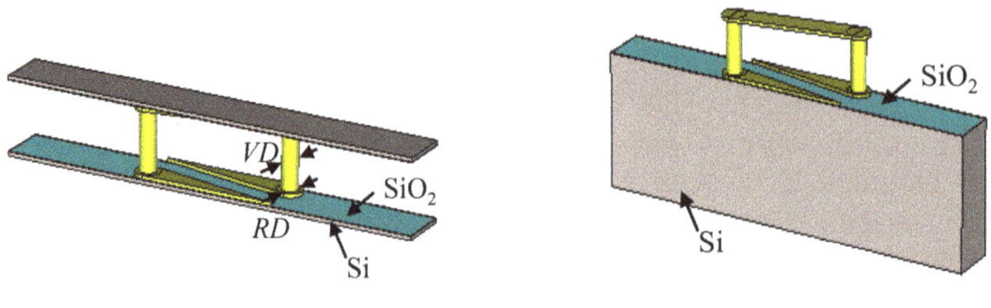

Figure 6.55. (a) one period of the symmetric PH-SEC, and (b) one period of the PH-SEC on a thick substrate. © [2014] IEEE. Reprinted, with permission, from [38].

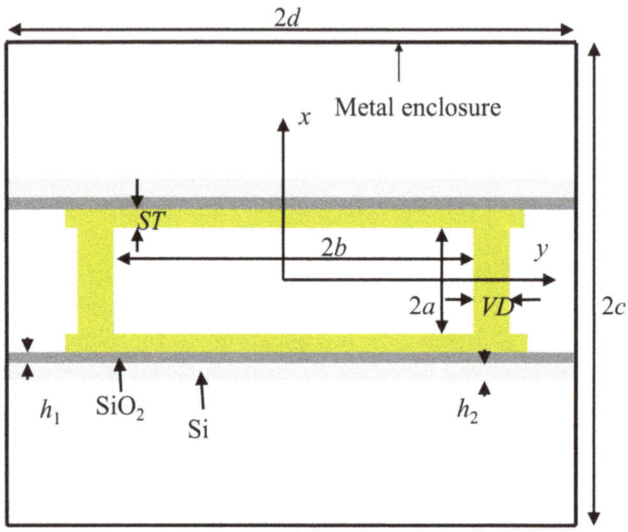

Figure 6.56. Cross-section of the symmetric PH-SEC. © [2014] IEEE. Reprinted, with permission, from [38].

6.5.1.1 Structural configuration

Figure 6.56 shows a cross section of the symmetric PH-SEC; $2a$ and $2b$ are the height and width of the electron-beam tunnel, respectively. The conductor strips have a width of *SW* and a thickness of *ST* and terminate in ring pads of diameter *RD*. The straight-edge connections of diameter *VD* connect two ring pads that face each other. The Si substrate has an ε_r of 11.9 and a thickness of h_2. There is a thin layer of SiO_2 on each Si substrate, with an ε_r of 3.9 and a thickness of h_1. The size of the metal enclosure is $2c \times 2d$. The period is L. The dimensions of the symmetric PH-SEC used in the simulations in CST MWS Eigenmode Solver are listed in table 6.4. A PH-SEC on a thick Si substrate (750 μm) is also simulated for comparison with the symmetric PH-SEC. The period of the latter structure is adjusted (450 μm) so that the two structures have similar values of phase velocity over the frequencies of

Table 6.4. Dimensions of the symmetric PH-SEC [38].

Parameter	Value (μm)
$2a$	300
$2b$	700
$2c$	1000
$2d$	2500
L	300
ST	20
SW	150
VD	100
RD	150
h_1	3
h_2	20

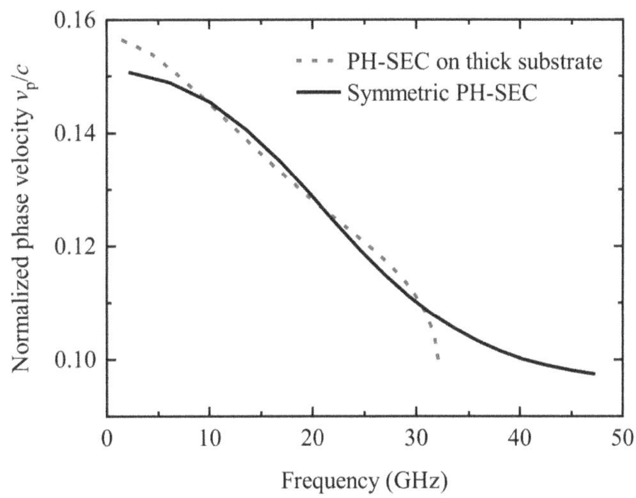

Figure 6.57. Normalized phase velocity values for the PH-SEC on a thick substrate and the symmetric PH-SEC. © [2014] IEEE. Reprinted, with permission, from [38].

interest. The metal enclosure of the PH-SEC with a thick substrate starts at the bottom of the substrate and extends 750 μm above the PH-SEC.

6.5.1.2 Results and discussion

The phase velocity values of both structures are presented in figure 6.57. These values for the two structures are quite close to each other over a wide range of frequencies, but the symmetric structure has a wider bandwidth.

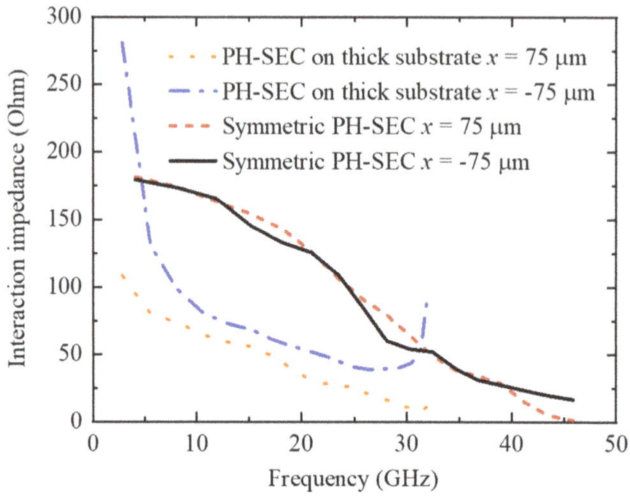

Figure 6.58. Interaction impedance for the PH-SEC on a thick substrate and the symmetric PH-SEC. © [2014] IEEE. Reprinted, with permission, from [38].

Figure 6.58 shows the interaction impedance of the PH-SEC with a thick substrate at two symmetric positions, $x = \pm 75$ μm. It can be seen that the interaction impedance values are not symmetric about the $x = 0$ plane; the values are higher for locations closer to the substrate, since the E field is stronger at these locations. This may lead to problems such as mode competition and gain loss in a TWT. The interaction impedance values of the symmetric PH-SEC are also shown in figure 6.58. As less dielectric material is present in this configuration, the E field is stronger in the region of the electron-beam tunnel. Consequently, the interaction impedance values are much higher than those for the PH-SEC with a thick substrate. In addition, the interaction impedance values at $x = \pm 75$ μm are quite close to each other for the symmetric configuration.

6.5.2 Dielectric charging in the PH-SEC

Due to the presence of dielectric substrates in the PH-SEC, there is an increased risk of dielectric charging. For instance, consider the Ka-band symmetric PH-SEC described in the previous section. As shown in figure 6.59(a), the symmetric PH-SEC has two Si substrates, each with a thin isolation layer of SiO_2.

Since the SiO_2 layer has very poor electrical conductivity, the voltage can build up between the metal strips and the SiO_2 layer due to dielectric charging. Figure 6.59(b) shows the same structure with some modifications to avoid dielectric charging and will be discussed later.

6.5.2.1 Dielectric charging models

One-layer model

The properties of the dielectric material (namely the dielectric constant and conductivity) and the dimensions of the dielectric substrate affect the dielectric

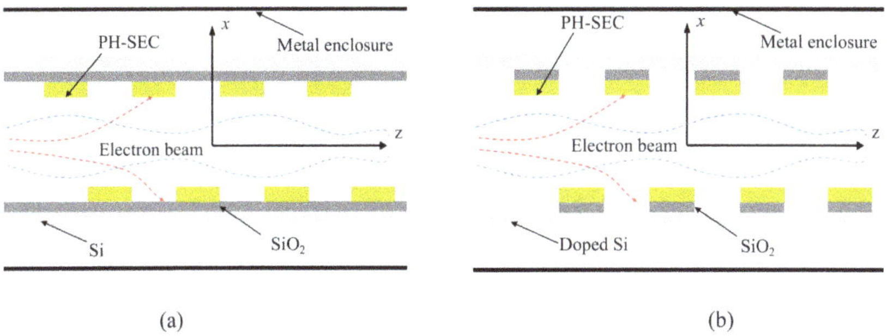

Figure 6.59. (a) Side view of the unmodified symmetric PH-SEC. (b) Side view of the modified symmetric PH-SEC. © [2015] IEEE. Reprinted, with permission, from [39].

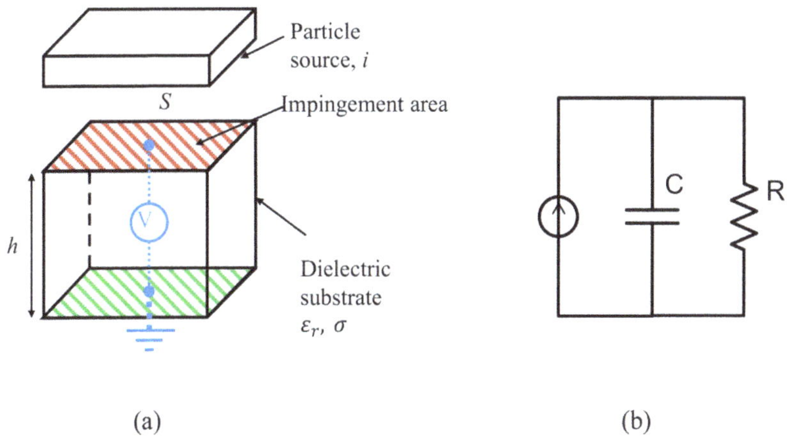

Figure 6.60. (a) Single-layer dielectric charging and discharging model. (b) Equivalent circuit of the model. © [2015] IEEE. Reprinted, with permission, from [39].

charging. In order to understand the impact of these parameters, a simple single-dielectric model is proposed first. The behavior of the model is analyzed using a simple equivalent circuit and also simulated using CST Particle Studio.

As shown in figure 6.60(a), we consider a dielectric substrate with a dielectric constant ε_r, a conductivity σ, a cross-sectional area S, and a thickness h. The substrate is grounded at the bottom and has a particle source with current i placed above the substrate with the same emission area as the cross-sectional area of the substrate. The electron beam generated by the particle source impinges on the top of the substrate, causing a buildup of charge and voltage at the top surface. At the same time, current flows from the top surface to the bottom surface of the substrate. The twin phenomena of charging and discharging can be represented by an equivalent circuit, as shown in figure 6.60(b). In this circuit, the particle source is represented as a constant current source which is connected to a parallel combination of a capacitor C and resistor R. The values of C and R are calculated as follows:

$$C = \frac{\varepsilon_0 \varepsilon_r S}{h}(\mathrm{F}), \ R = \frac{h}{\sigma S}(\Omega). \tag{6.38}$$

The voltage V_c on the capacitor is the solution of the simple differential equation:

$$C\frac{dV_c}{dt} = -\frac{V_c}{R} + i \tag{6.39}$$

$$V_c = Ri(1 - e^{-\frac{t}{T}})(V). \tag{6.40}$$

In equation (6.40), $T = RC = \varepsilon_0 \varepsilon_r / \sigma$ is the well-known relaxation time of the dielectric material. From equation (6.40), when $t \ll T$, using:

$$V_c \approx Ri\left\{1 - \left[1 - \frac{t}{T} + \frac{1}{2}\left(\frac{t}{T}\right)^2 - \cdots\right]\right\} \approx Ri\frac{t}{T}(V) \tag{6.41}$$

it can be seen that the voltage increases linearly with time for small values of t. Then, as the time t increases, the voltage increases less rapidly. Finally, when $t \gg T$, the voltage reaches the steady-state value of Ri. For a low-conductivity material, the relaxation time T and resistance R are large and the steady-state voltage can build up to a high level. On the other hand, for a high-conductivity material, the steady-state voltage is low and is reached sooner.

The behavior of the model in figure 6.60(a) is also examined through simulations by monitoring the time-varying voltage on the top surface of the substrate. Si is chosen as the substrate material with a dielectric constant of 11.9. The conductivity (σ) of Si can vary according to the dopant concentration. Four different cases with different combinations of S, h, and σ, resulting in different values of the resistance R, have been calculated and simulated for a particle current equaling 0.02 A. The parameters for the four cases are shown in table 6.5.

The voltage vs. time results for both calculations (based on equation (6.40)) and simulations are presented in figure 6.61. The curves represent simulation results, and the markers represent the calculation results. The simulated and calculated results match very well. It should be noted that the current here is negative. As a result, the corresponding voltage is also negative. Since the steady-state voltage is proportional to the resistance R, this voltage is different for each of the four cases. From the results for cases 3 and 4, which have the same dimensions but different conductivity values, it can be noted that the magnitude of the steady-state voltage is inversely

Table 6.5. Four cases of the one-layer model [40].

Cases	1	2	3	4
h (mm)	4	2	4	4
S (mm^2)	400	100	100	100
σ (S m^{-1})	0.01	0.01	0.015	0.01
ε_r	11.9	11.9	11.9	11.9

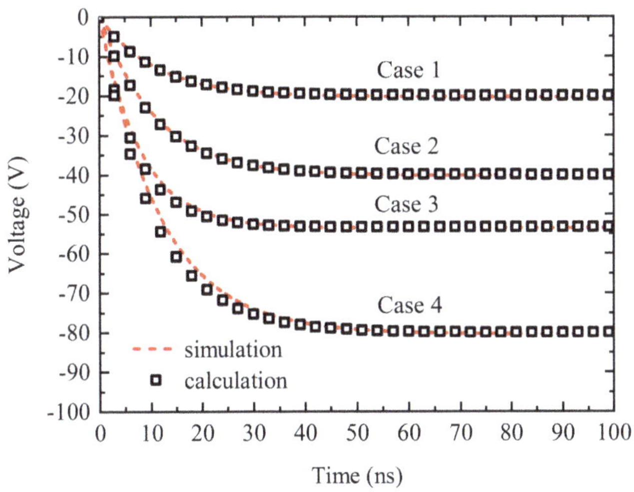

Figure 6.61. Dielectric charging voltage vs. time for four different cases. © [2015] IEEE. Reprinted, with permission, from [39].

proportional to conductivity. Thus, it can be concluded that, with other parameters held constant, the conductivity of the dielectric material of the substrate can be used to restrict the buildup of voltage due to dielectric charging. In this context, it is very useful that the conductivity of Si can be precisely varied by controlling the dopant concentration.

In an actual TWT, the situation is more complex than that represented by the above model. For instance, the electrons do not hit the dielectric material at normal incidence. There may also be secondary emission from the dielectric material [35]. The secondary electron yield (SEY) depends on the material (including surface coatings), the beam energy, and the angle of incidence of the primary electrons. The SEY is first considered when high-energy electrons of several keV impact the Si layer. According to [42], the SEY of Si is below 0.2 for electron energies in the range of 3–10 keV. We believe such a low value of the SEY will not cause a significant change in our results. On the other hand, for Si or SiO_2-on-Si, SEY can be greater than one for low impact energies [43] and may lead to changes in the extent and polarity of dielectric charging at some locations in a TWT. If the dielectric charge is allowed to accumulate, it may also lead to multipacting [44]. However, if the dielectric charge is dissipated through higher conductivity of the dielectric material —as proposed by us—such problems are not likely to occur. Therefore, we do not consider the effect of secondary electrons in this work. This helps us to examine more realistic and complex TWT structures through relatively simple simulations.

Two-layer model

Instead of the single layer considered in the previous model, the proposed symmetric PH-SEC (figure 6.59(a)) has two layers of dielectric materials—Si and SiO_2. These layers are not grounded on the opposite side of the impingement area. Further, the

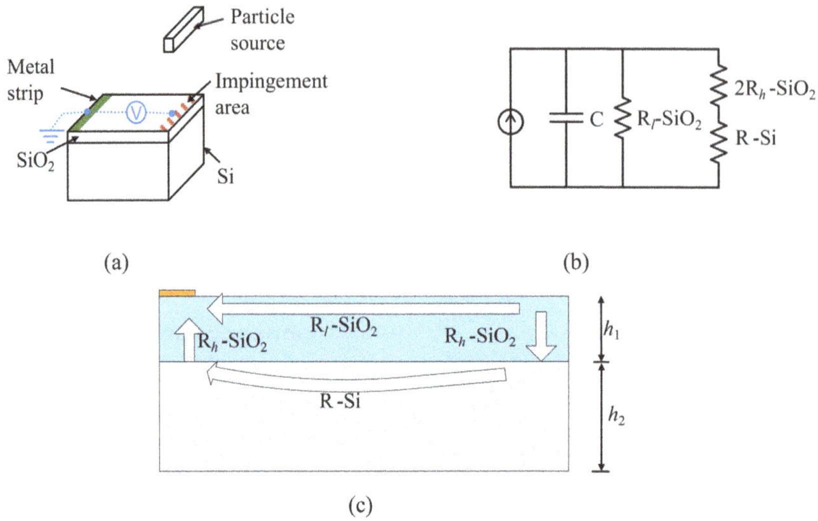

Figure 6.62. (a) Charging/discharging model with two layers of dielectrics. (b) Equivalent circuit of the model. (c) Current paths in the two layers. © [2015] IEEE. Reprinted, with permission, from [39].

maximum voltage difference due to dielectric charging is expected to develop between the metallic helix and locations midway between the helix turns. Therefore, we now study another model that considers these differences.

In the two-layer model shown in figure 6.62(a), the upper layer is SiO_2 and the lower layer is Si. SiO_2 has a dielectric constant of 3.9 and an extremely low conductivity of 1×10^{-14} S m^{-1}. The strip-shaped particle source generates electrons that hit a narrow strip on the right edge of the top surface of the SiO_2 layer. There is a thin metal strip on the left side of this surface that is connected to ground. When the electrons impact the SiO_2 layer, the charge and voltage build up, causing a current flow at the grounded metal strip.

An equivalent circuit for the two-layer model is presented in figure 6.62(b). As shown in figure 6.62(c), the current can flow either directly through the SiO_2 layer or through a path involving both SiO_2 and Si layers. The current through the first path encounters a resistance labeled as R_l–SiO_2. The second path goes down through the SiO_2 layer, covers the length of the Si layer, and finally returns to ground, so the corresponding resistances for this path are labeled as $2R_h$–SiO_2 and R–Si. Equations (6.40) and (6.41) still apply for the voltage V_c, but the expressions for the capacitor and resistors are complicated by the fact that the charge and current distribution may not be uniform [45]. In general, assuming that the thickness of the SiO_2 layer is quite small compared to the distance between the impingement area and the grounded strip, the resistance of the second path is much smaller than that of the first path.

The two-layer model is simulated using CST Particle Studio to study the effect of the thickness of the two layers and the conductivity of the Si layer. The surface area of the two layers is 20×5.5 mm^2, the size of the particle source is 20×0.5 mm^2, and the current is 1 mA.

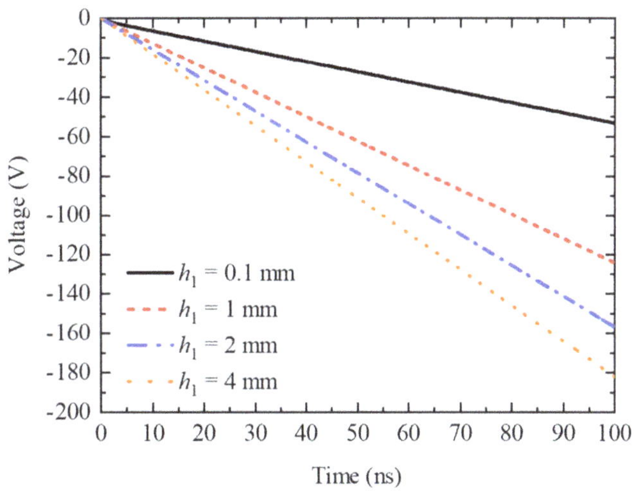

Figure 6.63. Dielectric charging voltage vs. time for different thicknesses of SiO_2. © [2015] IEEE. Reprinted, with permission, from [39].

For a Si layer thickness of 4 mm and a conductivity of 0.1 S m^{-1}, the effect of the thickness h_1 of the SiO_2 layer is shown in figure 6.63. It can be seen that the voltage buildup between the impingement area and the grounded strip is quite sensitive to h_1; as this thickness decreases, the voltage also decreases. This happens because the resistor value R_h–SiO_2 decreases as the thickness of the SiO_2 layer decreases. It should be pointed out that in these results, the simulation time falls within the $t \ll T$ regime; the steady-state voltage ($t \gg T$) for some of these cases can be on the order of 1000 V.

The dielectric charging voltage for different conductivity and thickness values of the Si layer has also been studied, keeping the SiO_2 layer thickness at 1 mm. As can be seen in figure 6.64, the voltage decreases as the conductivity increases; this is caused by a decrease in the value of the resistor R–Si. Figure 6.65 indicates that the value of the voltage is not very sensitive to the thickness of the Si layer. Figures 6.63–6.65 indicate that the SiO_2 layer dominates the dielectric charging effect in the two-layer model. This is caused by a large relaxation time T associated with the high resistivity of SiO_2. Thus, one may conclude that dielectric charging in such a two-layer structure can be reduced by removing the SiO_2 layer where possible and by increasing the conductivity of the Si layer while keeping an eye on the RF insertion loss.

6.5.2.2 Dielectric charging in the symmetric PH-SEC

In this section, the dielectric charging effect is studied for the symmetric PH-SEC SWS (figure 6.59(a)) for application in a TWT. Most of the dimensional parameters are shown in figure 6.60. The values of all the parameters used in the following simulations are the same as those used for the symmetric PH-SEC listed in table 6.4.

As mentioned at the beginning of section 6.5, to avoid the dielectric charging problem, two modifications are applied to the SWS. First, according to the

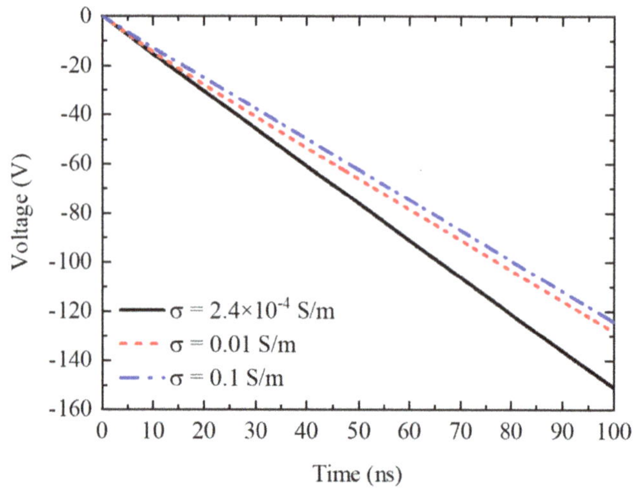

Figure 6.64. Dielectric charging voltage for different conductivities of Si. © [2015] IEEE. Reprinted, with permission, from [39].

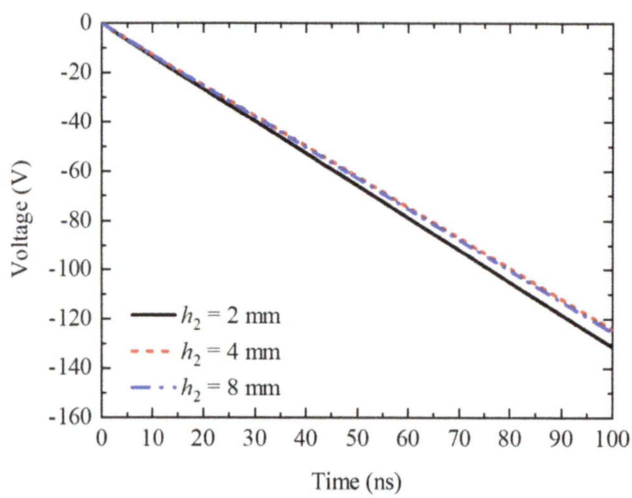

Figure 6.65. Dielectric charging voltage for different thicknesses of Si. © [2015] IEEE. Reprinted, with permission, from [39].

conclusion of the previous subsection that the SiO_2 layer dominates the dielectric charging effect, the SiO_2 layer is removed except beneath the metal strips. This largely prevents the electrons from hitting the SiO_2 layer directly. In addition, the RF performance of the circuit is unaffected, since the SiO_2 layer is very thin. Second, doped Si with appropriate conductivity is used instead of intrinsic Si. These modifications are expected to reduce the relaxation time of the two-layer structure

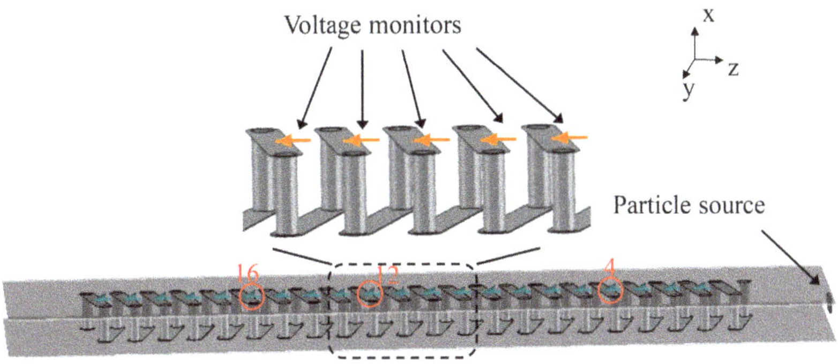

Figure 6.66. Simulation model used for the symmetric PH-SEC; the enlarged view shows the locations of the voltage monitors. © [2015] IEEE. Reprinted, with permission, from [39].

and restrict the voltage buildup due to dielectric charging. The resulting modified symmetric PH-SEC is shown in figure 6.59(b).

Simulations have been carried out for the symmetric PH-SEC, both with and without the modifications mentioned above. We assume that the DC voltage of the SWS is maintained at 0 V. Therefore, when the electrons hit the metallic SWS, the charge is conducted away. On the other hand, when the electrons hit the dielectric substrate, the charge accumulates if the substrate has poor conductivity.

As shown in figures 6.66, 22 periods of the symmetric PH-SEC are simulated, placing 21 voltage monitors that measure the voltage between each metal strip and the nearest center point between two metal strips. Voltage monitor 1 is closest to the particle source and voltage monitor 21 is farthest. The size of the particle source is $350 \times 150\ \mu\text{m}$, and it generates a sheet beam. The voltage and current of the particle source are set to 3700 V and 0.15 A, respectively. To incorporate, to a certain extent, the effects of misalignment between the electron gun and the SWS and the angular spread of the electron beam, we assume that the particle source has an inclination of 2° with respect to the x-axis and emits a beam with an angular spread of 2°. The Brillouin magnetic field for a circular beam with the same charge density is 0.32 T. We use this value as a reference and set the focusing magnetic field to 0.3 T.

The PIC simulations are carried out without any RF input. However, the voltages measured by the voltage monitors show high-frequency oscillations (results not shown here). To explain these oscillations, we refer to figure 6.67, which shows the dispersion characteristics of the symmetric PH-SEC (without modifications) together with the beam line. The beam line intersects mode 2 at about 54 GHz, which is very close to the π point at 48 GHz. The Fourier transform of the signal from voltage monitor 4 presented in figure 6.68 shows a peak at 48 GHz; this frequency can be considered to be backward-wave oscillation but not band-edge oscillation, since this oscillation frequency value changes with the beam velocity [33, 46]. Since the focus here is on dielectric charging, in the following discussion, we filter out the high-frequency components so that we are only comparing the DC voltage values.

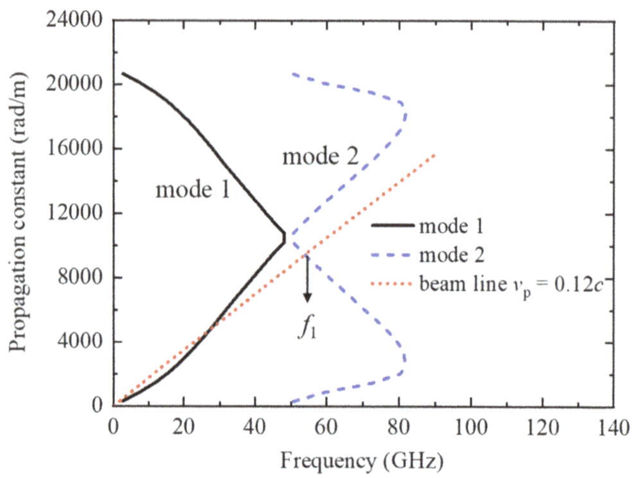

Figure 6.67. Dispersion diagram of the unmodified symmetric PH-SEC. © [2015] IEEE. Reprinted, with permission, from [39].

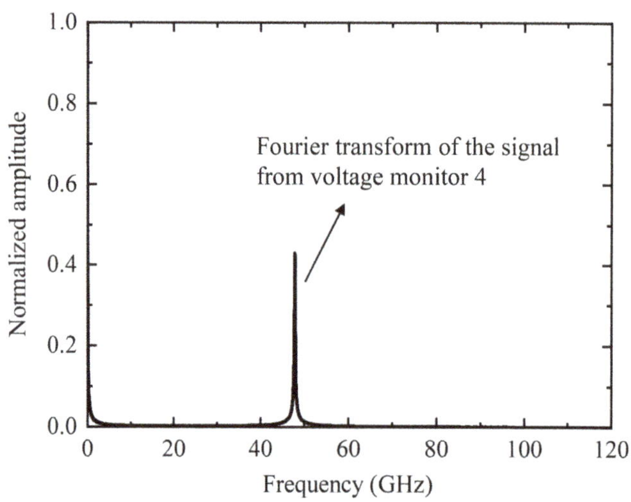

Figure 6.68. Spectrum of the dielectric charging voltage captured from voltage monitor 4. © [2015] IEEE. Reprinted, with permission, from [39].

The voltages corresponding to some of the voltage monitors for the unmodified structure are shown in figure 6.69. It can be seen that at some of the locations, where more electrons hit the dielectric substrate, the magnitude of the voltage keeps increasing for the simulation time considered here. This is the case for the voltages from monitors 4, 6, and 20. As mentioned in the previous section, the steady-state voltage values can reach 1000 V and therefore approach dielectric breakdown (the dielectric strength of Si is 4×10^6–10^7 V m^{-1}). In any case, such high voltages can cause defocusing of the electron beam. On the other hand, at some other locations

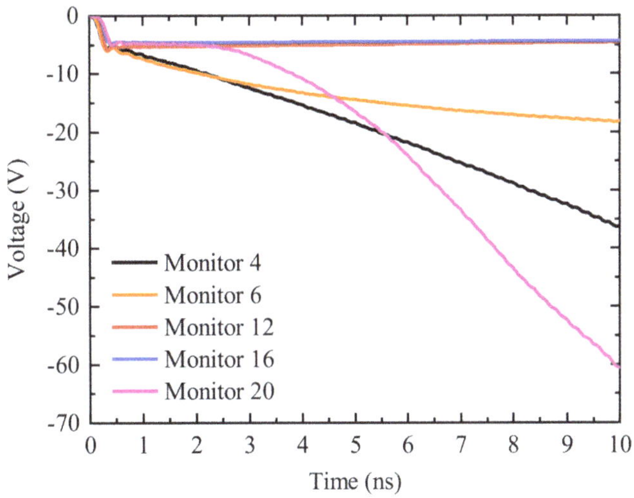

Figure 6.69. Dielectric charging voltage obtained from different voltage monitors for the unmodified structure. © [2015] IEEE. Reprinted, with permission, from [39].

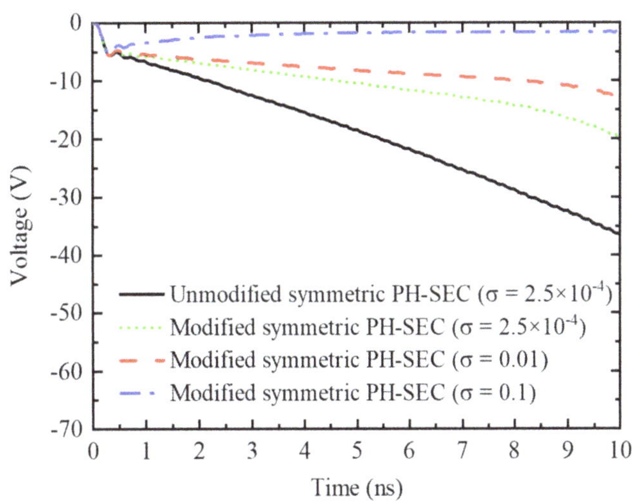

Figure 6.70. Voltage vs. time for different values of Si conductivity (from voltage monitor 4). © [2015] IEEE. Reprinted, with permission, from [39].

where fewer electrons hit the dielectric substrate, the voltage is very low; for example, for monitors 12 and 16.

The voltages obtained from monitor 4 for the unmodified and modified symmetric PH-SECs with different Si conductivity values are compared in figure 6.70. We can see that the voltage magnitude for the unmodified structure increases rapidly with time. The modified structure has a lower voltage magnitude, which decreases further as the conductivity increases. When the conductivity of Si is

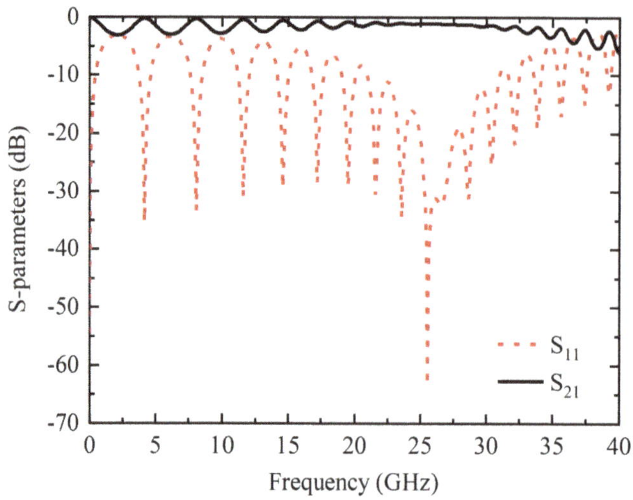

Figure 6.71. S-parameters of the modified symmetric PH-SEC with 22 periods. © [2015] IEEE. Reprinted, with permission, from [39].

Table 6.6. Loss per period at 27 GHz for different conductivities. © [2011] IEEE. Reprinted, with permission, from [40]

Conductivity (S m^{-1})	Loss (dB/period)
2.5×10^{-4}	0.035
0.01	0.038
0.1	0.064
0.25	0.106

0.1 S m^{-1}, the voltage reaches a steady-state value of only -1.5 V. These results show that the modified symmetric PH-SEC with increased Si conductivity can avoid the dielectric charging problem.

The additional loss caused by the higher conductivity of Si has also been examined. To estimate the loss per period, two identical symmetric PH-SEC structures, differing only by the numbers of periods (22 and 37), are simulated using CST MWS. The structures are fed with matched discrete ports that have an impedance of 80 ohms. As an example, the resulting S-parameters of the 22-period structure are shown in figure 6.71. The loss per period at 27 GHz for the proposed structure with four different conductivity values is calculated from the difference in the magnitude of S_{21} and the number of periods. The results are presented in table 6.6. It can be seen that even for 0.1 S m^{-1}, the loss is only 0.064 dB/period, which may be acceptable for a TWT operating in the Ka band. PIC simulations for a TWT consisting of 136 periods of the modified SWS show that increasing the conductivity of Si to 0.1 S m^{-1} reduces the TWT gain by ~2 dB compared to that for intrinsic Si. In addition, simulations using the CST MWS Eigenmode Solver show that the increased loss due to increasing the conductivity of Si does not have a significant effect on the phase velocity and interaction impedance values.

The modifications proposed here to avoid dielectric charging are compatible with microfabrication and may also be applicable to other microfabricated SWSs that have been proposed for application in TWTs, such as meander-line and biplanar interdigital structures.

6.6 Summary

This chapter describes some variations of the planar SWS that can be used to solve some of the issues that arise in practical TWT applications. As variations of the meander-line SWS, the logarithmic spiral and the angular log-periodic meander-line SWS are described for small TWTs operating at low voltage. Specifically, the dispersion properties, electromagnetic properties, and beam–wave interaction of the angular log-periodic meander-line SWS are analyzed and studied, confirming that it is a miniaturized TWT that operates at low voltage while offering wide bandwidth and high efficiency. In addition, we describe the V-shaped MML SWS, symmetric double V-shaped MML, 3D U-shaped MML SWS, and MML SWS with shaped substrates, which are used to enhance the interaction impedance. As variations of the PH-SEC SWS, we describe planar ring–bar SWSs and coupled helices (an unconnected pair of PH-SECs and a coaxial pair of PH-SECs), which reduce backward-wave oscillation. We show that the planar ring–bar SWS and the unconnected pair of PH-SECs have higher interaction impedance for the forward wave and a reduced risk of backward-wave oscillation. Finally, the dielectric charging effects in a PH-SEC fabricated using a Si substrate coated with a thin SiO_2 layer are examined. A simple single-layer dielectric slab model and a two-layer model are used to study the effects of the conductivity and dimensional parameters of the dielectric substrate on dielectric charging. Based on the results of these models, modifications to a Ka-band symmetric PH-SEC SWS are proposed to prevent dielectric charging in a TWT. We demonstrate that these modifications greatly reduce the voltage buildup due to dielectric charging.

References

[1] Chodorow M and Chu E L 1955 Cross-wound twin helices for traveling-wave tubes *J. Appl. Phys.* **26** 33–43

[2] Birdsall C K and Everhart T E 1956 Modified contra-wound helix circuits for high-power traveling-wave tubes *IRE Trans. Electron Devices* **3** 190–204

[3] Lopes D T and Motta C C 2008 Characterization of ring-bar and contrawound helix circuits for high-power traveling-wave tubes *IEEE Trans. Electron Devices* **55** 2498–504

[4] Cook J S, Kompfner R and Quate C F 1956 Coupled helices *Bell Syst. Tech. J.* **25** 127–78

[5] Singh V N, Basu B N, Pal B B and Vaidya N C 1983 Equivalent circuit analysis of a system of coupled helical transmission lines in a complex environment *J. Appl. Phys.* **54** 4141

[6] Pchelnikov Y N and Vlasov A N 2012 Novel slow-wave structure for high voltage TWTs *IVEC 2012* (Piscataway, NJ: IEEE) 273–4
[7] Savel'yev V S and Kushcenko G I 1970 Experimental investigation of a TWT with a radial electron stream *Radio Eng. Electron. Phys.* **15** 2267–72
[8] Putz J L and Scott A W 1974 *Development of a planar equiangular spiral amplifier* AD-787 339 National Technical Information Service, U.S. Department of Commerce (Varian Associates) 1974vara.rept.....P
[9] Wang S *et al* 2013 Study of a log-periodic slow wave structure for Ka-band radial sheet beam traveling wave tube *IEEE Trans. Plasma Sci.* **41** 2277–82
[10] Wang S, Gong Y, Wang Z, Wei Y, Duan Z and Feng J 2016 Study of the symmetrical microstrip angular log-periodic meander-line traveling-wave tube *IEEE Trans. Plasma Sci.* **44** 1787–93
[11] Solntsev V A 1994 Planar spiral systems with waves of constant radial phase velocity *J. Commun. Technol. Electron.* **39** 552–9 (Varian Associates) 1994RaEl...39..552S
[12] Wang S, Gong Y, Wei Y and Duan Z 2012 Study on the radial-sheet-beam electron optical system *IEEE Trans. Plasma Sci.* **40** 3442–8
[13] Shen F *et al* 2012 A novel V-shaped microstrip meander-line slow-wave structure for W-band MMPM *IEEE Trans. Plasma Sci.* **40** 463–9
[14] Shen F *et al* 2012 Symmetric double V-shaped microstrip meander-line slow-wave structure for W-band traveling-wave tube *IEEE Trans. Electron Devices* **59** 1551–7
[15] Sengele S, Jiang H, Booske J H, Kory C L, van der Weide D W and Ives R L 2009 Microfabrication and characterization of a selectively metallized W-band meander-line TWT circuit *IEEE Trans. Electron Devices* **56** 730–7
[16] Chua C and Aditya S 2013 A 3-D U-shaped meander-line slow-wave structure for traveling-wave tube applications *IEEE Trans. Electron Devices* **60** 1251–6
[17] Chua C S 2012 Studies on Planar Helical Slow-Wave Structures for Travelling-Wave Tube Applications (Singapore: Nanyang Technological University)
[18] Bates C D and Hartley J H 1978 *Low-Cost, Crossed-Field Amplifier Meanderline Circuit Concepts* ADA061147 Defense Technical Information Center https://apps.dtic.mil/sti/citations/ADA061147
[19] Wang Z *et al* 2020 Study on an X-band sheet beam meander-line SWS *IEEE Trans. Plasma Sci.* **48** 4149–54
[20] Wang Z *et al* 2020 Investigation on a Ka band diamond-supported meander-line SWS *J. Infrared, Millimeter, Terahertz Waves* **41** 1460–8
[21] Wang Z *et al* 2022 A Ka-band angular log-periodic meander-line SWS supported by diamond rods *IEEE Trans. Electron Devices* **69** 1374–9
[22] Chua C, Aditya S, Tsai J M, Tang M and Shen Z 2011 Microfabricated planar helical slow-wave structures based on straight-edge connections for THz vacuum electron devices *Int. J. Terahertz Sci. Technol.* **4** 208–29
[23] Naidu V B, Datta S K and Kumar L 2012 Novel multi helix structure for high frequency application *IVEC 2012* 449–50
[24] Zhao C 2016 *Planar Helix-Based Slow-Wave Structures for Millimeter Wave Traveling-Wave Tubes* (Singapore: Nanyang Technological University)
[25] Zhao C, Aditya S and Chua C 2013 Analysis of coupled planar helices with straight-edge connections for application in millimeter-wave TWTs *IEEE Trans. Electron Devices* **60** 1244–50

[26] Zhao C, Aditya S and Chua C 2014 Connected pair of planar helices with straight-edge connections for application in TWTs *IEEE Trans. Electron Devices* **61** 1692–8

[27] Arora R 1966 Surface waves on a pair of parallel undirectionally conducting screens *IEEE Trans. Antennas Propag.* **14** 795–7

[28] Aditya S and Arora R K 1979 Guided waves on a planar helix *IEEE Trans. Microw. Theory Tech.* **27** 860–3

[29] Fink H J and Whinnery J R 1982 Slow waves guided by parallel plane tape guides *IEEE Trans. Microw. Theory Tech.* **30** 2020–3

[30] Sinha M P 1986 Coupled planar helices *IETE J. Res.* **32** 38–41

[31] Chua C, Aditya S and Shen Z 2010 Planar helix with straight-edge connections in the presence of multilayer dielectric substrates *IEEE Trans. Electron Devices* **57** 3451–9

[32] Gilmour A S 1994 *Principles of Traveling Wave Tubes* (Boston, MA: Artech House)

[33] Hung D M H *et al* 2015 Absolute instability near the band edge of traveling-wave amplifiers *Phys. Rev. Lett.* **115** 124801

[34] Leou K C, McDermott D B and Luhmann N C 1992 Dielectric-loaded wideband gyro-TWT *IEEE Trans. Plasma Sci.* **20** 188–96

[35] Dallos A, Smith B H and Bowness C 1989 Simulation of rod charging in TWT helix structures *Int. Technical Digest on Electron Devices Meeting* (Piscataway, NJ: IEEE) 199–202

[36] Smith B H, Bowness C and Dallos A 1989 *Slow wave delay line structure having support rods coated by a dielectric material to prevent rod charging* US US-5038076-A

[37] Mikijelj B and Abe D K 2002 AlN-based lossy ceramics for high power applications *3rd IEEE Int. Vacuum Electronics Conf. (IEEE Cat. No.02EX524)* (Piscataway, NJ: IEEE) 32–3

[38] Zhao C, Aditya S and Chua C 2014 Symmetric planar helix slow-wave structure with straight-edge connections for application in TWTs *IEEE Int. Vacuum Electronics Conf.* (Piscataway, NJ: IEEE) 291–2

[39] Zhao C, Aditya S and Chua C 2015 A microfabricated planar helix slow-wave structure to avoid dielectric charging in TWTs *IEEE Trans. Electron Devices* **62** 1342–8

[40] Chua C *et al* 2011 Microfabrication and characterization of W-band planar helix slow-wave structure with straight-edge connections *IEEE Trans. Electron Devices* **58** 4098–105

[41] Chua C, Aditya S, Tsai J and Shen Z 2012 PIC simulation for W-band planar helix with straight-edge connections *IVEC 2012* 459–60

[42] Ciappa M, Ilgünsatiroglu E and Illarionov A Y 2012 Monte Carlo simulation of emission site, angular and energy distributions of secondary electrons in silicon at low beam energies *Microelectron. Reliab.* **52** 2139–43

[43] Yu S *et al* 2002 Secondary electron emission for layered structures *J. Vac. Sci. Technol. A: Vac., Surf. Films* **20** 950–2

[44] Lau Y Y, Verboncoeur J P and Valfells A 2000 Space-charge effects on multipactor on a dielectric *IEEE Trans. Plasma Sci.* **28** 529–36

[45] Zhang P, Lau Y Y and Timsit R S 2012 On the spreading resistance of thin-film contacts *IEEE Trans. Electron Devices* **59** 1936–40

[46] Ang L K and Lau Y Y 1998 Absolute instability in a traveling wave tube model *Phys. Plasmas* **5** 4408–10

IOP Publishing

Chen Zhao and Sheel Aditya

Chapter 7

Backward-wave oscillators (BWOs) and oscillator–amplifiers

7.1 Introduction

Previous chapters have described planar slow-wave structures (SWSs), namely the meander-line (ML) SWS and the planar helix with straight-edge connections (PH-SEC), along with their variations, which are intended for use in traveling-wave tubes (TWTs). Studies have demonstrated that these structures are conducive to micro-fabrication and exhibit commendable performance, specifically for millimeter-wave TWT amplifiers. In this chapter, we investigate the potential of planar SWSs for application in backward-wave oscillators (BWOs). BWOs are important, since they offer output signals with high spectral purity. Also, their signal frequency can be electronically tuned over a wide tunable bandwidth by varying the voltage of the electron beam. Due to these features, BWOs are attractive sources of RF signals in the millimeter-wave and terahertz frequency ranges.

While the ML SWS has gained popularity for millimeter-wave TWTs due to its simple structure and wide bandwidth, it exhibits certain drawbacks, rendering it unsuitable for BWO applications. First, it suffers from a high dielectric loading effect, leading to a decrease in interaction impedance and consequent low output power in a BWO. This is especially true when circuits are fabricated on thick substrates with high permittivity. Additionally, the input impedance of the micro-strip line is highly dependent on the thickness and permittivity of the substrate [1]. For a thick substrate, the input impedance becomes quite high; this may cause additional loss due to matching transitions. Moreover, to achieve a specific thickness of the substrate and to add the ground plane, additional steps of back-side processing, such as thinning, via etching, back-side metallization, etc. are required. The performance of the circuit may degrade if any step of the process is not handled well. Considering these constraints, two planar SWSs that are suitable for BWO

doi:10.1088/978-0-7503-5764-7ch7

applications, namely the interdigital SWS and the coplanar SWS, are introduced in this chapter. These two structures, while retaining the simple structure and wide bandwidth properties of the ML SWS, effectively address the drawbacks encountered when utilizing the ML SWS for BWO purposes.

The following sections describe the design of a BWO intended to operate in the W band and produced using a microfabrication-compatible PH-SEC. The chapter ends with the description of a new technique to enhance the efficiency of conventional nonrelativistic BWOs without increasing the length of the SWS. This technique is demonstrated for a PH-SEC-based BWO.

7.2 Interdigital SWS for BWO applications

7.2.1 Structural configuration

The interdigital SWS is one of the transmission structures used in microwave engineering. The configuration of the interdigital SWS is shown in figure 7.1. As depicted, the interdigital SWS consists of two comb-like metal strips whose fingers interlock with each other. This structure is usually positioned inside a metal enclosure. It can be considered a complementary structure to the ML SWS that interchanges the metal and the gaps. Consequently, the dispersion properties of the interdigital SWS closely resemble those of the ML SWS. However, unlike the ML SWS, the physical structure of the interdigital SWS introduces an additional phase shift of 2π per period, thereby constituting a fundamental backward-wave circuit, which is particularly suitable for BWO applications due to its high interaction impedance. With its planar configuration, it can be easily realized with microfabrication techniques, an important advantage for millimeter-wave applications.

7.2.2 Dispersion characteristics

Field-theory-based analysis of the interdigital SWS can be carried out under the quasi-static condition [2]. The analysis process is very similar to the analysis method of the ML SWS described in section 2.2.2. Here, only a brief description of the analysis process is presented, as further elaboration is available in section 2.2.2.

First, the field expressions of the generic structure of an infinite array of parallel metal strips (figure 7.2) are obtained. Such a generic SWS supports both even and odd modes. By solving the Helmholtz equation with periodic boundaries and

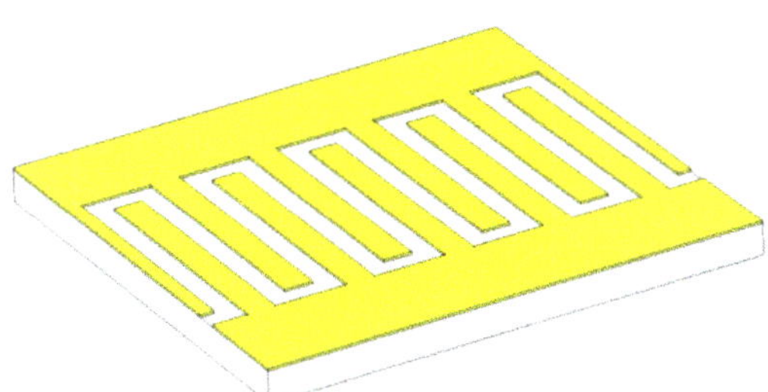

Figure 7.1. Configuration of the interdigital SWS.

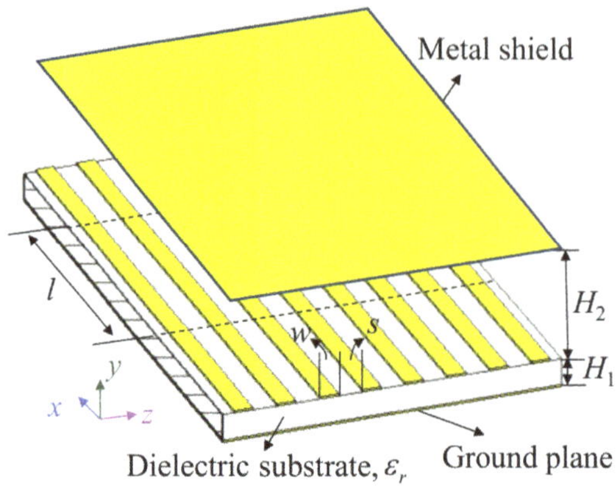

Figure 7.2. Configuration of an infinite array of parallel metal strips.

symmetric conditions, the electromagnetic properties, including the effective dielectric constant, the propagation constants (β_e and β_o), and the characteristic impedances (Z_e and Z_o), can be obtained. An interdigital SWS can then be realized by truncating the generic structure to a certain width (l) and connecting the ends of alternate strips. The boundary conditions for the interdigital SWS are:

$$\begin{cases} V_1(0.5l) = V_2(-0.5l) = 0 \\ I_1(-0.5l) = I_2(0.5l) = 0 \end{cases}. \tag{7.1}$$

The implementation of these boundary conditions yields the characteristic equation for the interdigital SWS [3]:

$$\frac{Z_e}{Z_o} = \begin{cases} \tan(\beta_e l/2)\tan(\beta_o l/2) \\ \cot(\beta_e l/2)\cot(\beta_o l/2) \end{cases} \tag{7.2}$$

where the two equations represent the backward and forward wave respectively. Moreover, according to Babinet's principle, the interdigital SWS can be expected to have the same dispersion properties as those of an ML SWS, which is complementary to the interdigital SWS. Accordingly, the dispersion characteristics of the interdigital SWS can be determined by examining the dispersion properties of the complementary ML SWS, which may be obtained using the field theory or equivalent circuit theory described in sections 2.2.2 and 2.2.3, respectively. However, as mentioned earlier, it is crucial to note that an additional phase shift of 2π per period should be considered for the interdigital SWS. This makes it an inherent backward-wave circuit.

7.2.3 Millimeter-wave and terahertz interdigital SWSs

The interdigital SWS has several advantages, including high interaction impedance, high tunable bandwidth, and ease of fabrication. When used for BWO applications, as an intrinsic backward-wave circuit, the SWS is much smaller than an intrinsic forward-wave circuit. This could lead to fabrication difficulties for millimeter-wave applications. However, the circuit has a simple planar configuration. Consequently, constructing the circuit on a substrate using modern microfabrication techniques is quite feasible. With the fast development of microfabrication techniques, there have been several successful attempts to produce millimeter-wave BWOs based on the interdigital SWS. Some of these works are mentioned in this subsection.

One of the early works to use an interdigital SWS in a millimeter-wave BWO was a study at the University of Utah, supported by NASA in the 1980s. This project aimed to produce a BWO capable of operating in the frequency range from 300 GHz to 2 THz. A milestone achievement was reported in [4], in which two circuits operating at 400–650 and 200–265 GHz were fabricated using lithographic techniques and tested. The former circuit was fabricated on a diamond substrate, while the latter was fabricated on a quartz substrate. Hot-test experiments were carried out. Although no output power was observed for the former circuit because of high circuit loss, a power of 50–100 μW was obtained from the latter circuit at a beam voltage of 2 kV and a beam current density of 20 A cm^{-2}.

Another interdigital SWS-based BWO was studied by James A. Dayton from Genvac Aerospace Corporation [5, 6]. A biplanar interdigital SWS operating at 650 GHz was used as the SWS. Unlike the traditional interdigital SWS, there was some space between the left and right parts of the interdigital SWS, forming an electron-beam tunnel between the two halves. At a beam voltage of 6 kV and a beam current of 2 mA, this BWO provided an output power of 26 mW at the output port.

More recently, a low-voltage V-band interdigital SWS was investigated by Saratov State University in Russia in 2017 [7]. The structure was fabricated using a photolithographic technique on a quartz substrate and placed in a rectangular waveguide. The cold-test measurements showed good reflection and transmission properties. Particle-in-cell (PIC) simulations showed that a BWO with an interdigital SWS of 50 periods could be tuned over the frequency range from 65 to 75 GHz when the beam voltage changed from 1 to 3 kV. The output power was 1 to 2 W for a beam current of less than 50 mA.

7.3 Coplanar SWS for BWO applications

In this section, the coplanar SWS for BWO applications is described. A perspective view of the SWS is shown in figure 7.3. This structure has a much higher interaction impedance and simpler input/output couplers than the ML SWS. Additionally, the proposed structure exhibits a notable reduction in dielectric charging issues and improved heat dissipation, which is primarily attributed to the extensive metal coverage on the top surface of the substrate and good metal contact with the metal enclosure. Furthermore, with a single metal layer configuration, the SWS is readily compatible with printed circuit or microfabrication techniques without requiring any back-side processing. The proposed SWS can also accommodate two pencil electron beams. The contents of this section closely follow the authors' work in [8] and [9].

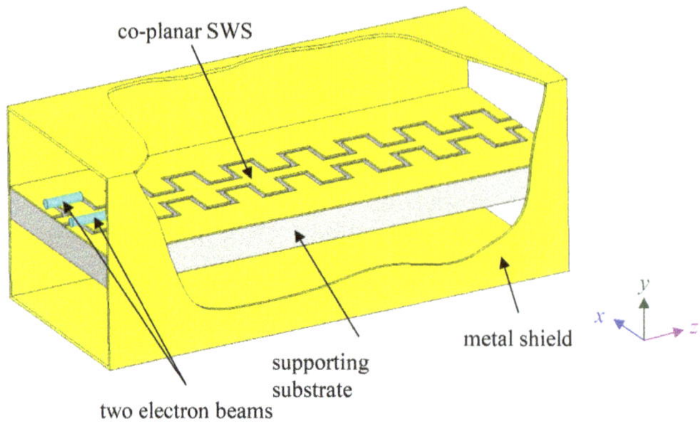

Figure 7.3. Perspective view of the coplanar SWS. © [2021] IEEE. Reprinted, with permission, from [8].

7.3.1 Structural configuration

The configuration of one period of the proposed SWS is presented in figure 7.4(a). The structure consists of three thin metal patterns printed on the top side of a dielectric substrate. The two patterns on the left and right have a comb-like structure. The central pattern has fins on both sides. Two pencil electron beams are accommodated at a height h' above the surface of the SWS, located above the gaps. As shown in figure 7.3, the entire structure is assumed to be placed inside a metal enclosure. Figure 7.4 also includes one period of a double microstrip meander-line (MML) SWS. The metal pattern of the proposed SWS is complementary to that of the double MML SWS. Of course, the ML SWS has a metal ground plane on the bottom surface of the substrate. In this structure, unlike the MML, the thickness of the substrate does not have much effect on the input impedance or the field distribution of the proposed SWS, as the field is mostly concentrated on the surface of the metal patterns. In addition, the structure can easily be microfabricated on substrate materials such as silicon, quartz, diamond, etc. All the structural parameters of the structure are indicated in figures 7.4(a), (c), and (e). For the targeted operation in the V band, the initial dimensional parameters of the proposed SWS are as follows: $L = 0.375$ mm, $p = 0.6$ mm, $h = 0.254$ mm, $w = 0.24$ mm, $g = 0.06$ mm, $e = 0.15$ mm, $d' = 0.3275$ mm, $d = 0.11$ mm, and $H_s = 1$ mm.

7.3.2 Dispersion characteristics

The proposed SWS has a complementary structure to that of the double ML SWS. These two structures have nearly identical dispersion properties. Consequently, the dispersion characteristics of the proposed SWS can be predicted from the characteristics of the double ML SWS.

The equivalent circuit analysis proposed in this section involves certain approximations. First of all, in general, the double ML SWS can support both symmetric and antisymmetric modes of propagation, depending on the different combinations

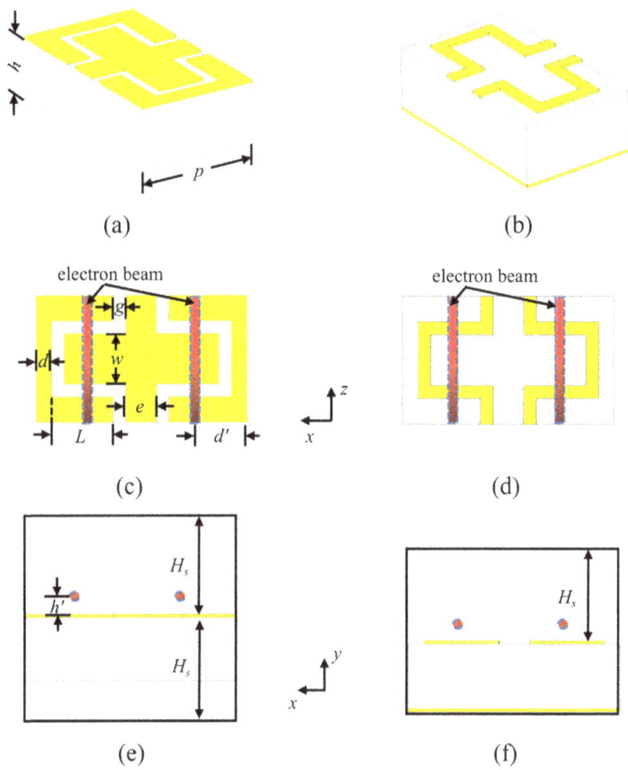

Figure 7.4. (a) Perspective view, (c) top view, and (e) side view of one period of the proposed SWS; (b) perspective view, (d) top view, and (f) side view of one period of the double MML SWS. © [2021] IEEE. Reprinted, with permission, from [8].

of potential on the meandering metal strips. These two modes exhibit different propagation constants compared to the single ML SWS because of the coupling between the double ML SWSs. However, the separation between the two meandering strips is large for most of the SWS period, except at the beginning and the end; in other words, the parameter *e* is not excessively small. It is therefore assumed that the coupling effect is low and that the propagation constants of the SWS for both propagating modes closely resemble those of a single ML SWS. Second, as the electric field is concentrated on the surface of the SWS and decays rapidly with the distance above the surface, the metal enclosure has very little impact on the electrodynamics of the SWS when the size of the metal enclosure is sufficiently large. As a result, the effect of the metal enclosure is not considered in the analytical model. Furthermore, this model treats the metal strips as perfect conductors with infinitesimal thickness. It is also important to note that the equivalent circuit analysis does not consider the field distributions. As a result, certain parameters such as the interaction impedance and the ohmic losses cannot be directly calculated using this method. To obtain these parameters, the field analysis method or numerical simulations are required.

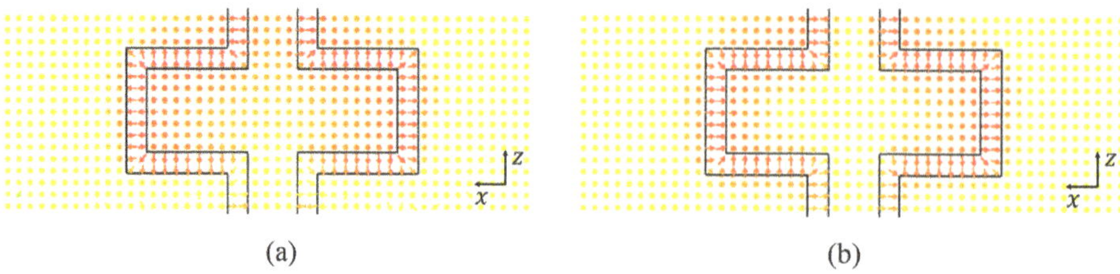

Figure 7.5. Field distribution of the (a) symmetric mode and (b) antisymmetric mode of the coplanar SWS at 72 GHz. © [2023] IEEE. Reprinted, with permission, from [9].

The equivalent circuit analysis method for a single ML SWS has been described in detail in section 2.2.3. The structure can be divided into nine sections, including five sections of uniform transmission lines and four sections of discontinuities. The dispersion characteristics can be obtained by cascading the transmission matrices of all nine sections.

The electric field of the SWS is simulated using the CST Microwave Studio (MWS) Eigenmode Solver. Figure 7.5 illustrates the electric field distributions of both the symmetric mode and the antisymmetric mode at 72 GHz over a single period of the SWS. For the symmetric mode, the electric fields in the gaps are symmetrically distributed with respect to the *yz* plane and point in opposite directions. For the antisymmetric mode, the electric fields in the two gaps point in the same direction. As the metal shield electrically connects the two comb-like metal patterns, only the symmetric mode is excited in the proposed SWS. In this mode, the electrons experience two phase shifts. The first occurs in the coplanar transmission line, while the second is an additional phase shift of 2π caused by the physical structure. This behavior is comparable to the behavior observed in an interdigital SWS. As such, the propagation constant of the proposed coplanar SWS can be calculated as:

$$\beta p = \cosh^{-1}\left(\frac{A + D}{2}\right) + 2\pi \tag{7.3}$$

where β is the phase constant, and A and D are the elements in the overall transmission matrix of one period of the SWS.

In order to compare and validate the dispersion characteristics of the SWS, simulations were performed using the CST MWS Eigenmode Solver. The results were compared with the numerical results obtained from the equivalent circuit model. The dispersion diagram of the coplanar SWS, with varying period lengths, is shown in figure 7.6. It is interesting to observe that, unlike the MML SWS, the fundamental mode of the proposed SWS is the backward-wave mode, making it very suitable for application in a BWO. It can be seen that as the period p increases, the operating frequency of the proposed SWS becomes lower. The numerical calculation results are in good agreement with the simulations.

A dispersion diagram comparing the simulation and numerical calculation results for different dielectric substrates is presented in figure 7.7. The substrate materials

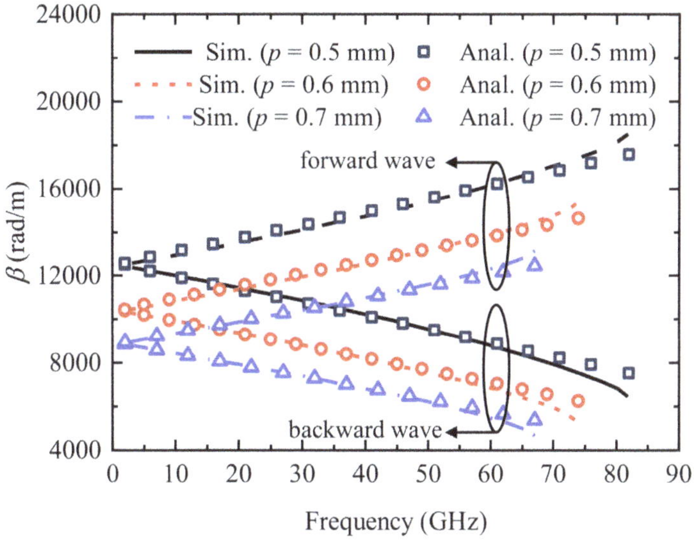

Figure 7.6. Simulated and calculated (from analysis) phase propagation constants of the coplanar SWS for different period lengths. © [2023] IEEE. Reprinted, with permission, from [9].

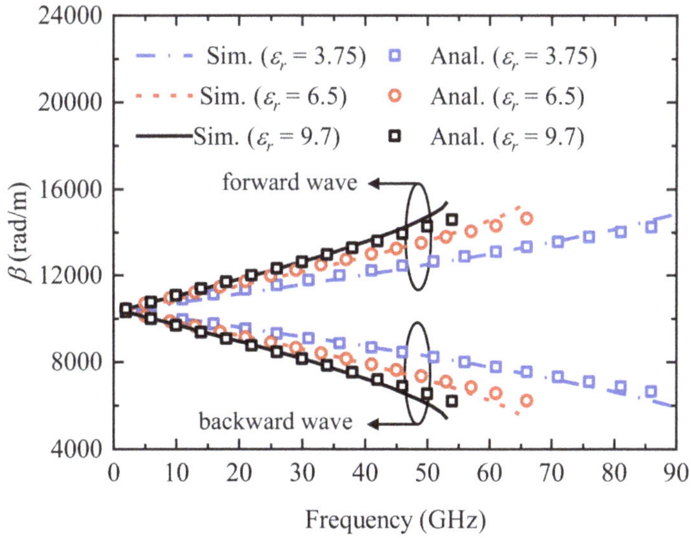

Figure 7.7. Simulated and analyzed phase propagation constants of the coplanar SWS for different dielectric permittivities. © [2023] IEEE. Reprinted, with permission, from [9].

include quartz ($\varepsilon_r = 3.75$), BeO ($\varepsilon_r = 6.5$), and Al_2O_3 ($\varepsilon_r = 9.7$). The analysis results are a good match for the simulations. For the backward-wave mode, a decrease in the propagation constant and the cutoff frequency can be observed as the value of the relative permittivity (ε_r) increases. This is because larger values of ε_r result in

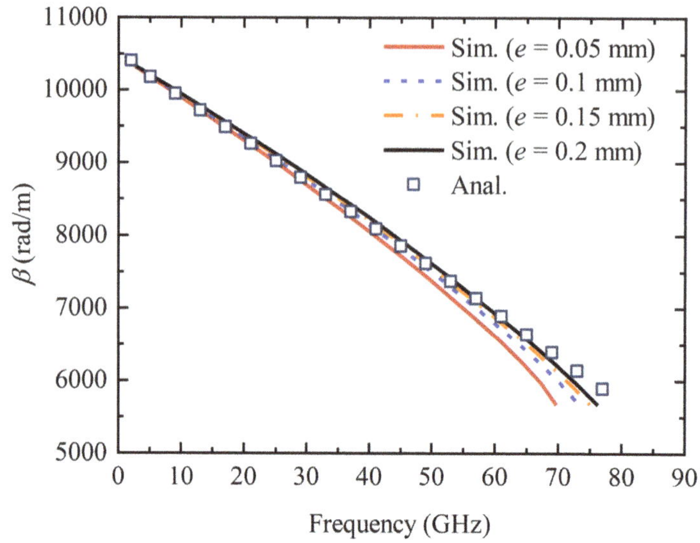

Figure 7.8. Dispersion diagram of the coplanar SWS with different values of *e* for the backward-wave mode. © [2023] IEEE. Reprinted, with permission, from [9].

shorter wavelengths of the electromagnetic wave, leading to a higher phase shift over one period of the SWS with the same dimensions.

In figure 7.8, the changes in the dispersion curve with parameter e are illustrated. It is evident that as e changes from 0.05 mm ($e/h = 0.197$) to 0.2 mm ($e/h = 0.787$), the simulated dispersion curve gradually converges towards the dispersion curve based on the analysis, exhibiting a higher π point frequency. This behavior arises due to the reduction in coupling between the two sides of the structure as 'e' becomes larger, thereby bringing the dispersion curve closer to that of a single ML SWS. When e is 0.05 mm, the maximum error between the analyzed and simulated dispersion diagrams is approximately 10%. When e becomes excessively small, this deviation increases further. In such cases, the couplings between the two sides of the double MML SWS need to be considered in the equivalent circuit model to improve the accuracy of the method.

For comparison, the dispersion characteristics for the backward wave of the double ML SWS with corresponding dimensions are also studied. As shown in figure 7.9(a), the two structures have very similar dispersion characteristics and slowing-down factors. Figure 7.9(a) also shows the beam line for a beam voltage of 18 kV, which intersects with the dispersion curve at about 73 GHz.

The interaction impedance values of the two structures are also calculated and compared for BWO applications. As shown in figure 7.4, two symmetrically arranged pencil electron beams with a radius of 0.06 mm are placed at the locations of maximum axial electric field in the gaps. The interaction impedance is calculated at the location of the axis of each electron beam, i.e., $h' = 0.1$ mm. The interaction impedance values for BWO are calculated using equation (1.2) in chapter 1. As both structures in figure 7.4 have symmetry with respect to the yz plane, the interaction

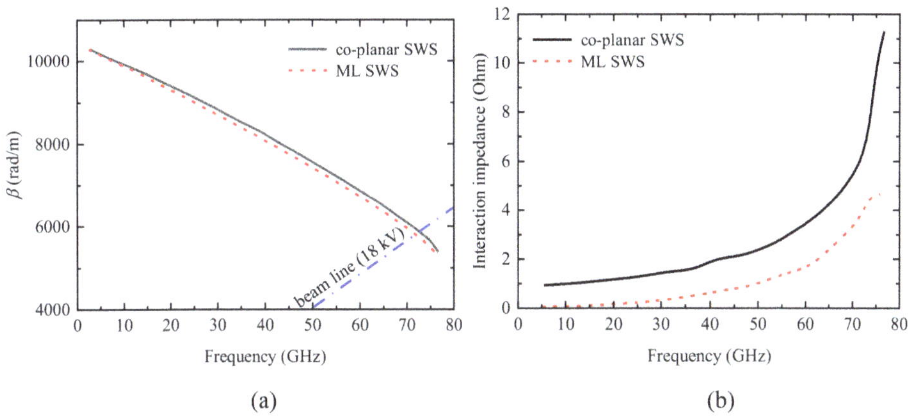

Figure 7.9. (a) Propagation constant and (b) interaction impedance of the backward wave of the ML SWS and the coplanar SWS.

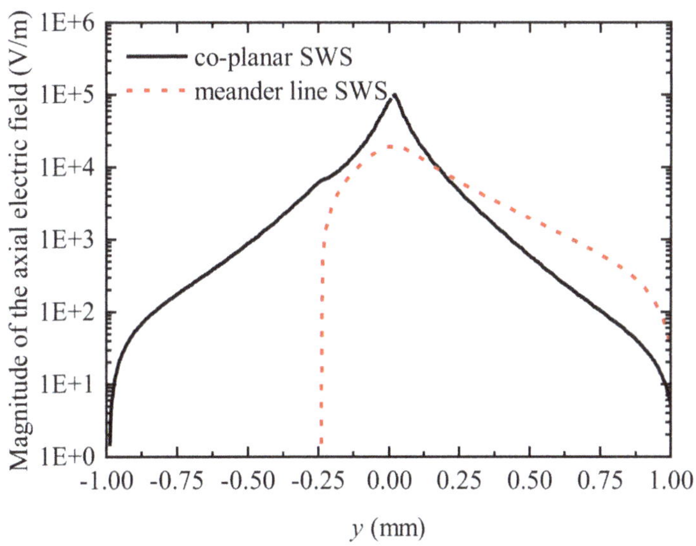

Figure 7.10. Magnitude of the axial electric field vs. *y*.

impedances at symmetric positions on the left and right sides of the *yz* plane should be identical. As a result, only the interaction impedance on the right side of the structure is calculated. The results for the calculated interaction impedance are presented in figure 7.9(b). The interaction impedance of the proposed SWS is 1.8–20 times higher than the interaction impedance of the ML SWS. This should lead to a higher output power and efficiency for a BWO based on the proposed SWS.

The electric field distributions of both structures are also examined. The variation of the axial electric field at 60 GHz with *y* is plotted in figure 7.10 ($d' = 0.3275$ mm). The thin metal patterns are at $y = 0$, and the input power is 1 W. It can be seen that the electric field of the coplanar SWS decays exponentially in both the $+y$ and $-y$ directions. It can also be noted that at locations close to the substrate surface, the axial electric field of the coplanar SWS is larger than that of the MML SWS. This is because the field is concentrated within the gaps between the coplanar patterns. Moreover, unlike the microstrip SWS, in which the presence of the ground plane in close vicinity does not allow the field to build up to a large value, there is no ground plane in the coplanar SWS. Higher values of the axial electric field result in higher interaction impedances of the proposed SWS at locations close to the surface. At $y = -0.254$ mm, i.e. at the ground plane of the microstrip line, the field of the microstrip SWS drops to zero. The axial electric field of the proposed SWS drops to zero at $y = \pm 1.0$ mm, i.e. at the surface of the metal enclosure.

7.3.3 Input and output couplers

Input/output couplers are important for good wave transmission along an SWS. To achieve good wave transmission in an SWS, it is important to ensure that the characteristic impedance of the couplers matches that of the SWS. For a periodic structure, the effective characteristic impedance, namely the Bloch impedance, is represented by:

$$Z_0 = \sqrt{\frac{B}{C}} \tag{7.4}$$

where B and C are elements of the transmission matrix of one period of the SWS. Given information about the characteristic impedance, designers can choose suitable types of couplers and match them to the SWS with specific impedance-matching techniques.

A design procedure for the input/output couplers of the coplanar SWS operating in the V band is described here, and the transmission characteristics are presented. The dielectric substrate is quartz with a thickness of 0.254 mm. The material for the metal layer is gold with a conductivity of 2×10^7 S m^{-1}. The thickness of the metal layer is 4 μm. The other dimensions are the same as the initial parameters.

The real and imaginary parts of Z_0 are calculated using equation (7.4) and shown in figure 7.11. It can be seen that the imaginary part of Z_0 is close to 0 Ω, and the average of the real part is about 54 Ω over the operating frequencies. The coplanar configuration of the proposed SWS naturally fits the coplanar waveguide (CPW). In this design, the SWS is first connected to a 54 Ω CPW, which tapers to a 50 Ω CPW. The widths of the two gaps of the CPW transition from 0.037 to 0.032 mm. Here, we consider an SWS with 45 periods. The configuration of the SWS with the input/output couplers is shown in figure 7.12. To facilitate the connection of the proposed SWS to the electron gun and collector, the CPW feed lines are bent at right angles. The S-parameters of the SWS with the input–output coupler are shown in figure 7.13. The S_{11} value is below −15 dB, and the S_{21} value is higher than −5.9 dB over the frequency range of 0–70 GHz.

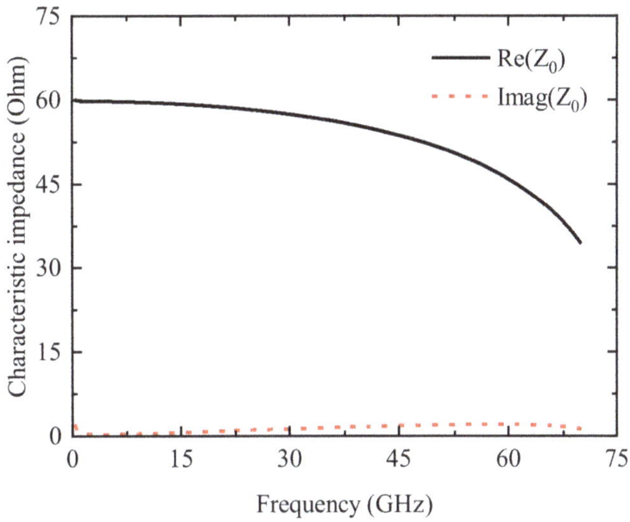

Figure 7.11. The real and imaginary parts of the characteristic impedance. © [2023] IEEE. Reprinted, with permission, from [9].

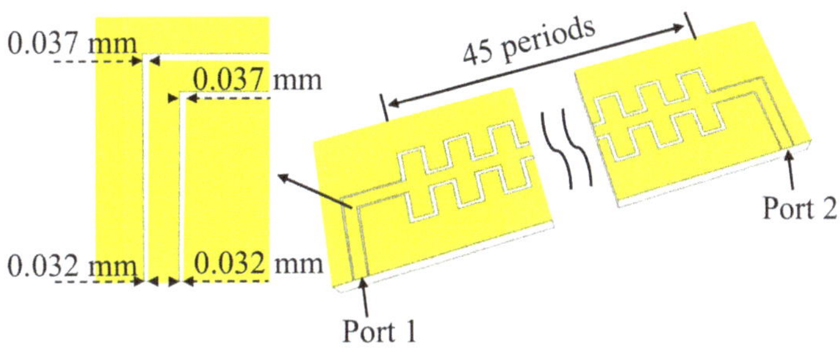

Figure 7.12. Perspective view of the proposed SWS with input and output couplers. © [2023] IEEE. Reprinted, with permission, from [9].

7.3.4 Simulation results for hot-test parameters

In this section, the hot-test parameters of a BWO based on the proposed 45-period SWS are studied using the CST Particle Studio simulator. The distance between the two circular electron beams and the surface of the SWS, h', is 0.1 mm. The electron-beam radius is 0.06 mm, and the current is 0.045 A in each electron beam; i.e. the total current is 0.09 A. A uniform axial magnetic field of 1 T is applied. Figure 7.14 illustrates the variation of the output frequency and power of the BWO with beam voltage. The operating frequency changes from 56 to 73 GHz when the beam voltage changes from 8 to 22 kV, corresponding to a tunable frequency range of 17 GHz. The corresponding frequency of the intersection point between the beam

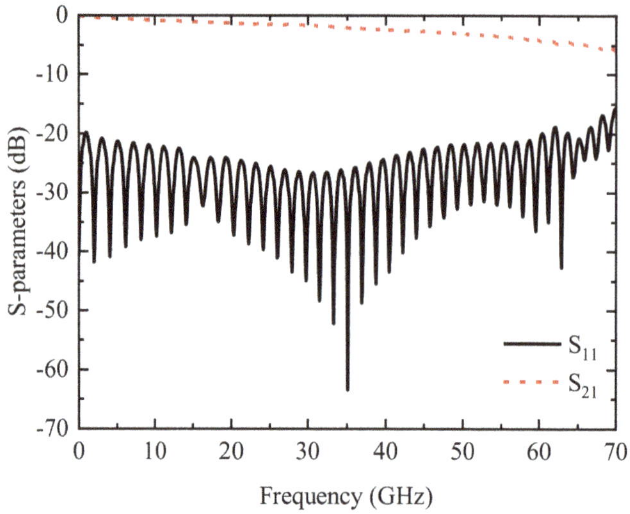

Figure 7.13. Simulated S-parameters of the proposed SWS. © [2023] IEEE. Reprinted, with permission, from [9].

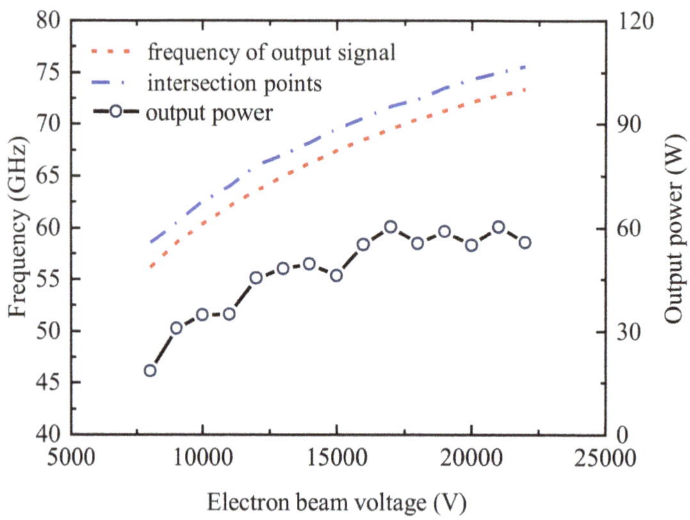

Figure 7.14. Variation of output frequency and power with electron-beam voltage. © [2023] IEEE. Reprinted, with permission, from [9].

line and the dispersion curve is also shown in this figure. The frequency of the output signal is quite close to that indicated by the intersection point. A maximum output power of 60 W is obtained at 69.5 GHz when the beam voltage is 17 kV. The corresponding RF efficiency is 4%. Figure 7.15 shows the port signals for beam voltages of 17 and 22 kV. Stable output signals are obtained at port 1 for both beam voltages after about 5 ns.

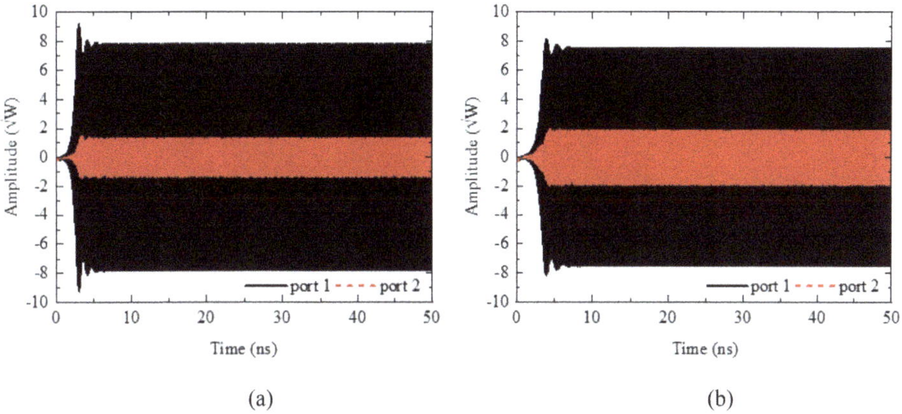

Figure 7.15. BWO output signals vs. time at electron-beam voltages of (a) 17 kV and (b) 22 kV. © [2023] IEEE. Reprinted, with permission, from [9].

Figure 7.16. Top view of the fabricated X-band 15-period coplanar SWS. © [2021] IEEE. Reprinted, with permission, from [8].

7.3.5 Fabrication and measurement

To verify the simulation results presented in the previous sections, two proof-of-concept structures are fabricated and measured. One X-band coplanar SWS is fabricated using the printed circuit technique on an alumina substrate. A second V-band coplanar SWS is fabricated using microfabrication techniques on a quartz substrate. This subsection presents the fabrication and measurement results for these SWSs.

7.3.5.1 X-band coplanar SWS

The X-band coplanar SWS is fabricated using printed circuit techniques on a 0.508 mm thick alumina substrate with a permittivity of 9.7 and a loss tangent of 2×10^{-4}. The metal layer consists of Ti/Pt/Au. The thickness of the Au is 3–4 μm, which ensures the overall high conductivity of the metal layer. Following the symbols for the dimensional parameters shown in figure 7.4, the dimensions of one period of the scaled SWS are as follows: $p = 2.4$ mm, $w = 0.96$ mm, $h = 0.508$ mm, $L = 1.2$ mm, and $g = 0.24$ mm. To achieve good impedance matching to 50 Ω, the SWS is connected to a 44.3 Ω CPW, which gradually tapers to a 50 Ω CPW. A top view of the fabricated SWS is shown in figure 7.16.

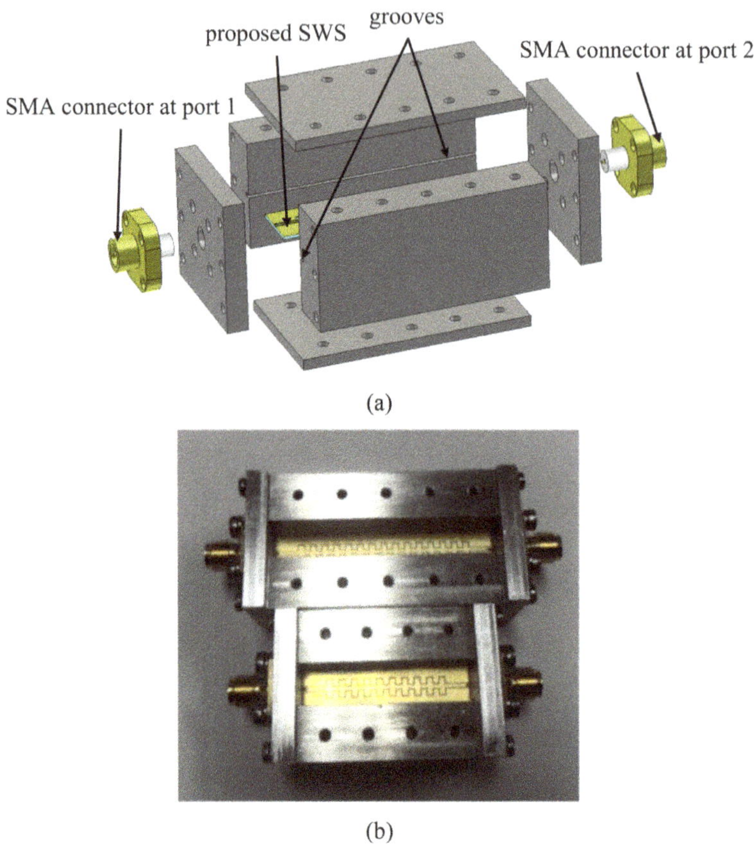

Figure 7.17. (a) Assembly of the coplanar structure. (b) Photo of the assemblies for SWSs of 10 and 15 periods.

To facilitate measurements of the fabricated SWS, a metal fixture-cum-enclosure is designed and fabricated. As shown in figure 7.17(a), grooves are made on the left and right side plates so that the substrate fits tightly in the grooves to provide good electrical and thermal contact. In order to obtain the phase velocity from the measured results, two circuits are fabricated with different numbers of periods, namely 10 and 15 (figure 7.17(b)).. The S-parameters of the structure are measured using a vector network analyzer (VNA). Figure 7.18 shows both the simulated and measured S-parameters of the SWS with 15 periods. The S_{11} value of the SWS is below -10 dB, and the insertion loss is less than 2 dB over the frequency range from 0 to 12 GHz. The difference between the simulated and measured results might be caused by tolerances in the fabrication and assembly processes of the structure.

The propagation constant of the forward wave is calculated using:

$$\beta_0 = \frac{\varphi_2 - \varphi_1}{(n_2 - n_1) \times p} \tag{7.5}$$

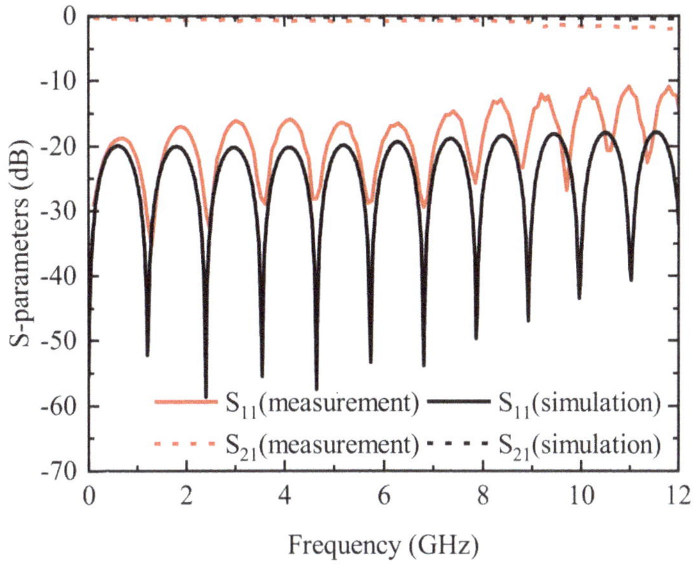

Figure 7.18. Simulated and measured S-parameters of the SWS with 15 periods. © [2021] IEEE. Reprinted, with permission, from [8].

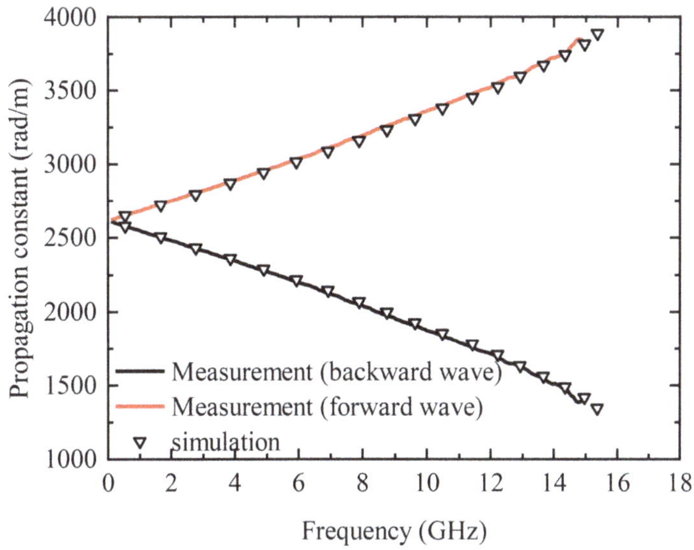

Figure 7.19. Simulated and measured propagation constants.

where φ_1 and φ_2 are the phases of the measured S_{21} values of the two circuits and n_1 and n_2 are the numbers of periods of the two circuits ($n_2 > n_1$). The measured propagation constants of the fundamental and the first negative space harmonic are presented in figure 7.19. The measured results show very good agreement with the simulation results.

7.3.5.2 V-band coplanar SWS

The designed V-band coplanar structure described in subsection 7.3.3 is fabricated on a quartz substrate using a simple photolithographic process. The main steps in the fabrication process are presented in figure 7.20 and are described in the following.

(a) Deposit a 100 nm thick titanium film onto the surface of a quartz substrate. Then sputter a 300 nm thick gold coating, which forms the main metal layer for the SWS.
(b) Spread a layer of positive photoresist over the gold coating layer. Expose the surface of the photoresist to UV light through a photomask.
(c) Dissolve the photoresist in the exposed areas using a developer, transferring the comb-like pattern from the photomask onto the remaining photoresist.
(d) Remove the metal layer using a wet etching process, creating a symmetrical U-shaped groove.
(e) Electroplate gold onto the etched metal layer to increase the metal thickness to 4 μm. Remove the remaining photoresist.
(f) Use laser cutting to dice the quartz wafer into individual samples.

A photograph of the fabricated structure is shown in figure 7.21(a). To facilitate testing, the fabricated input/output couplers are not bent. To measure the dispersion characteristics of the SWS, structures of two different lengths (20 and 25 periods) are fabricated. The propagation constant can be calculated using equation (7.5). To prevent radiation loss, the SWS is placed in a metal enclosure. As illustrated in figure 7.21(b), the designed metal enclosure consists of upper and lower metal blocks. The lower metal block has thin grooves on the inner side to hold the SWS. The thickness of the grooves is consistent with the thickness of the SWS. The two ends of the SWS are extended by 1 mm and have a greater width to prevent the SWS from moving inside the metal enclosure and to maintain its position and alignment

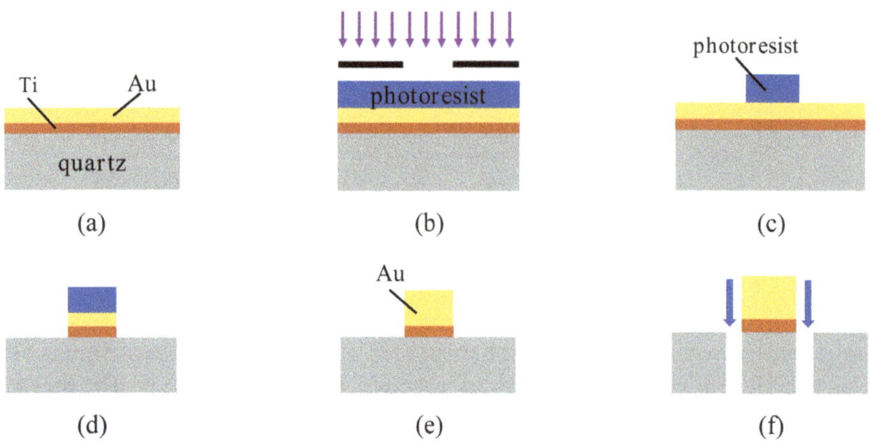

Figure 7.20. Fabrication process of the coplanar SWS. © [2023] IEEE. Reprinted, with permission, from [9].

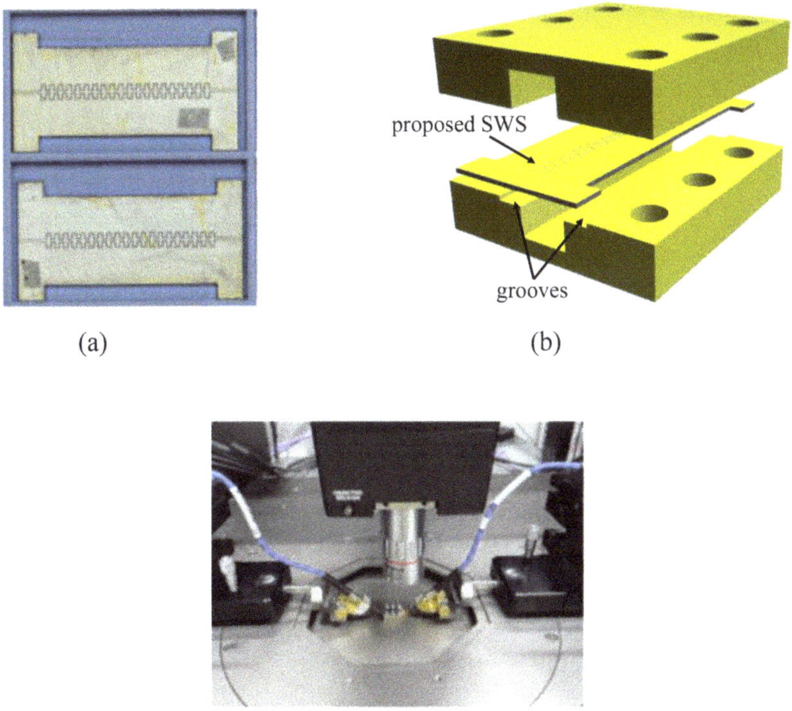

(a) (b)

(c)

Figure 7.21. (a) Photograph of the microfabricated SWS. (b) Assembly of the coplanar SWS. (c) Photograph of the probe station measurement. © [2023] IEEE. Reprinted, with permission, from [9].

during operation. It should be noted that the additional grooves in the metal shield and the extended structures at the two ends slightly affect the S-parameters of the SWS, but they do not significantly influence its impedance-matching properties.

The transmission characteristics of the fabricated SWS are measured with a probe station which is connected to the VNA shown in figure 7.21(c). The measured S-parameters of the SWS with 20 and 25 periods are shown in figures 7.22(a) and (b). It can be seen that the measurement results are in very good agreement with the simulation results. Over the frequency range from 0 to 65 GHz, the reflection coefficient S_{11} is better than -15 dB. Additionally, the insertion loss, represented by S_{21}, is less than 2 dB throughout the frequency range from 0 to 70 GHz, demonstrating low power loss during transmission over a wide frequency range. The propagation constant is calculated from the phases of the measured S_{21} values. The obtained results are presented in figure 7.23 and compared with the simulated and analytical results. The measured propagation constant values are consistent with both the simulation results and the equivalent circuit analysis results.

As in other millimeter-wave SWSs, the dimensions of many features of the fabricated coplanar SWS are only tens of micrometers. The smallest gap of 0.032 mm occurs at the CPW couplers. The high electric field at this position may

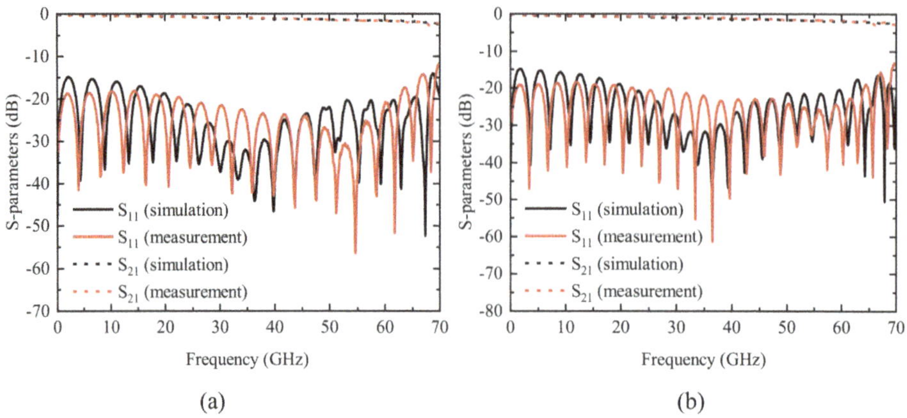

Figure 7.22. Simulated and measured S-parameters of the SWSs with (a) 20 periods and (b) 25 periods. © [2023] IEEE. Reprinted, with permission, from [9].

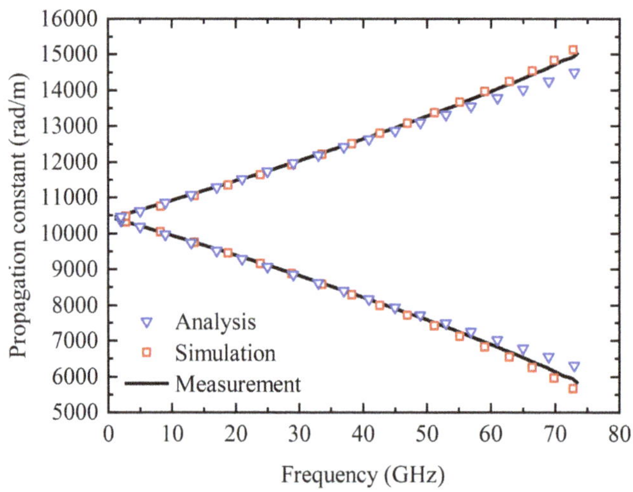

Figure 7.23. Phase constant of the V-band coplanar SWS. © [2023] IEEE. Reprinted, with permission, from [9].

lead to electric breakdown, adversely affecting the power handling capability of the BWO. The average electric field across the smallest gap is evaluated using CST MWS for an input power of 1 W. The results are presented in figure 7.24. It can be seen that the electric field is less than 2.5e5 V m^{-1} over the operating frequencies. For an output power of 60 W, the corresponding RF electric field is less than 1.9e6 V m^{-1}, which is much lower than the breakdown field strength value of vacuum (~1e11 V m^{-1}) and the breakdown field strength value of quartz (~5e7 V m^{-1}).

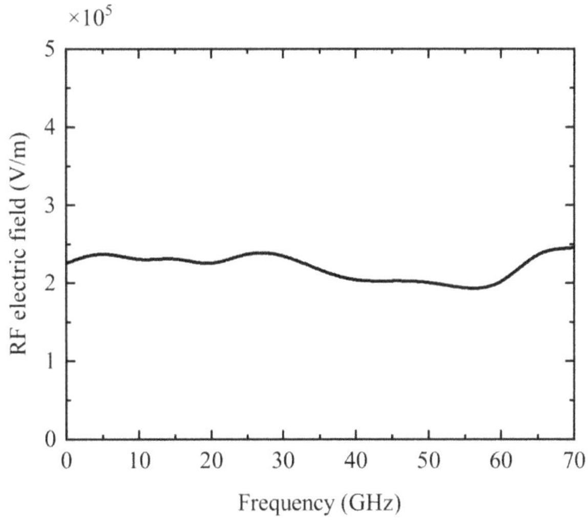

Figure 7.24. Average RF electric field value at the minimum gap of 0.032 mm for an input power of 1 W. © [2023] IEEE. Reprinted, with permission, from [9].

7.4 A W-band BWO based on a PH-SEC

Chapter 4 describes the PH-SEC SWS and its application in TWT amplifiers. Chapter 8 describes examples of the fabrication of PH-SEC SWSs intended for use in various frequency bands. In view of some unique features of BWOs, as mentioned at the beginning of this chapter, it is also desirable to explore the application of PH-SECs in BWOs. This section describes the design of a W-band BWO based on a PH-SEC. The description closely follows that reported by the authors' research group in [10, 11].

Parts of sections 7.3.2–7.4 have been reprinted, with permission, from [9]. © [2023] IEEE.

7.4.1 Design of PH-SEC-based BWO

A BWO based on PH-SEC and operating in the frequency range from 15.5 to 20 GHz has been reported in [12]. Here, we present the design of a W-band BWO which uses a microfabrication-compatible PH-SEC. Figure 7.25(a) shows a simplified 3D view of the PH-SEC, and figure 7.25(b) shows a cross-sectional view of the PH-SEC used for the W-band BWO. It has the same structure as that used in [13] for the design of a PH-SEC-based amplifier. The dimensions of the various geometrical features are also shown in figure 7.25(b). The pitch or the period L of the SWS is chosen as 375 μm to enable operation of the BWO with a beam voltage of around 10 kV.

The materials used for the SWS are also kept the same as those in [13]. Quartz substrates with a dielectric constant of 4.43 and a loss tangent of 5.1×10^{-5} are used as supports for the PH-SEC. The PH-SEC and the substrates are bonded together using a low-loss nonconductive B-staged bisbenzocyclobutene (BCB) layer with a dielectric constant of 2.65 and a thickness of 5 μm. Silicon with a dielectric constant

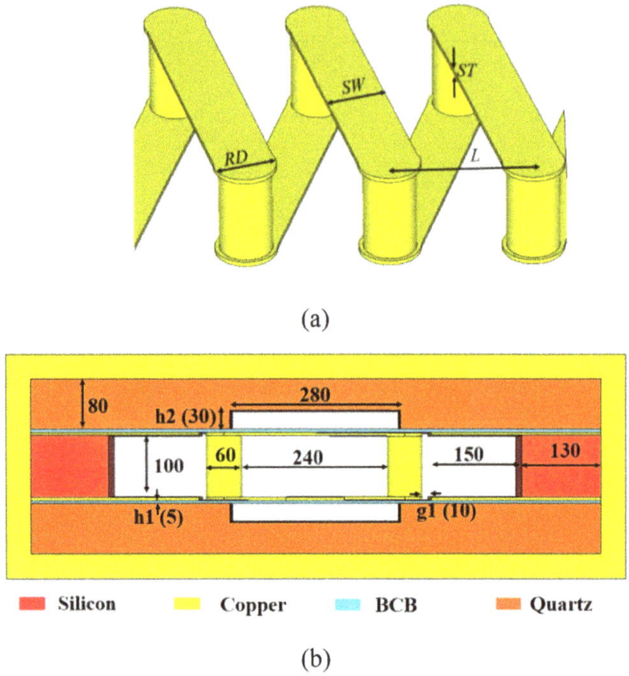

Figure 7.25. (a) Simplified 3D view of the PH-SEC. (b) Cross-sectional view of the PH-SEC SWS with dimensions (all dimensions are in μm). © [2018] IEEE. Reprinted, with permission, from [11].

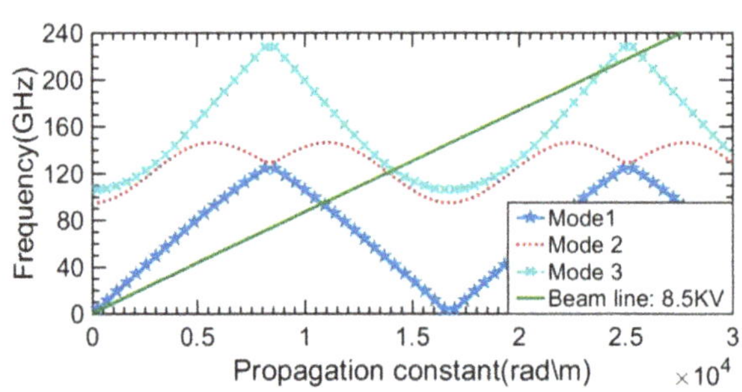

Figure 7.26. Dispersion characteristics of the PH-SEC SWS shown in figure 7.25(b). © [2018] IEEE. Reprinted, with permission, from [11].

of 11.9 and a loss tangent of 2.5×10^{-4} is used as the spacer between the quartz substrates. The planar helix and the metal enclosure are considered to be made of copper with an assumed RF conductivity of 2×10^7 S m^{-1}.

The dispersion characteristics of the SWS together with an 8.5 kV beam line are shown in figure 7.26. Simulations are carried out using the CST MWS Eigenmode Solver. Mode 1 is the fundamental mode, while modes 2 and 3 are the first and second higher-order modes of the SWS, respectively. The 8.5 kV beam line intersects

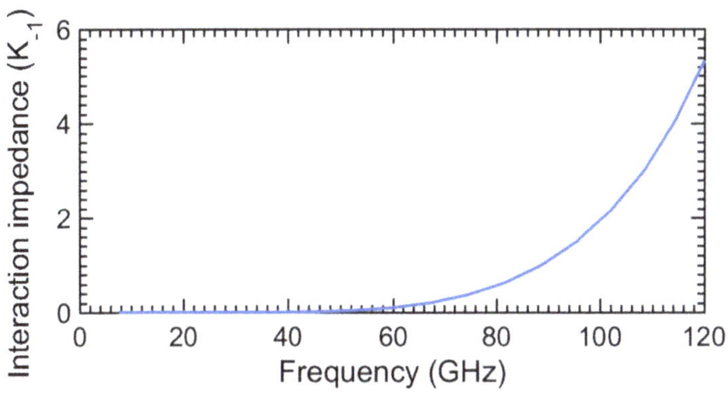

Figure 7.27. On-axis interaction impedance of the first backward-wave space harmonic of the PH-SEC. © [2018] IEEE. Reprinted, with permission, from [11].

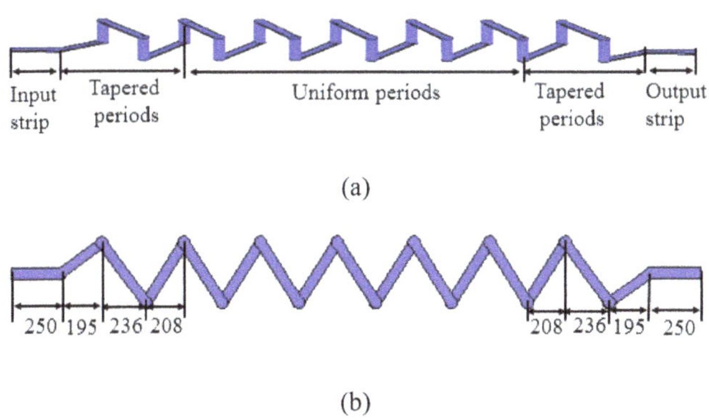

Figure 7.28. Structure of the SWS with four periods and strip-line feeds. The metal enclosure and dielectric supports are not shown. (a) 3D view, and (b) 2D view with the main dimensions (all dimensions are in μm). © [2018] IEEE. Reprinted, with permission, from [11].

the first backward-wave space harmonic of the first mode at 94 GHz. The beam line also intersects the backward-wave harmonics of the second and third modes at about 120 and 225 GHz, respectively, indicating the possibility of unwanted oscillations around these frequencies. Figure 7.27 shows the on-axis interaction impedance for the first backward-wave space harmonic in the SWS, indicating a value of around 1.5 Ω at the design frequency.

Sixty-two periods of the PH-SEC described above are used to form the SWS for the BWO. The overall length of the SWS is 24.78 mm. Strip-line feeds are used at the input and output ports of the SWS. Figure 7.28(a) shows the SWS with the strip-line feeds and tapered periods at both ends. As shown in the figure, pitch tapering is used at both ends to improve the matching of the feed. The dimensions of the pitch taper are included in figure 7.28(b). The strip-line feeds and the horizontal strips of the PH-SEC are in the same plane; hence, both can be fabricated together in the same

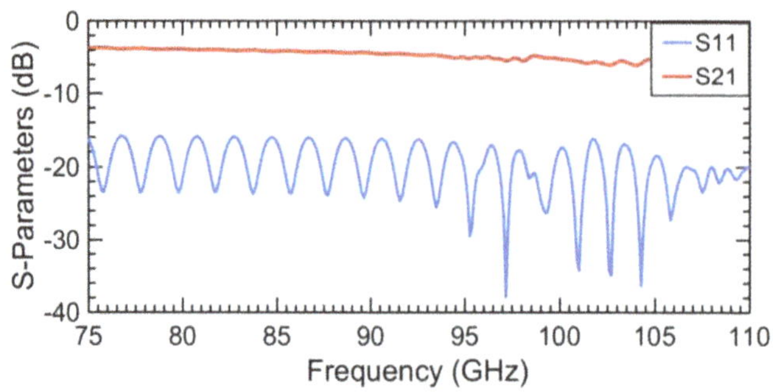

Figure 7.29. S-parameters of the PH-SEC SWS designed for a BWO. © [2018] IEEE. Reprinted, with permission, from [11].

process step. Figure 7.29 shows the S-parameters of the SWS; S_{11} is less than -16 dB and S_{21} is better than -6.5 dB over the entire W band.

With reference to figure 7.25(b), the thickness of BCB (h_1), the depth of the trench in the quartz substrates (h_2), and the gap between the PH-SEC and the ground layer (g_1) are likely to vary due to fabrication tolerances and thus can affect the properties of the SWS. The variation in the dispersion characteristics and interaction impedance of the SWS due to the variation in these parameters has been studied [11]. An increase in h_1, h_2, or g_1 slightly increases the value of the propagation constant for a given frequency, leading to a higher slowing-down factor for the EM wave. A smaller value of h_1 and g_1 can increase the interaction impedance of the SWS, whereas the effect of h_2 on the interaction impedance is negligible. Even though a smaller value of g_1 is more desirable for the operation of the BWO, we fix it at 10 μm for ease of fabrication. Similarly, the value of h_1 is chosen as 5 μm to facilitate realization using the normal photolithographic process. A larger value of h_2 can provide a higher slowing-down factor for the first backward-wave space harmonic of the SWS, but this reduces the mechanical stability of the quartz substrate; hence, the h_2 value chosen is 30 μm.

7.4.2 PIC simulation results

The performance of the BWO based on the PH-SEC described in the previous subsection is estimated using CST PIC simulations. A centrally located sheet electron beam with dimensions of $160 \times 40\ \mu m^2$ and a beam current of 20 mA is used. The beam voltage is varied from 7 to 11 kV. A magnetic field of 1 T is used to confine the flow of the electron beam through the SWS. Simulations indicate that 100% of the beam is transmitted through the SWS.

Figure 7.30(a) shows the time evolution of the output signal from the BWO for a beam voltage of 8.5 kV. A strong output signal is generated at the port near the electron gun. Oscillations start to build up at around 20 ns, and the amplitude of the signal grows to a steady value by around 27 ns. As shown in figure 7.30(b), the spectrum of the output signal is strongest at 92.46 GHz, which is very close to the frequency of 94 GHz indicated by the dispersion curve in figure 7.26. There are small traces of noise in the frequency spectrum. The strongest noise is close to 235 GHz,

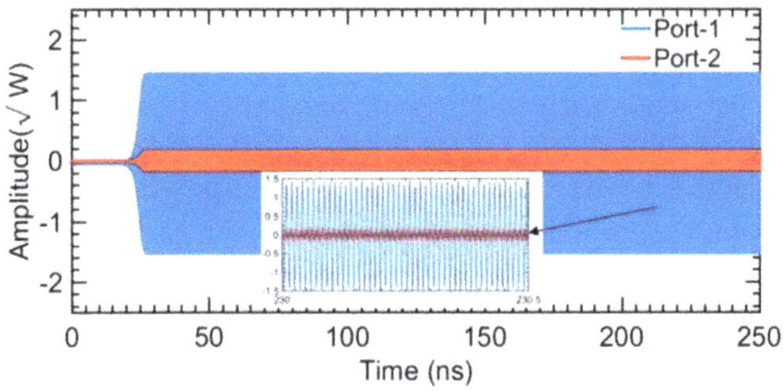

(a)

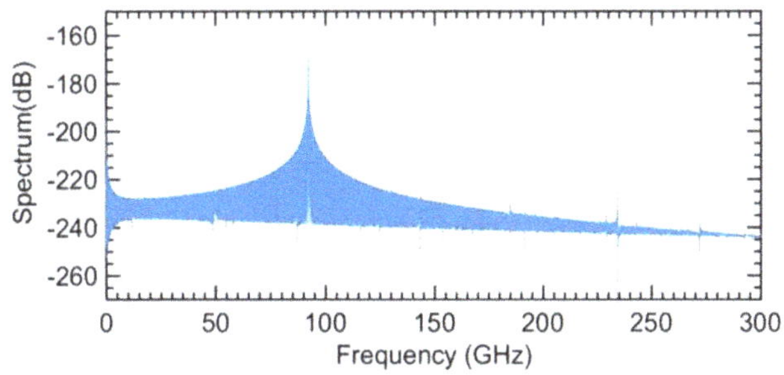

(b)

Figure 7.30. PIC results for the BWO at a beam voltage of 8.5 kV. (a) The output signal versus time with a magnified version in the inset. (b) The spectrum of the output signal. © [2018] IEEE. Reprinted, with permission, from [11].

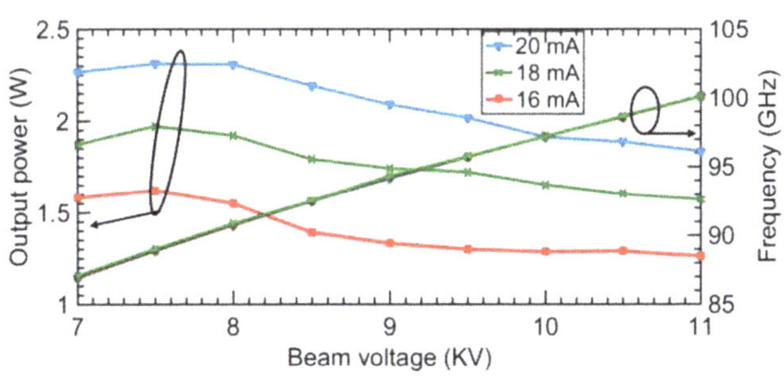

Figure 7.31. Variation of output power and frequency with beam voltage for three different beam currents. © [2018] IEEE. Reprinted, with permission, from [11].

and it is 59 dB below the main signal. This unwanted signal corresponds to the intersection of the third mode and the 8.5 kV beam line, with a shift in frequency due to beam loading.

As shown in figure 7.31, the frequency of the output signal can be tuned from 86.9 to 100.07 GHz by tuning the beam voltage from 7 to 11 kV. Thus, the BWO provides a tunable bandwidth of 14% around the central frequency of 93.48 GHz.

This figure includes the results for beam currents of 16 and 18 mA. A magnetic field of 0.75 T is used for simulations with a beam current of 16 mA, and a magnetic field of 0.9 T is used for simulations with a beam current of 18 mA. The figure indicates that the frequency tuning of the output signal is independent of the beam current. The figure also shows the variation of peak output power with beam voltage (or frequency) and beam current. The output power of the oscillator increases with the beam current for a given frequency. The maximum peak output power of 2.3 W at 103 GHz is obtained for a beam voltage of 7.5 kV and a beam current of 20 mA. The maximum peak output powers of the oscillator are reduced to 1.97 and 1.62 W for beam currents of 18 and 16 mA, respectively. The corresponding peak RF efficiency values are 1.62%, 1.48%, and 1.41%, respectively.

7.4.3 Fabrication of the PH-SEC and measurement results

To verify the simulation results, a scaled and simplified version of the PH-SEC operating in the X band is designed with a CPW feed and fabricated using the printed circuit technique. Figure 7.32 shows the structure and the dimensions of the fabricated SWS. As shown in the figure, the top and bottom horizontal strips of the PH-SEC SWS are fabricated on two identical dielectric substrates. RO4003 substrates with a dielectric constant of 3.55, a loss tangent of 0.0027, and a thickness of 0.203 mm. The substrates are separated by 1 mm and are connected using copper wires to realize the straight-edge connections. A metal enclosure is used to hold and shield the PH-SEC. Two SWSs, which are identical except for having different numbers of periods (25 and 30), are fabricated to enable the measurement of phase velocity.

Figure 7.33 shows photos of the assembled SWS with thirty periods. As shown in figure 7.33(a), the top horizontal strips of the PH-SEC and the CPW feed lines are printed on the upper substrate. The bottom horizontal strips of the SWS are printed on the lower substrate shown in figure 7.33(b). As shown in figure 7.33(c), the straight-edge connections are realized using copper wires 0.4 mm in diameter, which connect the top and bottom horizontal strips. The copper wires are soldered to circular patches 0.7 mm in diameter on the ends of the horizontal strips. The widths of the CPWs are tapered to achieve wideband matching. Pitch tapering is also used

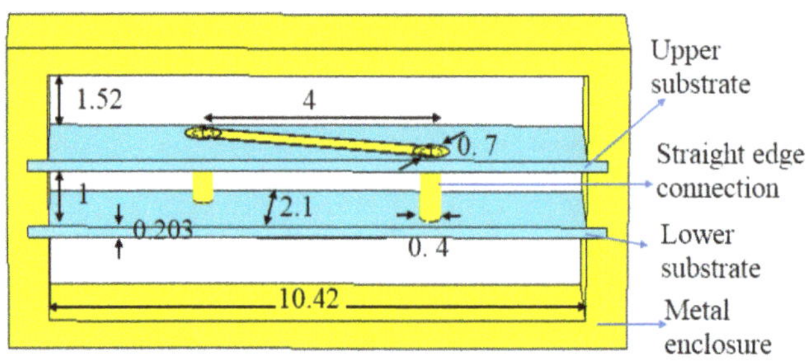

Figure 7.32. Structure and dimensions of the fabricated SWS (all dimensions are in millimeters). © [2018] IEEE. Reprinted, with permission, from [11].

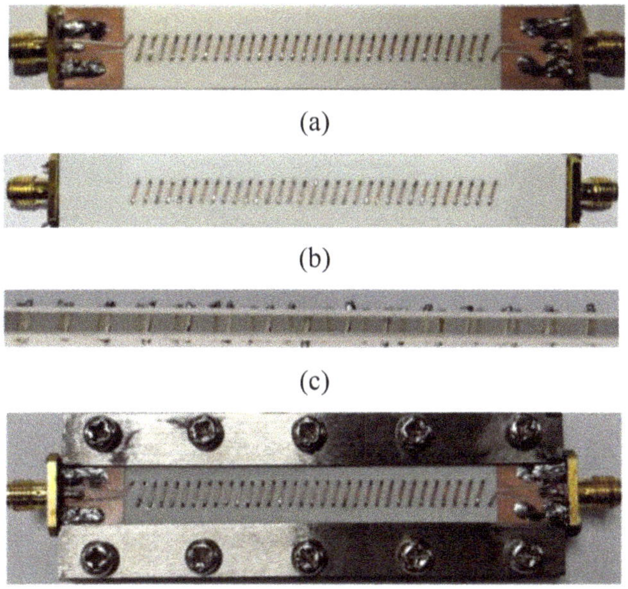

Figure 7.33. Fabricated prototype assembly of the PH-SEC SWS with 30 periods. (a) Top view. (b) Bottom view. (c) Side view. (d) SWS inside the metal enclosure (without the top cover of the metal enclosure). © [2018] IEEE. Reprinted, with permission, from [11].

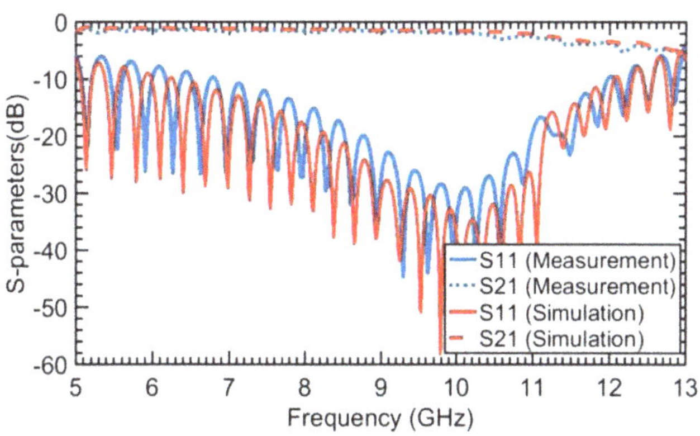

Figure 7.34. Measured and simulated S-parameters. © [2018] IEEE. Reprinted, with permission, from [11].

at both ends of the SWS for better matching. Figure 7.33(d) shows the assembled SWS with the top metal cover removed.

The CST MWS Transient solver is used to simulate the two SWSs with different numbers of periods. Dielectric and conductor losses are included in the simulation; copper is considered to have a conductivity of 2.9×10^7 S m^{-1}. The simulated and measured S-parameters of the SWS with thirty periods are shown in figure 7.34.

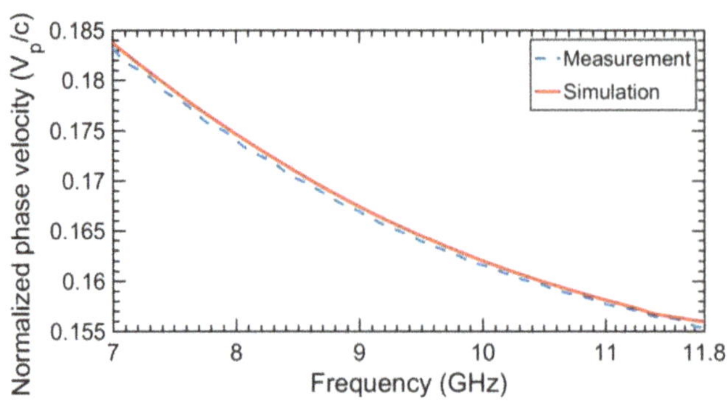

Figure 7.35. Comparison between the simulated and measured normalized phase velocities of the SWS. © [2018] IEEE. Reprinted, with permission, from [11].

The measured values of S_{11} and S_{21} are good matches with those obtained through simulation over a wide frequency range.

The propagation constant of the PH-SEC is calculated using the measured S_{21} phase values of the SWSs with two different lengths. The phase velocity of the PH-SEC is estimated from the propagation constant and is compared with that obtained through simulation in figure 7.35. The measured phase velocity values match closely with those obtained through simulation. The measured phase velocity values are slightly lower than the simulated values. This difference may be due to fabrication tolerances, the extra length of the straight-edge connections at solder joints, and the slight difference between the conductivity values for the fabricated structure and those used in simulation.

7.5 Two-beam oscillator–amplifier

While the advantages of BWOs include high spectral purity, electronic frequency tunability, and wide tunable bandwidth, many types of BWOs suffer from low DC-to-RF conversion efficiency (i.e. RF efficiency). Some examples supporting this observation, taken from [14], are mentioned below.

One commercial BWO operating in the frequency range of 33–54 GHz produces a maximum peak power of 180 mW and achieves a maximum electronic efficiency of 0.72% [15]. Another commercial BWO that operates in the frequency range of 36–55 GHz and uses a maximum beam voltage of 1.2 kV produces a maximum output power of 80 mW and achieves an RF efficiency of 1% [16].

Microfabricated TWTs are currently attracting a lot of attention due to their ability to operate at relatively high frequencies. Microfabricated BWOs operating at low beam voltages have been reported for high-frequency applications, but these too suffer from very low RF efficiency. An efficiency of 1.74% and an output power of 20 W at 35 GHz have been obtained from a BWO with a folded waveguide as the SWS [17]. A folded-waveguide-based BWO with a 12 kV beam has been designed to operate around 95 GHz, producing 5 W of output power and achieving an efficiency of 0.84% [18]. A BWO with a staggered double-vane SWS that requires an electron beam with an input DC power of 2 kW has been estimated to produce only watt-level output power at 90 GHz [19].

Some modified structures have been proposed in the past to overcome the low output power and efficiency of conventional BWOs. The orotron and the clinotron are two variants of the conventional BWO that were developed to improve the power level [20]. The orotron uses a Fabry-Perot cavity with two mirrors, in which a periodic structure is placed on one of the mirrors. However, the very low group velocity of the quasi-cutoff operating wave increases the ohmic losses in the orotron and impedes its application in the continuous-wave (CW) regime [21, 22]. Clinotrons use an inclined and thick electron beam to improve the interaction with the evanescent surface wave of the SWS. The electron beam in this device directly hits the SWS, increasing heat generation. The likelihood of noise generation and device failure is higher in this device. In addition, efficiency improvement using a depressed collector is not possible in the clinotron [23].

The rippled-wall waveguide (RWG) SWS and a relativistic electron beam are commonly used to generate high power outputs from relativistic BWOs. The interaction between the electron beam and the electromagnetic wave is stronger in relativistic BWOs; hence, they have the potential to operate at very high efficiency (15%–40%) and in the megawatt range of output power. Sometimes, a plasma background is used for space-charge neutralization in relativistic BWOs [24, 25]. However, relativistic BWOs require very high beam voltages.

Conventionally, for nonrelativistic BWOs, high power is achieved by feeding the output of a BWO to a separate traveling-wave tube amplifier (TWTA) [26], but this increases the size and cost. Some other attempts to improve BWO efficiency have also been reported in the literature [27–29]; however, efficiency improvement has come at the cost of a reduction in tunable bandwidth in these attempts.

This section describes a novel technique to improve the efficiency of the conventional nonrelativistic BWO by including an additional electron beam which synchronizes with the fundamental forward-wave space harmonic and amplifies the signal generated by the BWO. This technique, first reported by the authors' research group in [14], preserves the electronically tunable bandwidth of the BWO without increasing its overall size.

The generation of two beams with different beam voltages is challenging. However, in the past, two electron beams with different beam voltages have been proposed and realized in electron tubes; this work has been reviewed in [30].

7.5.1 The proposed technique

The schematic of a TWT is shown in figure 7.36(a). In a conventional BWO, the output signal is generated at port 1, and port 2 has matched termination. The electron beam of the conventional BWO is velocity synchronized with the first backward-wave space harmonic of the SWS. On the other hand, to operate the TWT as a TWTA, one feeds the input signal into port 1, and the output signal is obtained

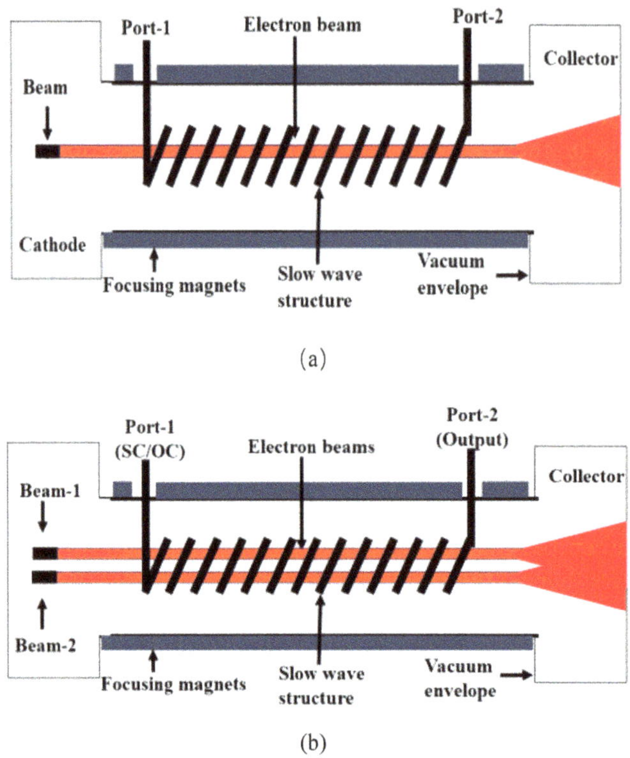

Figure 7.36. Schematic diagram of (a) the TWT; (b) the proposed vacuum electronic two-beam oscillator–amplifier. © [2017] IEEE. Reprinted, with permission, from [14].

from port 2. The electron beam of the TWTA is velocity synchronized with the fundamental forward-wave space harmonic of the SWS. Figure 7.36(b) shows the schematic of the proposed two-beam oscillator–amplifier. The structure of the proposed oscillator–amplifier has two changes compared to that of a conventional BWO. The first is the inclusion of a second electron beam in the BWO; the second electron beam is velocity synchronized with the phase velocity of the fundamental space harmonic of the SWS. In the second change, port 1 of the SWS is terminated in an open circuit (or a short circuit).

The operation of the proposed device can be explained by combining the operating principles of both the BWO and the TWTA. Beam 1 in figure 7.36(b), which is synchronized with the backward wave, generates an RF signal at port 1 according to the usual method of BWO operation. This signal is fully reflected toward port 2. The reflected signal acts as an input signal applied at port 1 of a TWTA; it interacts with the second beam, which is synchronized with the fundamental forward-wave space harmonic and is amplified. Thus, the operations of both the BWO and the TWTA are combined in a single SWS without increasing the overall length. The efficiency of the new device is a combination of the efficiencies of the BWO and the TWTA.

The technique is first tried using a circular helix SWS operating in the Ku band [14]. Both a conventional BWO and an oscillator–amplifier are designed using the same SWS interaction length. Three different two-beam configurations are considered to study the performance of the oscillator–amplifier using the PIC Solver. The results show that the oscillator–amplifier improves the DC-to-RF conversion efficiency of the conventional BWO by a factor ranging from three to six, depending on the beam configuration. The results show that the oscillator–amplifier achieves a maximum DC-to-RF conversion efficiency of 16.77% and a peak output power of 624.5 W.

7.5.2 Design of BWO and oscillator–amplifier using PH-SEC

The design of a two-beam oscillator–amplifier using the PH-SEC SWS was first reported in [31] and has been described in detail in [31]. A conventional BWO is also designed to enable a comparison of the performance of the two devices. This subsection presents the structure, dimensions, dispersion characteristics, and interaction impedance of the PH-SEC SWS. The S-parameters of the SWS are also presented for both the BWO and the oscillator–amplifier. The description here closely follows that in [10].

Similar to the circular helix in [14], the PH-SEC is also designed to operate in the Ku band. As shown in figure 7.37, the PH-SEC is supported by two identical dielectric substrates inside a metal enclosure. The straight-edge connections of the PH-SEC consist of copper wires that have a diameter of 0.5 mm. The top horizontal strips of the PH-SEC are printed on the upper dielectric substrate, while the bottom horizontal strips are printed on the lower dielectric substrate. The design considers RO4003 substrates with a thickness of 0.203 mm, a dielectric constant of 3.55, and a loss tangent of 0.0027. The dimensions of the SWS are tabulated in table 7.1.

The dispersion characteristics of the SWS obtained using the CST MWS Eigenmode Solver are shown in figure 7.38. This figure includes the propagation constant of the fundamental mode and the first higher-order mode of the SWS. Three beam lines corresponding to beam voltages of 5, 7, and 15.75 kV are also included in the figure. The beam lines corresponding to the 5 and 7 kV beam voltages intersect the first backward-wave space harmonic of the fundamental mode

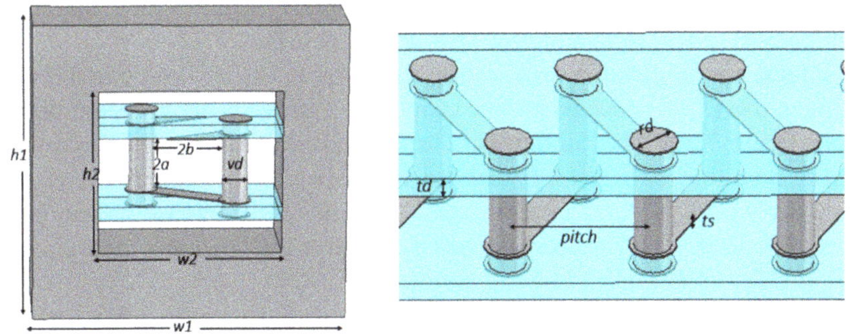

Figure 7.37. Cross-sectional view and a perspective view of the PH-SEC. © [2019] IEEE. Reprinted, with permission, from [31].

Table 7.1. Dimensions of the PH-SEC. © [2019] IEEE. Reprinted, with permission, from [31].

Parameter	Value (mm)	Parameter	Value (mm)
$2a$	1	Pitch	1.5
$2b$	1	h_2	2.44
rd	0.5	h_1	4.44
vd	0.4	w_2	2.8
td	0.203	w_1	4.8
ts	0.017		

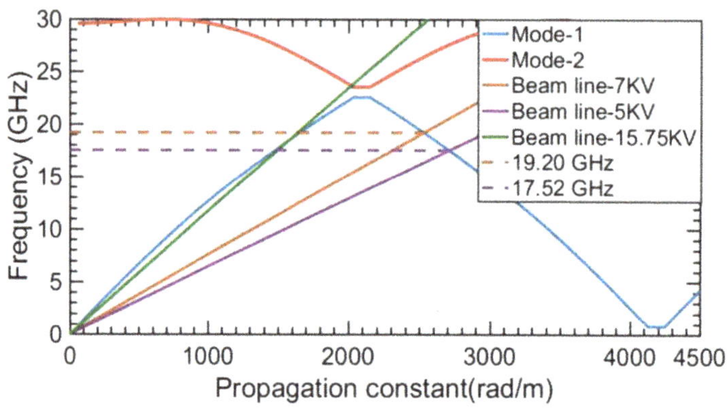

Figure 7.38. Dispersion characteristics of the SWS; also shown are three different beam lines. © [2019] IEEE. Reprinted, with permission, from [31].

at 17.52 and 19.20 GHz, respectively. These intersections indicate that electron beams with voltages of 4 and 6 kV can generate EM signals at frequencies of around 16.7 and 18.9 GHz, respectively. The beam line corresponding to the 15.75 kV beam voltage runs very close to the curve for the fundamental forward-wave space harmonic of mode 1. This indicates that the 15.75 kV beam voltage can amplify a wide range of frequencies, including 16.7 and 18.9 GHz.

The interaction impedance versus frequency for both the fundamental (K_0) and the first negative (K_{-1}) space harmonic of mode 1 at three different beam positions is shown in figure 7.39. The fundamental space harmonic has a higher interaction impedance than that of the first backward-wave space harmonic. For a given beam position, the values of K_0 decrease, while those of K_{-1} increase as the frequency increases. The values of K_{-1} decrease when we move the beam position closer to the horizontal strips or the straight-edge connections.

Fifty periods of the PH-SEC shown in figure 7.37 form the SWS for both the BWO and the oscillator–amplifier. We use a similar interaction length for both devices. Both port 1 and port 2 of the SWS are considered to be matched in the

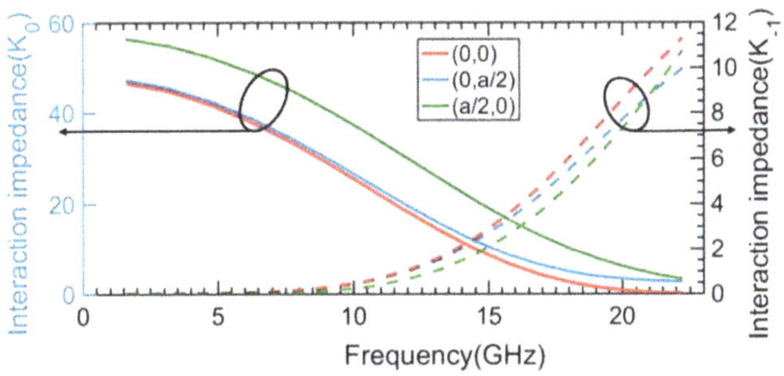

Figure 7.39. Interaction impedances of the fundamental (K_0) and first backward-wave (K_{-1}) space harmonic of the PH-SEC at three different beam positions; (x, y) are the coordinates of the beam position. © [2019] IEEE. Reprinted, with permission, from [31].

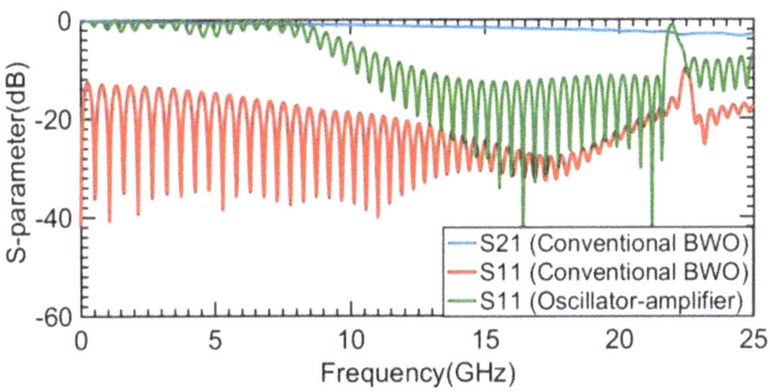

Figure 7.40. S-parameters of the PH-SEC SWS used as a BWO and as an oscillator–amplifier. © [2019] IEEE. Reprinted, with permission, from [31].

BWO simulations. Port 1 is considered to be a perfect open circuit, and port 2 is matched in the simulations for the oscillator–amplifier. CPW feed lines are designed for both ports of the BWO and one port of the oscillator–amplifier. In addition, pitch tapering is used at both ends of the SWS to obtain a better match.

The simulated S-parameters of both devices are shown in figure 7.40. The BWO has an S_{11} value of less than −22 dB and an S_{21} value better than −2.5 dB over the 14–20 GHz frequency range. This frequency range represents a bandwidth of 35.3% and can be considered the expected operational bandwidth of the BWO. The proposed oscillator–amplifier is a one-port device; it has an S_{11} value only, which is better than −11 dB for frequencies from 14–20 GHz. The complete mismatch at port 1 is the reason for the poor reflection characteristic of the oscillator–amplifier.

7.5.3 PIC simulation results

The beam configuration assumed for PIC simulations is shown in figure 7.41. We use two solid cylindrical beams inside the SWS. The beams have a radius of 0.1 mm and are separated by a distance of 0.2 mm. As shown in the figure, the upper beam is

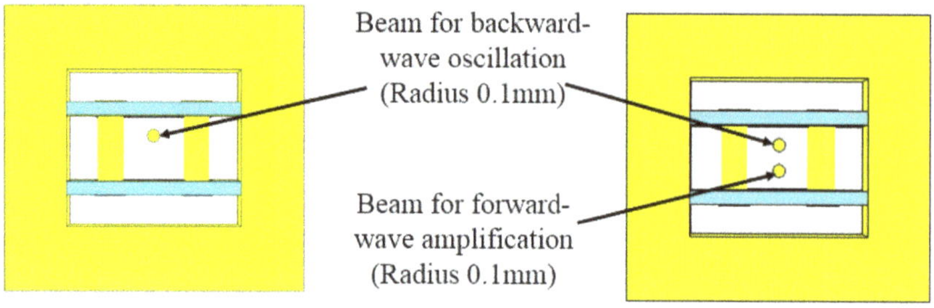

Figure 7.41. Positions of the e-beams in the BWO and the oscillator–amplifier. Reprinted with permission from [10].

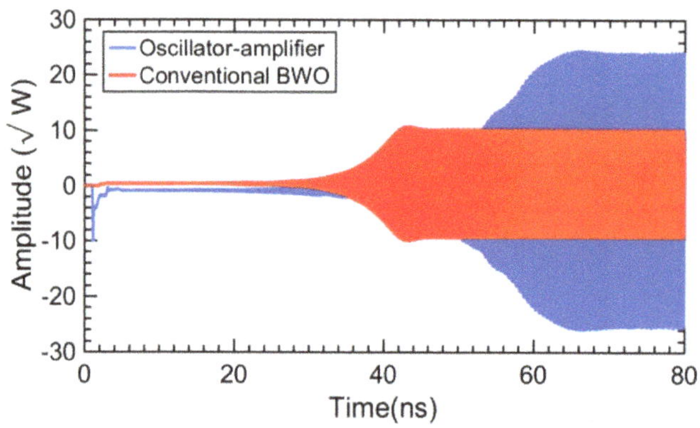

Figure 7.42. Output signals produced by the oscillator–amplifier using 5 and 15.75 kV solid beams and the output signal produced by the BWO with a 5 kV solid beam. Reprinted with permission from [10].

used for the backward-wave interaction, while the lower beam is used for the forward-wave interaction. Simulations of the BWOs use only the beam for backward-wave interaction. The shape, position, and size of this beam are kept the same for the oscillator–amplifier. The beam current for the BWO is taken to be 0.25 A. Simulations of the oscillator–amplifier use an additional solid beam with a beam current of 0.15 A for forward-wave amplification. Both devices use a focusing magnetic field of 0.55 T.

The simulated time-domain output signal from the oscillator–amplifier is shown in figure 7.42, based on a 5 kV beam for the backward-wave interaction and a 15.75 kV beam for the forward-wave interaction. The time-domain output signal for the BWO with a 5 kV beam is also shown in this figure. The output signals from both devices grow with time and settle to a steady value. The amplitude of the output signal obtained from the oscillator–amplifier is 2.5 times higher than that obtained from the conventional BWO. Fourier transforms of the output signals from the oscillator–amplifier at beam voltages of 5 and 7 kV for backward-wave interaction are shown in figure 7.43.

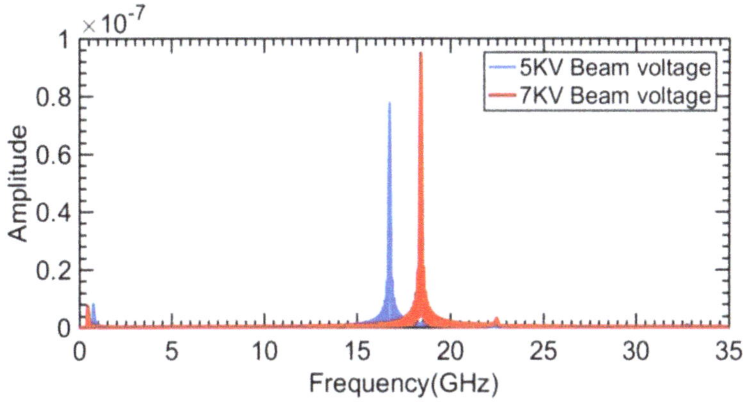

Figure 7.43. Fourier transforms of the output signals produced by the oscillator–amplifier at two different beam voltages for backward-wave interaction. Reprinted with permission from [10].

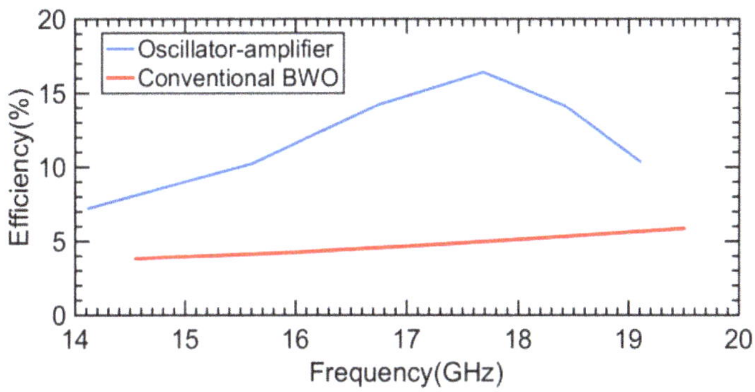

Figure 7.44. Comparison between the RF efficiency of the oscillator–amplifier and that of the conventional BWO. © [2019] IEEE. Reprinted, with permission, from [31].

The tunable bandwidth of the oscillator–amplifier is studied by varying the beam voltage used for the backward-wave interaction from 3 to 8 kV while keeping the beam voltage for the forward-wave interaction constant. Similarly, the tunable bandwidth of the BWO is also studied by varying the beam voltage from 3 to 8 kV. The frequency of the output signal produced by the oscillator–amplifier varies from 14.1 to 19.1 GHz, while the output frequency of the BWO varies from 14.5 to 19.5 GHz. Both devices have a bandwidth of 5 GHz, clearly showing that the oscillator–amplifier does not compromise the tunable bandwidth of the BWO. For a given beam voltage and backward-wave interaction, the frequency of the output signal produced by the oscillator–amplifier has a frequency deviation of 0.4 GHz with respect to that of the conventional BWO. This deviation occurs because of the change in the dispersion characteristics of the oscillator–amplifier caused by the addition of the beam for forward-wave interaction.

A plot of RF efficiency versus frequency for both devices is given in figure 7.44. As shown in the figure, the oscillator–amplifier achieves more than double the

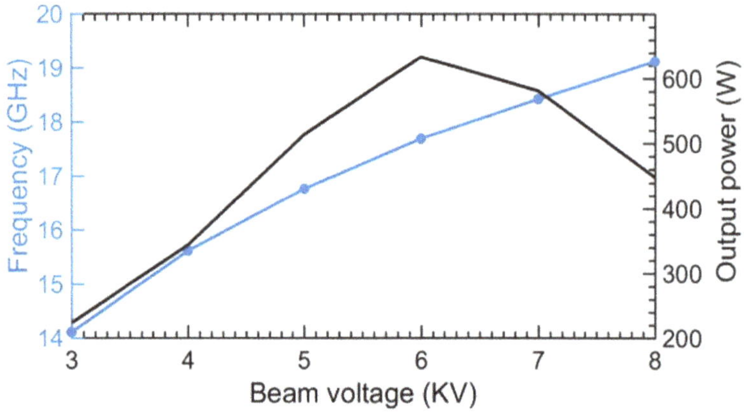

Figure 7.45. Variation of the frequency and power of the output signal with beam voltage for the backward-wave interaction in the oscillator–amplifier.

efficiency of the BWO over the entire frequency range. The oscillator–amplifier has a maximum efficiency improvement of three times compared to that of the BWO. The power and frequency of the output signal produced by the oscillator–amplifier versus beam voltage for backward-wave interaction are shown in figure 7.45. The figure shows that the oscillator–amplifier can generate a maximum output power of 633 W.

7.6 Summary

BWOs offer special features such as high spectral purity and electronic tunability of the output frequency. While the MML SWS has gained popularity for millimeter-wave TWTs, it does exhibit certain drawbacks. This chapter describes two planar SWSs suitable for BWO applications, namely the interdigital SWS and the coplanar SWS, which address these issues and yet retain the advantage of a simple structure for fabrication.

The interdigital SWS has several advantages, including high interaction impedance, high tunable bandwidth, and ease of fabrication. We briefly describe several attempts to produce millimeter-wave BWOs based on interdigital SWSs. The design of a coplanar SWS-based BWO operating in the V band is described in some detail. Simulation results show that the operating frequency of the BWO changes from 56 to 73 GHz when the beam voltage changes from 8 to 22 kV, with a tunable frequency range of 17 GHz. A maximum output power of 60 W is obtained at 69.5 GHz when the beam voltage is 17 kV. The corresponding RF efficiency is 4%. The cold-test parameters have been verified by fabricating a scaled version intended for use in the X band, as well as a version for the V band. In both cases, the measured results match the simulation results very well.

This chapter includes the design and performance of a W-band BWO based on a microfabrication-compatible PH-SEC. The simulation results for the BWO show that for a beam voltage varying from 7 to 11 kV, the device can generate signals with frequencies varying from 86.9 to 100.07 GHz, corresponding to a tuneable bandwidth of 14%. For a beam current of 20 mA, the oscillator can generate a peak output power of 2.3 W, and it can achieve a maximum DC-to-RF efficiency of 1.62%. A scaled version of the PH-SEC SWS designed to operate at X-band frequencies has been fabricated. The measured S-parameters and the phase velocity of the SWS closely match the simulated values.

This chapter also describes a new technique for improving the efficiency and output power of conventional nonrelativistic BWOs. The new technique combines the operation of both an oscillator and an amplifier in a single SWS. Both an oscillator–amplifier and a conventional BWO are designed using a PH-SEC for operation in the Ku band. The simulation results show that the oscillator–amplifier improves the DC-to-RF conversion efficiency of the conventional BWO by a factor ranging from two to three. Both the conventional BWO and the oscillator–amplifier provide a tunable bandwidth of 5 GHz. The oscillator–amplifier achieves a maximum DC-to-RF conversion efficiency of 16.39% and a peak output power of 633 W.

References

[1] Collin R E 1992 *Foundations for Microwave Engineering* (New York: McGraw-Hill)

[2] Weiss J A 1974 Dispersion and field analysis of a microstrip meander-line slow-wave structure *IEEE Trans. Microw. Theory Tech.* **22** 1194–201

[3] Crampagne R and Ahmadpanah M 1977 Meander and interdigital lines as periodic slow-wave structure. II. Applications to slow-wave structures *Int. J. Electron.* **43** 33–9

[4] Barnett L R, Grow R W and Baird J M 1988 Backward-wave oscillators for the frequency range from 600 GHz to 1800 GHz *Technical Digest., Int. Electron Devices Meeting* (Piscataway, NJ: IEEE) 858–61

[5] Kory C L and Dayton J A 2008 Interaction simulations of two 650 GHz BWOs using MAFIA *2008 IEEE Int. Vacuum Electronics Conf.* (Piscataway, NJ: IEEE) 390–1

[6] Dayton J A *et al* 2008 Assembly and preliminary testing of the prototype 650 GHz BWO *2008 IEEE Int. Vacuum Electronics Conf.* (Piscataway, NJ: IEEE) 394–5

[7] Ryskin N M, Benedik A I, Rozhnev A G, Sinitsyn N I, Torgashov R A and Torgashov G V 2017 Study of low-voltage millimeter-wave tubes with planar slow-wave structures on dielectric substrates *2017 10th UK-Europe-China Workshop on Millimetre Waves and Terahertz Technologies (UCMMT)* (Piscataway, NJ: IEEE) 1–2

[8] Zhao C, Aditya S and Wang S 2021 A novel coplanar slow-wave structure for millimeter-wave BWO applications *IEEE Trans. Electron Devices* **68** 1924–9

[9] Zhao C, Tian S, Liu W, Liao X, Fang X and Wang S 2024 Design and RF characterization of the co-planar slow wave structure for millimeter-wave BWO applications *IEEE Trans. Electron Devices* **71** 833–9

[10] Ajith Kumar M M 2019 *Application of Planar Helix Slow-Wave Structure in Backward-Wave Oscillators* (Singapore: Nanyang Technological University)

[11] Ajith Kumar M M, Aditya S and Wang S 2018 A W-band backward-wave oscillator based on planar helix slow wave structure *IEEE Trans. Electron Devices* **65** 5097–102

[12] Ajith Kumar M M, Zhao C, Wang S and Aditya S 2016 Backward wave oscillator using a planar helix slow-wave structure with straight-edge connections *2016 IEEE Int. Vacuum Electronics Conf. (IVEC)* (Piscataway, NJ: IEEE) 1–2

[13] Wang S and Aditya S 2017 Wideband power combining of four microfabricated W-band traveling-wave tubes *IEEE Trans. Electron Devices* **64** 3849–56

[14] Ajith Kumar M M and Aditya S 2017 Vacuum electronic two-beam oscillator–amplifier *IEEE Trans. Plasma Sci.* **45** 2260–7

[15] *SGMW-xxx Millimetre Wave Oscillator (With external BWO Modules)* http://elva-1.com/data/files/Manuals/SGMWxxx_User_Man.pdf (accessed 22 Februay 2024)

[16] *MM-Wave Sources from 36 to 178 GHz Based on BWO (Backward Wave Oscillators)* http://insight-product.com/mmbwo3.htm (accessed 22 Februay 2024)

[17] Han S-T *et al* 2004 Low-voltage operation of Ka-band folded waveguide traveling-wave tube *IEEE Trans. Plasma Sci.* **32** 60–6

[18] Shin Y M *et al* 2006 Experimental investigation of 95GHz folded waveguide backward wave oscillator fabricated by two-step LIGA *2006 IEEE Int. Vacuum Electronics Conf.* (Piscataway, NJ: IEEE) 419–20

[19] Park Y *et al* 2016 Design and development of 90 GHz backward-wave oscillator with staggered double-vane slow-wave structure *2016 IEEE Int. Vacuum Electronics Conf. (IVEC)* (Piscataway, NJ: IEEE) 1–2

[20] Booske J H *et al* 2011 Vacuum electronic high power terahertz sources *IEEE Trans. Terahertz Sci. Technol.* **1** 54–75

[21] Bratman V L, Fedotov A E, Makhalov P B and Manuilov V N 2014 Design and numerical analysis of W-band oscillators with hollow electron beam *IEEE Trans. Electron Devices* **61** 1795–9

[22] Rusin F S and Bogomolov G D 1969 Orotron—an electronic oscillator with an open resonator and reflecting grating *Proc. IEEE* **57** 720–2

[23] Ponomarenko S S *et al* 2013 400-GHz continuous-wave clinotron oscillator *IEEE Trans. Plasma Sci.* **41** 82–6

[24] Goebel D M, Adler E A, Ponti E S, Feicht J R, Eisenhart R L and Lemke R W 1999 Efficiency enhancement in high power backward-wave oscillators *IEEE Trans. Plasma Sci.* **27** 800–9

[25] Lin A T and Chen L 1989 Plasma-induced efficiency enhancement in a backward wave oscillator *Phys. Rev. Lett.* **63** 2808–11

[26] Wharton C B and Butler J M 1990 Relativistic O-type oscillator-amplifier systems *Proc. SPIE Conf. Intense Microwave and Particle Beams* 1226 (Bellingham, WA: SPIE) 23–35

[27] Barreto G and Wharton C B 1992 Experimental results from a tandem BWO-TWT system used to generate high-power microwaves *IEEE Trans. Plasma Sci.* **20** 493–8

[28] Pchelnikov Y N and Solntsev V A 2004 BWO with an amplifying section *Fifth IEEE Int. Vacuum Electronics Conf.* (Piscataway, NJ: IEEE) 73–4

[29] Chipengo U, Zuboraj M, Nahar N K and Volakis J L 2015 A novel slow-wave structure for high-power Ka-band backward wave oscillators with mode control *IEEE Trans. Plasma Sci.* **43** 1879–86

[30] Phillips P M, Zaidman E G, Freund H P, Ganguly A K and Vanderplaats N R 1990 Review of two-stream amplifier performance *IEEE Trans. Electron Devices* **37** 870–7

[31] Ajith K M M and Aditya S 2019 Two-beam Ku-band oscillator-amplifier using a planar helix slow-wave structure *2019 Int. Vacuum Electronics Conf. (IVEC)* (Piscataway, NJ: IEEE) 1–3

Chapter 8

Design and fabrication of planar helix slow-wave structures for TWTs

8.1 Introduction

In previous chapters, we described the planar helix slow-wave structure (SWS) with straight-edge connections (PH-SEC) and its variations, such as the rectangular ring–bar SWS with straight-edge connections (RRB-SEC). We also investigated their application in TWTs. These chapters described the methods of analysis and simulation used to obtain the cold-test and hot-test results. Additionally, we discussed the experimental verification of cold-test results using scaled structures operating at relatively low frequencies (<10 GHz). PH-SECs and related SWSs have significant advantages. Like circular helices, they offer wide bandwidth and high interaction impedance. In addition, even SWSs intended for use at millimeter-wave frequencies can be fabricated relatively easily using photolithographic and micro-fabrication techniques. These techniques have enabled the low-cost mass production of semiconductor devices, and a similar impact is expected in the case of vacuum electron devices. For instance, as described in section 8.3, more than 2000 W-band (75–110 GHz) PH-SEC and RRB-SEC SWSs have been realized on an 8″ silicon wafer using microfabrication techniques. Another advantage of PH-SEC SWSs is their ability to accommodate a sheet electron beam more readily.

This chapter describes the design and fabrication of PH-SEC SWSs by the authors' research group. Described in the following sections are PH-SEC structures designed to operate in the C/X (7–10 GHz) and Ka bands (26–40 GHz) and PH-SEC and RRB-SEC structures intended to operate in the W band. Each of these structures used a different fabrication technique. For the C/X-band structure, we not only considered RF performance but also addressed issues such as the vacuum compatibility of materials, the ability to withstand high temperatures, and the ease and accuracy of assembly techniques. The aim was to realize a structure that was

doi:10.1088/978-0-7503-5764-7ch8

ready to mate with an electron gun and a collector. For the Ka-band structure, a number of dispersion-shaping techniques described in chapter 5 were applied to design a wideband PH-SEC SWS. In addition to achieving wideband performance, the design specifically considered the feasibility of microfabrication; also considered were the thermal properties and electric field values at some critical locations in the SWS to examine the possibility of dielectric breakdown. The W-band structures were microfabricated on a silicon substrate. A detailed assessment of the dimensions, shape, and surface roughness of the fabricated structures was carried out. The effects of surface roughness, contact resistance, and silicon resistivity were studied. Thermal and dielectric breakdown issues were also considered.

The sections of this chapter follow a chronological order for these structures. For the W-band structures, measurement of the cold-test S-parameters was carried out 'on-wafer' using a coplanar waveguide (CPW) probe station. This avoided the need to dice individual SWSs and fix each SWS in a precision-fabricated mechanical assembly, thus enabling a really fast evaluation of a large number of microfabricated SWSs. This aspect is one of the highlights of the work done by our research group.

8.2 Design and fabrication of a C/X-band PH-SEC SWS

This section describes the design and fabrication of a PH-SEC intended for use at C/X-band frequencies. In addition to the RF performance, various issues such as the vacuum compatibility of materials, the ability to withstand high temperatures, and the ease and accuracy of assembly techniques are specifically emphasized. The simulated and measured cold-test parameters of the fabricated structure are presented. The interaction impedance and hot-test parameters are also studied using simulations. It is shown that the proposed structure is able to provide a saturation output power of 100 W over the frequency range from 6 to 8 GHz. Such a structure is an attractive candidate for low-cost C- and X-band TWTs fabricated using printed circuit techniques. The description in this section closely follows [1], which reported the work done by the authors' research group in collaboration with the Microwave Tubes Research & Development Centre (MTRDC), India.

8.2.1 Configuration and fabrication

A possible basic configuration of the PH-SEC is shown in figure 8.1. Figure 8.2 shows a configuration that includes a metal enclosure together with feed couplers. These two figures indicate the dimensional parameters as well as the materials chosen for the different parts of the structure. As shown in figure 8.2(a), the PH-SEC is fabricated on a pair of alumina substrates ($\varepsilon_r = 9.9$, thickness $AT = 0.254$ mm). Alumina (Al_2O_3) is a dimensionally stable, vacuum-compatible material that can withstand temperatures over 1000 °C. Inclined metal strips of width SW are printed on the inner surface of the substrates; this helps to reduce the dielectric charging effect. The straight-edge connections or the vertical posts connecting the upper and lower strips have a diameter of VD and are realized by using short lengths of

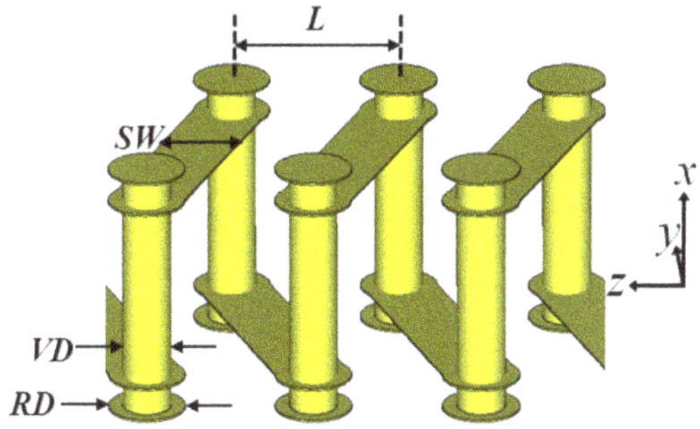

Figure 8.1. Basic configuration of the PH-SEC. © [2017] IEEE. Reprinted, with permission, from [1].

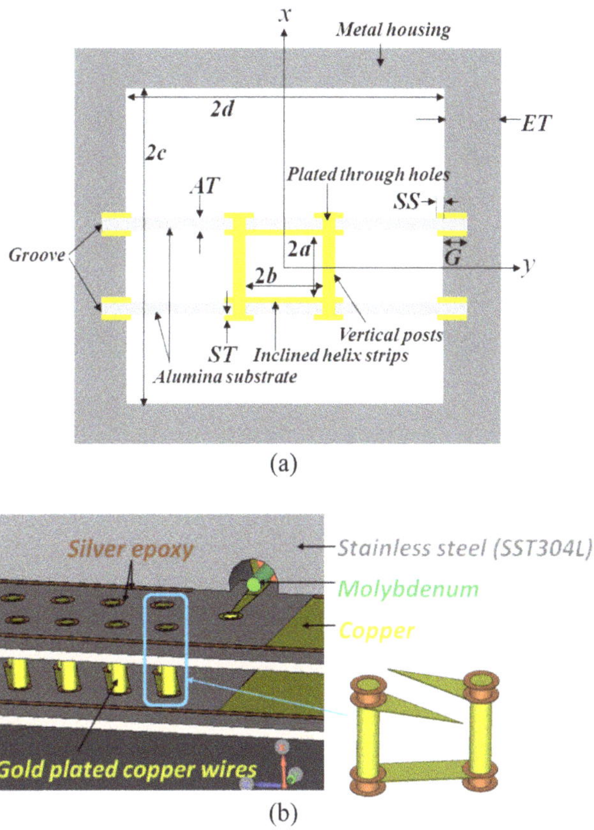

Figure 8.2. (a) Cross-sectional view and (b) perspective view of the proposed SWS. © [2017] IEEE. Reprinted, with permission, from [1].

gold-plated copper wires. Ring pads with a diameter of *RD* are applied to both ends of the vertical posts. These pads on the outer surface of the alumina substrates are connected to the inclined strips using plated-through holes (PTHs). These pads also act as the connection pads for soldering the copper wires and inclined strips.

Table 8.1. Dimensions of the proposed PH-SEC in mm [1].

$2a$	$2b$	AT	ET	G	SS	ST
1.5	1.5	0.254	2	1	0.25	0.017
L	SW	VD	RD			
1.4	0.65	0.41	0.65			

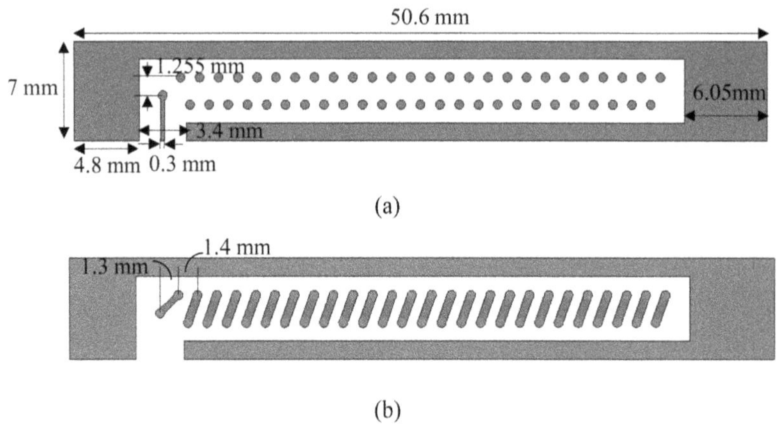

Figure 8.3. Configuration of the PH-SEC with 25 periods and feed design. (a) front side and (b) rear side. © [2017] IEEE. Reprinted, with permission, from [1].

The dimensions $2a$, $2b$, and L determine the operational frequency range, the phase velocity, and the interaction impedance. These dimensions are optimized to provide a wide bandwidth, a phase velocity synchronous with the electron beam, and high gain. The width of the metal strip SW and the diameters of the vias and ring pads VD and RD are determined by considering fabrication feasibility and the need to achieve high return loss and low insertion loss. The dimensions of the other parts, such as the metal housing and flanges, are chosen for compatibility with the electron gun and collector. The optimized dimensions of the structure are presented in table 8.1.

The SWS is fed with SubMiniature version A (SMA) coaxial couplers. Figure 8.3 presents the pattern for the PH-SEC and the feed for a design with 25 periods. The basic PH-SEC is surrounded by metallized areas that act as a ground plane. A metal strip comes out from the last turn of the PH-SEC to the edge of the substrate and forms a 50 Ω CPW feed. The dimensions of the feed are also presented in figure 8.3. For the proof of concept, a relatively short assembly with 25 periods is designed and fabricated. Similar design and fabrication steps should be applicable to longer assemblies.

The PH-SEC SWS is fixed in a metal housing made of nonmagnetic stainless steel (SST304L) so that it does not interfere with the focusing magnetic field. The metal housing consists of four separate panels that are laser welded together after fixing the

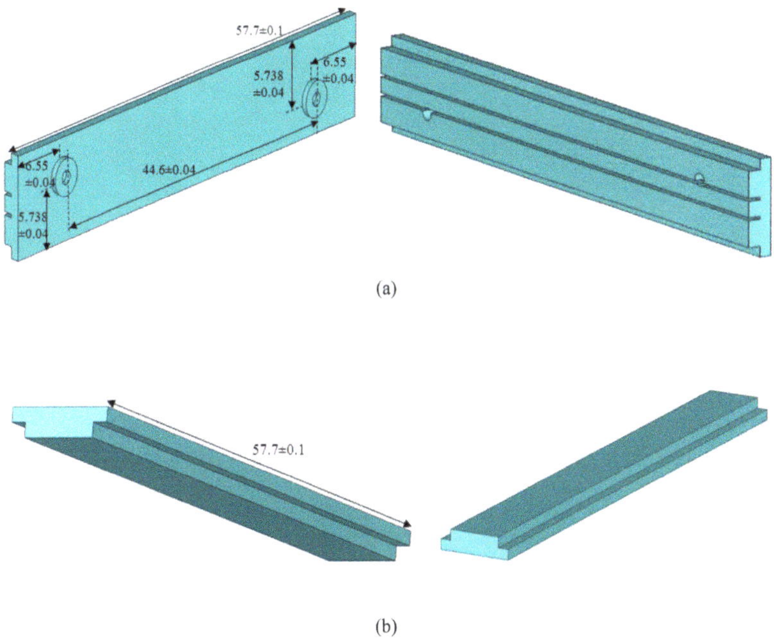

Figure 8.4. Design and dimensions in millimeters of (a) the right panel, (b) top and bottom panels of the metal housing. © [2017] IEEE. Reprinted, with permission, from [1].

SWS. In addition to these panels, flanges are also fabricated and attached at the two ends of the SWS to mate with the electron gun and collector. The design of the right, top, and bottom panels, with dimensions in millimeters, is presented in figure 8.4. The right panel (figure 8.4(a)) has two small buttons with holes in order to accommodate the center pin of the SMA coaxial coupler.

Figure 8.5(a) shows a schematic of the assembly in progress. The inner conductor pin of the SMA connector is soldered to the CPW feed, and the outer conductor of the SMA connector is soldered to the panel. The left panel has the same design as the right panel, except that there are no buttons. Grooves are made on both the left and right panels to accurately hold the alumina substrates in position and to provide good electrical and thermal connections. The vertical posts are realized by inserting gold-plated copper wires through the PTHs and soldering the wire ends to the ring pads. An assembly ready for the soldering of vertical posts is shown in figure 8.5(b).

Typical metal-to-metal solders have relatively low melting points (below 200 °C). In order for the SWS to withstand high-temperature vacuum treatment and for the TWT to work at high power levels, high-temperature solders such as silver epoxy or a gold eutectic need to be used. Silver epoxy (Duralco 124, usable up to 315 °C) is an economical and easy-to-apply conducting epoxy that is vacuum compatible and has good electrical and thermal conductivities. Gold eutectics also have high melting points and better vacuum compatibility than silver epoxy. We use a eutectic available from the Indium Corporation of America (Au 96.8%, Si 3.2%) with a melting point of 363 °C. To facilitate application, the eutectic is preformed into the

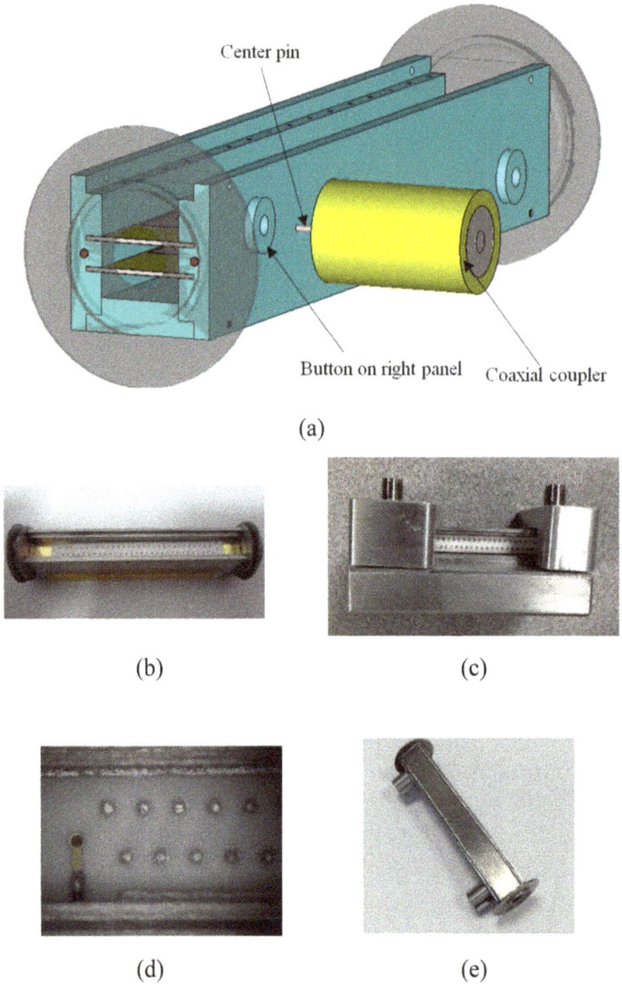

Figure 8.5. (a) Assembly of the proposed structure. Photos of the various stages of assembly: (b) assembly ready for the soldering of wires; (c) assembly showing Au–Si soldered points and couplers; (d) soldered points and coupler pin with silver epoxy; (e) whole assembly with caps and flanges. © [2017] IEEE. Reprinted, with permission, from [1].

shapes of 0.15 mm thick washers and rectangles. The washers sit on the ring pads around the vertical posts, and the rectangles are inserted into the grooves in the side panels. The use of the Au–Si eutectic required considerable optimization of the flux and cleaning process.

We used both silver epoxy and the Au–Si eutectic to solder the copper wires in the structure. The copper wires are gold plated for a good solder connection and to prevent oxidation. Figures 8.5(c) and (d) show the soldered points after successful soldering with gold eutectic and silver epoxy, respectively.

Several steps are required for assembly, soldering, and welding. These steps need to be carried out in a certain sequence. Suitable fixtures were designed and fabricated to assist in the execution of these steps. Figure 8.5(c) includes one such fixture. One also needs to attach flanges on either side of the structure to mate with the gun and collector. These flanges are made of CE-Core Iron, which is a special type of soft iron that has high permeability. A completed 25-period-long assembly is shown in figure 8.5(e). In this assembly, the RF connectors are replaced by caps as an interim step to facilitate leak checks for vacuum.

Compared to the fabrication and assembly of a conventional circular helix inside a metal barrel, this approach makes the fabrication of the metal housing and assembly much simpler. The conventional helix assembly requires precision fabrication of the helix, dielectric supports, and metal barrel, followed by hot or cold stuffing of the helix as well as supports into the metal barrel, which requires special equipment. The quality of contact between the helix, supports, and the barrel may vary from piece to piece; this greatly impacts the thermal resistance of the contacts and, hence, the power handling.

The approach presented here does not require precision fabrication of parts or special equipment and can yield excellent piece-to-piece uniformity. In addition, our study [2] shows that the presence of continuous substrates leads to very good thermal properties. Although soldering was done manually in this proof of concept, we used preformed washers and rectangles of Au–Si eutectic, and the SWS was designed with PTHs (see figure 8.2(a)). These measures minimize the need for operator skill and time. Furthermore, in mass production, one would adopt processes that are largely automated, e.g. wave soldering, PCB stacking/placement, solder reflow, etc. leading to low cost. The general fabrication process described here can also be applied to other planar SWSs, such as meander lines, in which little soldering is required.

8.2.2 Cold-test parameters and measurement results

The dispersion characteristics of the SWS are simulated using CST Microwave Studio (MWS). In order to measure the phase velocity, two assemblies with different lengths are fabricated (25 and 30 periods). The propagation constant β is calculated by taking the difference between the phase delays of the two assemblies divided by the difference in lengths. The phase velocity of the SWS is simulated using the CST MWS Transient solver. Both the simulated and measured normalized phase velocities from 6 to 9 GHz are presented in figure 8.6. The measurement results match the simulations very well. The simulated interaction impedance is also presented in figure 8.6. It has a relatively high value of 20 Ω at the central frequency of 7 GHz.

As shown in figure 8.2(b), silver epoxy is used at the ring pads and the side metal plates. The conductivity of silver epoxy is much lower than that of copper, inducing additional circuit loss. Moreover, some of the other metals used, such as stainless steel (metal housing) and molybdenum (coaxial coupler tip), also have relatively lower conductivities. Table 8.2 lists the conductivity values of the different metals used in the simulations. The return loss of the SWS simulated using the CST MWS Transient solver is shown in figure 8.7. This figure also includes the measured return loss. It can be seen that the return loss is less than -10 dB over the frequency range

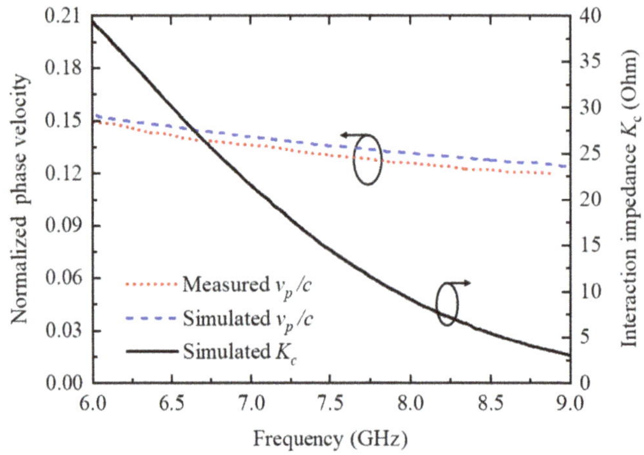

Figure 8.6. Simulated and measured normalized phase velocity and simulated interaction impedance. © [2017] IEEE. Reprinted, with permission, from [1].

Table 8.2. Conductivities of different materials [1].

Material	Conductivity (S m^{-1})
Copper	5.8×10^7
SST304L	1.39×10^6
Molybdenum	1.82×10^7
Silver epoxy	5.88×10^3

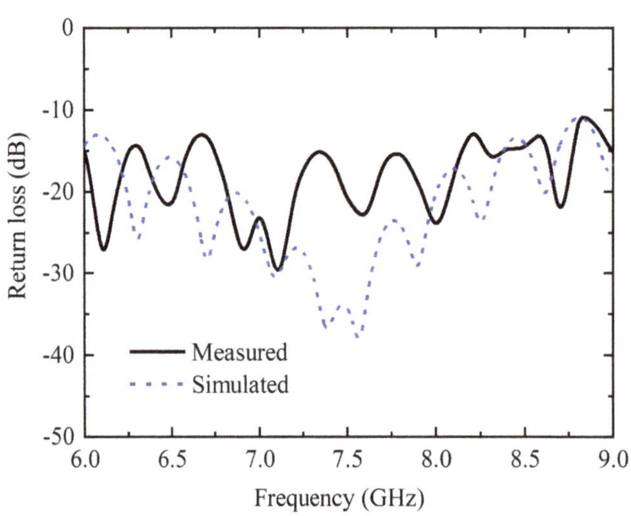

Figure 8.7. Simulated and measured return losses. © [2017] IEEE. Reprinted, with permission, from [1].

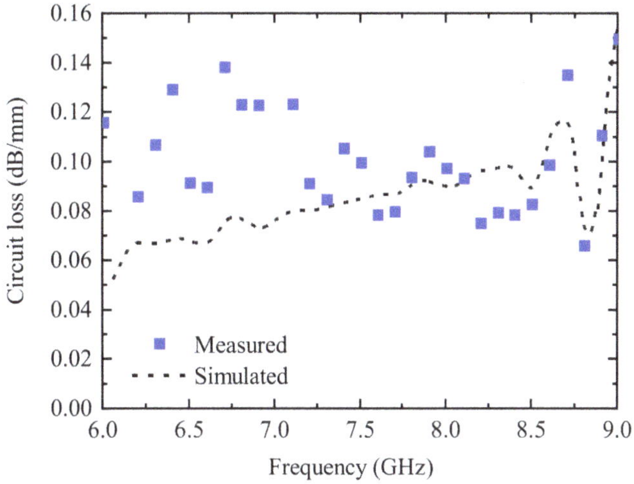

Figure 8.8. Simulated and measured circuit losses. © [2017] IEEE. Reprinted, with permission, from [1].

of 6–9 GHz. The measurement results show a good match with the simulations, with only a small frequency shift. This shift may be attributed to the tolerances in fabrication and assembly.

The simulated and measured circuit losses at different frequencies are presented in figure 8.8. The circuit loss per unit length is obtained by considering the difference between the insertion losses of assemblies of different lengths. The simulated circuit loss per unit length is 0.06 dB mm^{-1} at 6 GHz; it increases with frequency and reaches 0.15 dB mm^{-1} at 9 GHz. The measurement results match the simulations reasonably well in the 7–9 GHz frequency range. The rather high difference in loss toward the lower frequencies is likely caused by the RF feed coupler. The S-parameters for the standalone coupler show that the insertion loss is high toward the lower frequencies. Although the circuit loss per unit length is obtained by considering the difference between the insertion losses of assemblies of different lengths, due to usual fabrication tolerances, the effect of the high insertion loss of the coupler is not completely canceled. Au–Si eutectic is also used for soldering the vertical post wires, and the measured circuit loss is slightly lower than that for the circuit with silver epoxy.

8.2.3 Simulation results for hot-test parameters

The hot-test parameters are also studied for the proposed SWS using CST Particle Studio. A length of 80 periods is considered for these simulations. The SWS uses a centered electron beam with a radius of 0.35 mm and a circular cross section. The beam voltage and current are 5.5 kV and 150 mA, respectively. A focusing magnetic field of 0.22 T is assumed. The output power and efficiency versus input power at 7.25 GHz are presented in figure 8.9. The maximum output power is around 50 dBm (100 W) for an input power of 21.76 dBm (0.15 W). The corresponding RF efficiency is 12.2%. The total efficiency of the TWT can be enhanced using techniques such as pitch tapering and a depressed collector.

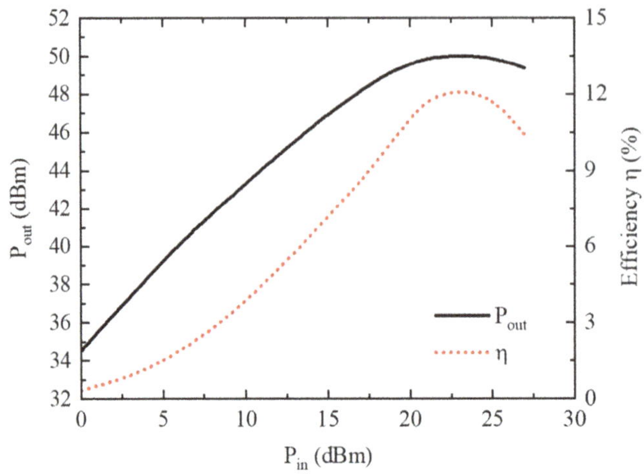

Figure 8.9. Output power and RF efficiency vs. input power at 7.25 GHz for an 80-period structure. © [2017] IEEE. Reprinted, with permission, from [1].

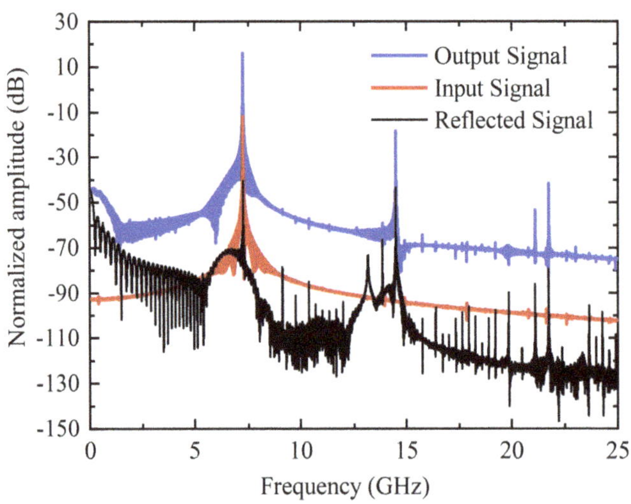

Figure 8.10. Frequency spectra of the input, output, and reflected signals at $P_{in} = 21.76$ dBm (saturated input power) at 7.25 GHz. © [2017] IEEE. Reprinted, with permission, from [1].

The frequency spectra of the signals at the input and output ports are also simulated. In order to obtain an accurate spectrum of the signals, the simulations are carried out for 300 ns with an input power of 21.76 dBm (saturating input power) at 7.25 GHz. The frequency spectrum is obtained using a Fourier transform of the signal in the time domain. The spectra of the input, output, and reflected signals with normalized amplitude are presented in figure 8.10. As can be seen, the amplitude of the reflected wave is 28 dB smaller than the input signal. In addition, the second harmonic power in the output signal is 34 dB smaller than that of the fundamental.

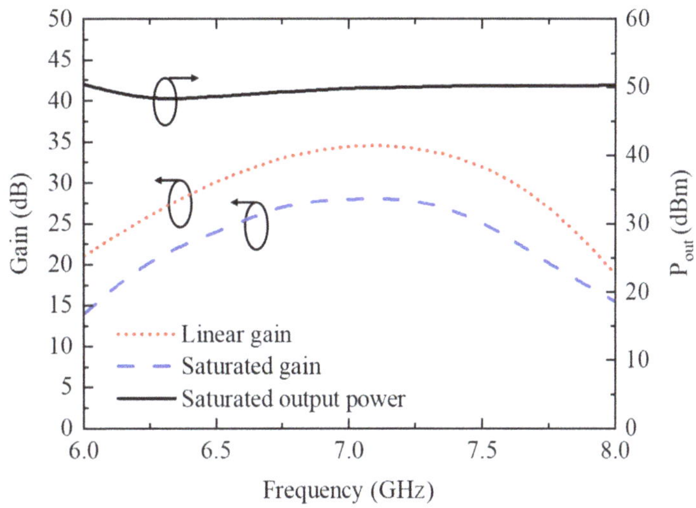

Figure 8.11. Linear gain, saturated gain, and saturated output power for frequencies from 6 to 8 GHz. © [2017] IEEE. Reprinted, with permission, from [1].

Figure 8.11 shows the output power and gain over the frequency range from 6 to 8 GHz. The maximum linear gain of 35 dB is obtained at 7.25 GHz. The maximum saturated gain also occurs at 7.25 GHz and is lower than the linear gain by 5–8 dB. A nearly constant saturated output power of 50 dBm is obtained over this frequency range.

8.3 Design and fabrication of a W-band PH-SEC SWS

To extend the operating frequency range of the PH-SEC to higher frequencies, such as those above the Ku band, microfabrication techniques are called for to achieve the required small feature sizes and three-dimensional structures. This section describes the microfabrication of PH-SEC SWSs operating at W-band frequencies. The simulated cold-test parameters for a PH-SEC on a thick silicon substrate are presented, and the effects of reducing the amount of silicon beneath the structure are discussed. A brief description of the microfabrication process and the difficulties encountered during fabrication are included. Cold-test results obtained by on-wafer measurement, covering the 70–100 GHz frequency range, are presented for PH-SEC structures fabricated on a thick silicon substrate. A detailed assessment of the dimensions, shape, and surface roughness of the fabricated structures is carried out. The effects of surface roughness, contact resistance, and silicon resistivity are studied. Also considered are the thermal and dielectric breakdown issues. The description in sections 8.3.1–8.3.5 closely follows [3, 4], which reported the work done by the research group of the authors of this book in collaboration with the Institute of Microelectronics, Singapore.

Parts of this section have been reprinted, with permission, from [3]. © [2011] IEEE.

8.3.1 Configuration and cold-test parameters

A perspective view of the basic PH-SEC SWS is shown in figure 8.12. A cross-sectional view of a shielded PH-SEC on a silicon substrate ($\varepsilon_r = 11.9$) with a thickness of h_2 is shown in figure 8.13(a). The PH-SEC consists of three metal layers, as shown in figure 8.12: the bottom horizontal layer, the top horizontal layer, and the vertical pillars (straight-edge connections or vias) that connect the two horizontal layers. The top and bottom layers have a thickness of ST and a width of SW. Each vertical pillar has a diameter of VD, and it connects top and bottom ring pads which have a diameter of RD ($>VD$). A layer of silicon dioxide ($\varepsilon_r = 3.9$) with a thickness of h_1 is assumed to be deposited on top of the silicon substrate. L is the period of the structure. The dimensions of the PH-SEC structure along the x- and y-directions are $2a$ and $2b$, respectively. The dimensions of the metal shield along the x- and y-directions are $2c$ and $2d$, respectively. The metal shield is only used in simulations that use the CST MWS Eigenmode Solver, and it is placed far away from the PH-SEC in order to reduce its effect on the properties of the PH-SEC. The center of the PH-SEC structure coincides with the origin (0, 0). Table 8.3

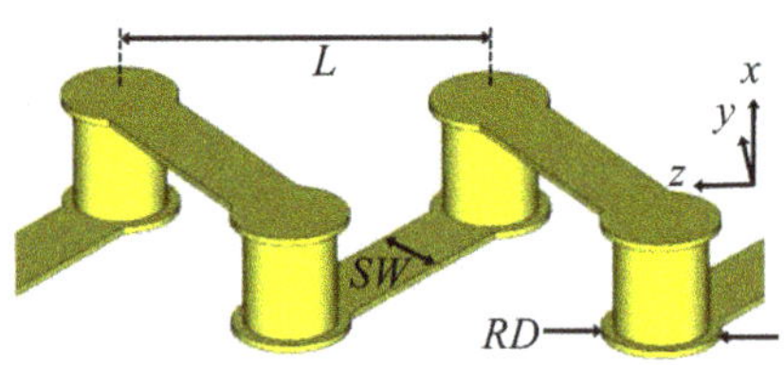

Figure 8.12. Perspective view of the PH-SEC, showing three metal layers: two horizontal layers, connected by a third metal layer comprising vertical pillars or vias. © [2011] IEEE. Reprinted, with permission, from [3].

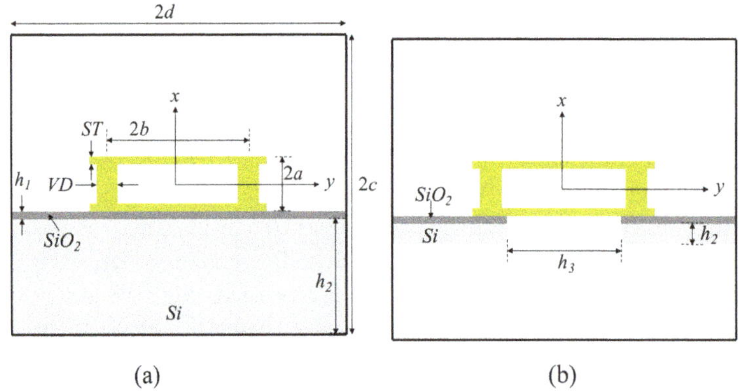

Figure 8.13. Cross-sectional view of (a) a PH-SEC on a thick silicon substrate, and (b) a PH-SEC on a thin silicon substrate with a trench in the silicon beneath the PH-SEC. © [2011] IEEE. Reprinted, with permission, from [3].

Table 8.3. Dimensions of the fabricated W-band PH-SEC.

Parameter	Value (μm)
h_1	3
h_2	750
$2a$	44
$2b$	70, 150, 170, 190, 270
L	92.85, 112.85, 152.85
ST	3
SW	40
VD	40
RD	50

summarizes the various dimensions for the PH-SEC fabricated on a thick silicon substrate.

A period of the PH-SEC configuration shown in figure 8.13(a) is simulated using the CST MWS Eigenmode Solver. Figure 8.14(a) shows the simulated dispersion diagram for the fundamental mode (mode 1) and the next higher-order mode (mode 2). The fundamental mode has a bandwidth of ~130 GHz. Mode 2 begins at 30 GHz and propagates as a slow wave in the same direction as the fundamental mode for frequencies below 170 GHz. The lower frequency limit of mode 2 is highly dependent on the size of the metal shield. In our simulations, relatively large shield dimensions are used to accommodate the 750 μm thick silicon wafer that is used in fabrication. If the dimensions of the shield are decreased, the lower frequency limit of mode 2 can be raised, and the frequency overlap with the fundamental mode can be avoided.

Figure 8.14(b) shows the interaction impedance K_c, a figure of merit used to characterize the strength of the beam–wave interaction, for both modes 1 and 2 at four locations along the x-axis, with reference to figure 8.13(a), with $y = 0$. Three of these locations lie within the helical structure ($x = -12.5$ μm, 0, and 12.5 μm), while the fourth is above the helical structure ($x = 37.5$ μm). The variation of interaction impedance along the y-axis is ignored, since this variation is small for relatively large widths of $2b$. The interaction impedance values for mode 2 are negligibly small over the frequency range that overlaps with the fundamental mode at those four points. Hence, mode 2 is insignificant for beam–wave interaction.

As shown in figure 8.14(b), due to the presence of the thick silicon substrate beneath the PH-SEC structure, the variation in the interaction impedance along the x-axis is asymmetric, with generally higher values for negative values of x. This is due to a higher field concentration closer to the silicon substrate. Such an asymmetry may cause problems such as mode competition and gain loss if the configuration shown in figure 8.13(a) is used for actual TWT operation. It is desirable to reduce the asymmetry as much as possible. One possible way to achieve this is by reducing the

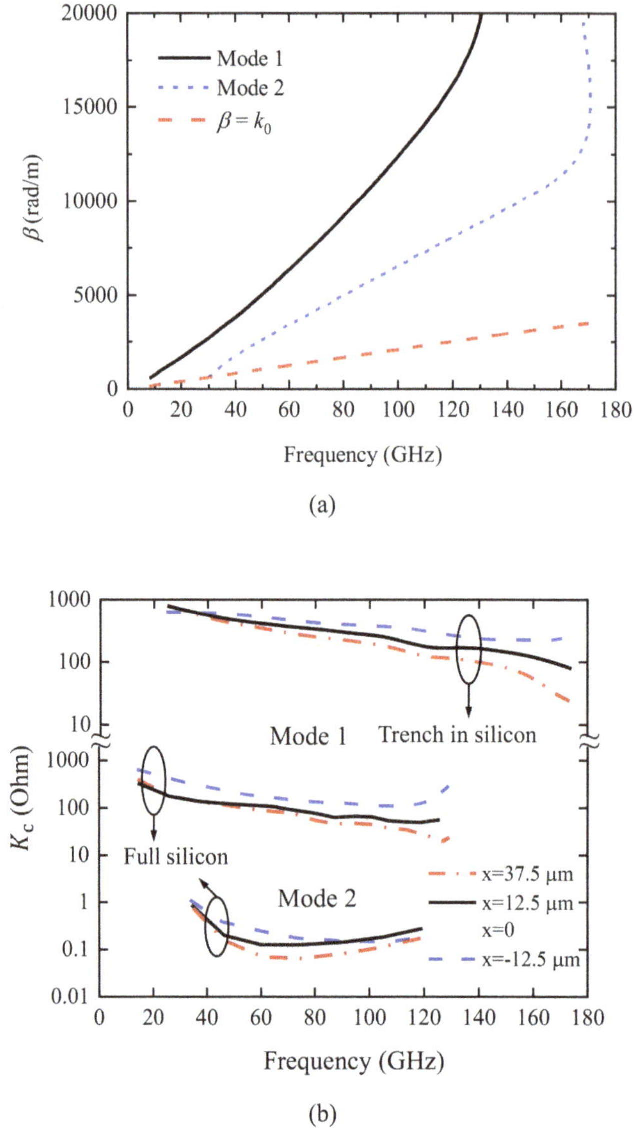

Figure 8.14. Simulated (a) propagation constant and (b) interaction impedance of the PH-SEC for the configuration shown in figure 8.13(a), with $2b = 230$ μm, $2c = 1556$ μm, $2d = 690$ μm, and $L = 152.85$ μm. Part (b) also includes the interaction impedance of the PH-SEC with the trench-in-silicon configuration shown in figure 8.13(b) with the same helical dimensions of $h_2 = 50$ μm and $h_3 = 180$ μm. © [2011] IEEE. Reprinted, with permission, from [3].

amount of silicon beneath the SWS. Figure 8.13(b) shows the configuration of a PH-SEC on a thin silicon substrate with a trench of width h_3 in silicon beneath the structure. The interaction impedance results for this modification are included in figure 8.14(b). It can be seen that the interaction strength at ±12.5 μm is more uniform, indicating a significant improvement in the symmetry of the field

Figure 8.15. Perspective view of the successful connections between the vertical posts and the suspended helical bridges. © [2011] IEEE. Reprinted, with permission, from [3].

distribution. In addition, the field strength and the interaction impedance at the center of the structure increase compared to the thick silicon case. Since several additional steps of silicon etching are required to produce a trench in the silicon wafer, we focus on the thick silicon configuration.

Several PH-SEC structures with the helix widths $2b$ varying from 110 to 310 μm and helix periods L varying from 92.85 to 152.85 μm have been designed for 22 and 32 helix turns. The helix height $2a$, via diameter VD, ring diameter RD, strip width SW, and strip thickness ST are fixed in these designs. A CPW feed has also been designed to couple the input and output microwave power to/from the helical structure. The silicon wafer is assumed to have a high resistivity (1000–10 000 Ω-cm). This design produces a simulated return loss of less than −20 dB and an insertion loss of less than −5 dB over a wide frequency range, using bulk material conductivity values in the CST MWS Transient solver. S-parameters for some of these structures are presented in figures 8.16 and 8.17, together with measured results.

8.3.2 Microfabrication process

The structures designed above are built on a high-resistivity 8-inch silicon wafer with a thickness of 750 μm. First, a 3 μm SiO_2 layer is deposited on top of the silicon wafer. In the next step, the bottom metal layer (Cu/Ti) of the helix strips and CPW feeds is formed using the negative-tone photoresist MaN-1440 lift-off process. A seed layer is then deposited for the electroplating of the vertical pillars, or vias. The vias are formed by lithography of a thick positive-tone photoresist layer (AZ 40XT-11D) and subsequent electroplating. A second seed layer is deposited on the thick photoresist for the electroplating of the third metallization layer (Cu). A layer of thick negative-tone dry film photoresist is laminated and patterned to form the electroplating mold for the suspended helix bridges (Cu). Finally, the structure is obtained through several wet immersions in photoresist strippers and Cu/Ti etchants. More details of the fabrication process can be found in [5]. More than 2000 W-band PH-SEC SWSs have been realized on a single wafer.

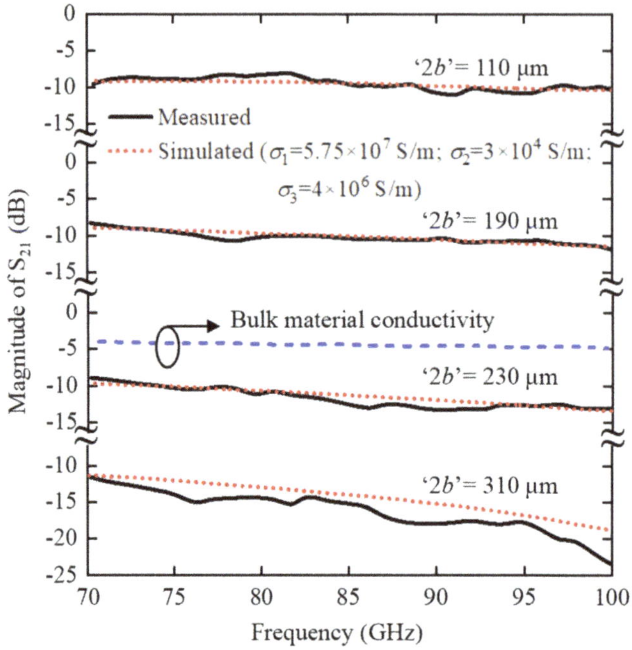

Figure 8.16. Comparison of measured and simulated S_{21} values for the microfabricated PH-SEC with different helix widths $2b$ for a fixed period $L = 152.85$ μm (22 periods). Different conductivity values for the different metal layers are considered in the simulations. © [2011] IEEE. Reprinted, with permission, from [3].

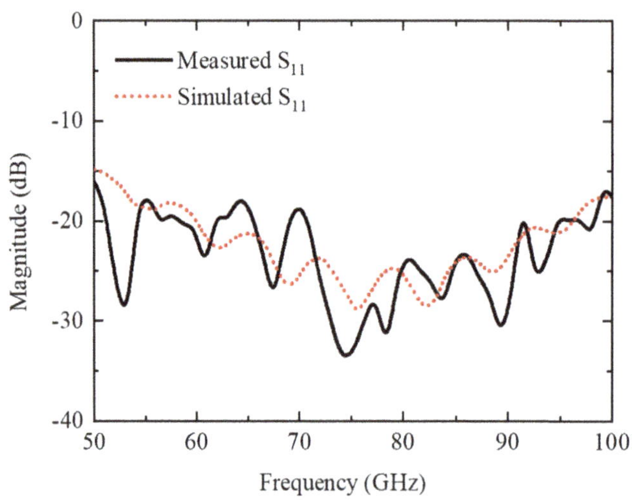

Figure 8.17. Comparison of measured and simulated S_{11} values for a PH-SEC with $2b = 230$ μm and $L = 152.85$ μm (22 periods). © [2011] IEEE. Reprinted, with permission, from [3].

The main challenges in this fabrication process include: (i) producing high-aspect-ratio vertical pillars, or vias, using a thick positive photoresist, (ii) building suspended helix bridges on top of the thick photoresist sacrificial layer, and (iii) achieving good connection between the vias and the helix bridges. Successful via-to-bridge connections, as shown in figure 8.15, were developed after several trials.

8.3.3 Measurement results for S-parameters and attenuation

The dimensions of the fabricated structures are measured using a profiler and a scanning electron microscope (SEM). The thickness of the thick photoresist layer has a nonuniformity of around 5–6 μm across the 8-inch wafer. The vertical pillars (or vias) have a 6–9 μm variation in diameter along the height. The height of the convex portion of the vias is 12 μm (see figure 8.15). The thicknesses of the bottom and top horizontal metal layers are 2.38 and 3.8 μm, respectively.

In the context of loss in high-frequency microfabricated structures, surface roughness is considered to play an important role. It is useful to characterize the additional power absorption due to surface roughness in terms of an increase in the resistivity of the material, since the surface roughness cannot be directly considered in three-dimensional RF simulators [6, 7]. Typically, in recent years, this has been done by using half the value of the bulk DC conductivity of the metal layer in the simulations. Here, we adopt a more detailed approach that considers the actual surface roughnesses of the different metal layers.

The surface roughness of the fabricated structures is measured using an atomic force microscope (AFM). The root-mean-square (RMS) roughness value of the bottom metal layer (created by the lift-off process) is 13.65 nm, while the RMS roughness of the top bridge layer (created by the electroplating process) is 143.40 nm. In the next step, the effective conductivities of the three metal layers are calculated, considering the surface roughness, following the models proposed in [6, 7]. Let σ_1, σ_2, and σ_3 represent the effective conductivity of the bottom horizontal metal layer, the via layer, and the top horizontal metal layer (the helix bridge layer), respectively. The DC conductivity of bulk copper is taken to be 5.8×10^7 S m^{-1}, corresponding to a skin depth δ_s of 0.21 μm at 100 GHz. σ_1 and σ_3 are determined based on the additional loss due to the surface roughness. According to [6], the absorption ratio of a rough surface to a smooth surface is modeled as:

$$\frac{P_{a,\mathrm{rough}}}{P_{a,\mathrm{smooth}}} = 1 + \frac{2}{\pi}\tan^{-1}\left[1.4\left(\frac{h}{\delta_s}\right)^2\right] \tag{8.1}$$

where h is the RMS height of the rough surface profile. The increase in resistivity of the bottom metal layer due to the surface roughness of 13.65 nm is ≈1.004 based on (8.1); this corresponds to an effective conductivity of $\sigma_1 \approx 5.75 \times 10^7$ S m^{-1} for this layer. For the helix bridge layer, with a surface roughness of 143.40 nm, significantly more power absorption occurs due to the additional effect of oxidation of the rough electroplated copper surface. According to [7], the increase in resistivity for protrusions due to foreign objects is

$$R_{max}(TE) = \left(\frac{4h}{\delta_s}\right) f_{bump} \tag{8.2}$$

where f_{bump} is the fraction of the surface area covered by the rough surface. In our case, the 'foreign objects' correspond to oxidized copper. The surface area covered by the rough surface is taken to be 100%, leading to an effective conductivity of $\sigma_3 \approx 4 \times 10^6$ S m^{-1} for this layer.

The value of effective conductivity σ_2 for the via layer is obtained by fitting the simulated S_{21} to the measured S_{21}. We find that $\sigma_2 = 3 \times 10^4$ S m^{-1} best fits the measured results for the various fabricated structures. A comparison of the measured and simulated S_{21} values is shown in figure 8.16 for structures with different widths $2b$ and a fixed period L of 152.85 μm. A good match is seen between the measured and simulated values for most of the cases covered in this figure. A relatively poor match is seen around 100 GHz for $2b = 310$ μm. The reason for this discrepancy lies in the fact that the π-point for this case lies close to 100 GHz. The S_{21} values vary rapidly with frequency around the π-point. Due to small dimensional discrepancies between the fabricated and simulated structures, the π-point in the two cases is slightly different, leading to a relatively large difference in the S_{21} values. For smaller values of $2b$, the π-point is well above 100 GHz; therefore, such an effect is not observed for these values of $2b$ in this figure.

It is difficult to explain which factors could result in such a low value for σ_2. Some important factors, such as the contact resistance between the vias and the helix bridges, uniformity, process alignment, and oxidation of the metal layers, are not fully captured in the models used to estimate the increase in resistivity. However, the σ_2 value obtained here is of the same order of magnitude as the conductivity of electroplated copper, 0.07×10^6–58×10^6 S m^{-1} reported in [8]. By improving the microfabrication process, one could expect to achieve effective conductivity values that are closer to the value for the bulk material and, hence, achieve lower insertion loss. Figure 8.16 also includes the results for simulated S_{21} values obtained by assuming a DC conductivity for bulk copper of 5.8×10^7 S m^{-1}, for a width $2b$ of 190 μm. The surface roughnesses of the three metallization layers of the PH-SEC are different from each other; in this respect, it is unlike the conventional circular helix [9]. Therefore, it is difficult in this case to find a simple representation of the change in attenuation as a function of surface roughness for the overall structure.

Figure 8.17 shows the comparison of measured and simulated S_{11} values for a helix width of $2b = 230$ μm and a helix period of $L = 152.85$ μm. The designed PH-SEC structure with a CPW feed has a wide −20 dB S_{11} bandwidth, covering the frequency range from 60 to 95 GHz. A fairly good match is seen between the simulated and measured results. The deviation from the simulated results is mainly attributed to dimensional errors arising during fabrication.

The phase velocity, as shown in figure 8.18, is obtained using the difference in the phase of S_{21} for two structures that are identical except for having different numbers of helical turns (22 and 32). The measured results are presented for three different values of the period L: 92.85, 112.85, and 152.85 μm. The simulation results in

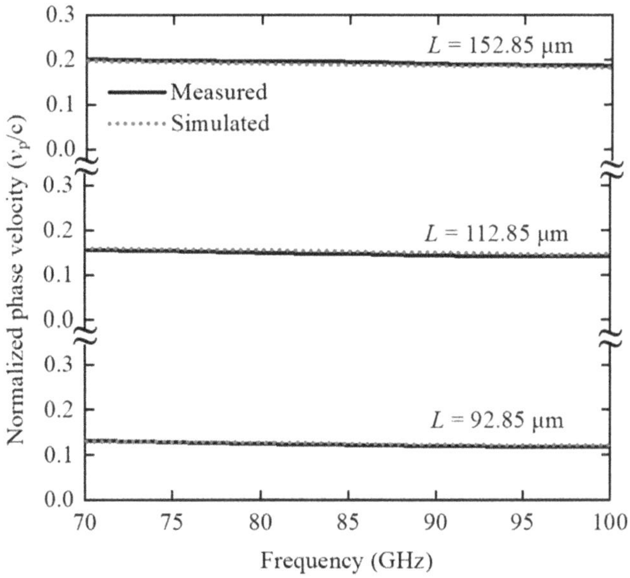

Figure 8.18. Comparison of measured and simulated phase velocities for a helix width of $2b = 230$ μm and different periods L. © [2011] IEEE. Reprinted, with permission, from [3].

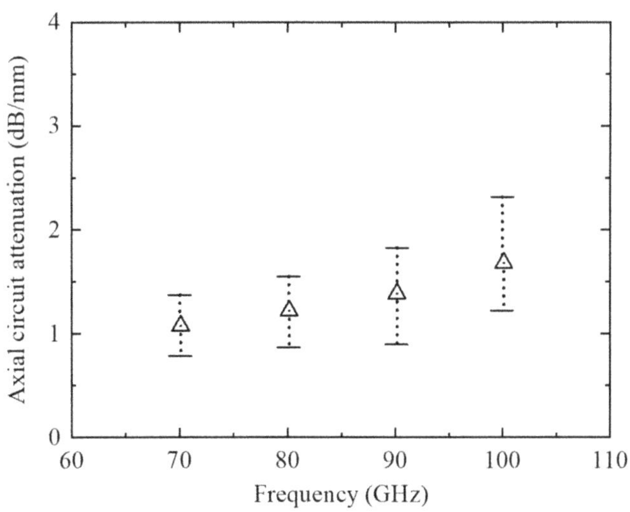

Figure 8.19. Measured circuit attenuation for $2b = 230$ μm with $L = 152.85$ μm. © [2011] IEEE. Reprinted, with permission, from [3].

figure 8.18 are obtained using the CST MWS Transient solver. There is a very good match between the measured and simulated results.

The measured axial circuit attenuation for a width $2b$ of 230 μm and a period L of 152.85 μm is shown in figure 8.19. These attenuation values are obtained using the difference in the magnitude of S_{21} for the fabricated structures with different numbers of helical turns. Due to small random fluctuations in the measured S_{21} values, the accuracy of the measured axial attenuation values is estimated to be better than ±0.5 dB. In general, the circuit attenuation increases as the frequency increases. It may be noted that the circuit attenuation of the PH-SEC on high-resistivity silicon is comparable to that of the raised meander-line SWS, which is reported to be 0.7 dB mm^{-1} in the W band [10].

8.3.4 Impact of loss tangent of silicon substrate

The various simulation results presented so far are based on high-resistivity silicon wafers with a DC resistivity of 10 000 Ω-cm (corresponding to a loss tangent of 0.000 17 at 90 GHz). Figure 8.20 shows the impact of variation in the loss tangent of the silicon wafer on the insertion loss for PH-SECs with different helix widths $2b$ at 90 GHz. The high-resistivity silicon wafer used in fabrication (1000–10 000 Ω-cm) exhibits very low insertion loss compared to the low-cost and low-resistivity silicon wafers. As an example, compared to the high-resistivity silicon wafer, the insertion loss is doubled if one uses a silicon wafer with 20 Ω-cm resistivity. In addition, based on the gradient of each loss tangent versus S_{21} curve shown in figure 8.20, the resistivity of the silicon wafer has a greater impact on helix structures with larger widths.

8.3.5 Helix temperature and dielectric breakdown

The helix temperature is important piece of information that determines the power handling capability in a corresponding TWT. The helix temperature depends on the construction details of the actual device, and an accurate determination of the temperature is quite complicated. Here, we present simplified simulation results

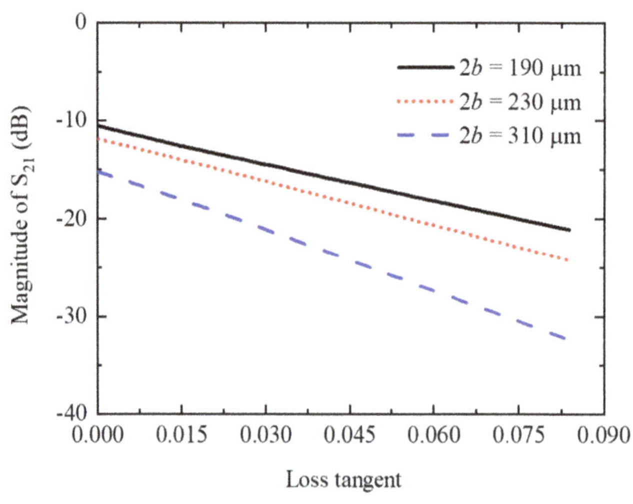

Figure 8.20. Simulated insertion loss (S_{21}) as a function of the loss tangent of the silicon wafer at 90 GHz for different helix widths $2b$ and $L = 152.85$ μm (22 periods). © [2011] IEEE. Reprinted, with permission, from [3].

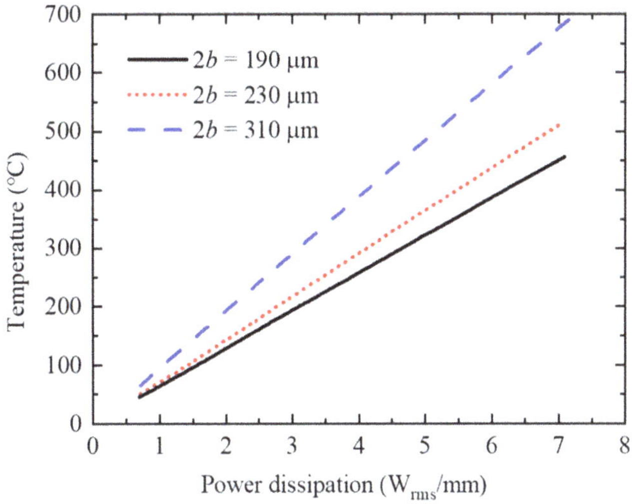

Figure 8.21. Simulated highest helix temperature as a function of dissipated RMS power at 90 GHz for different helix widths $2b$ and $L = 152.85$ μm (22 periods). © [2011] IEEE. Reprinted, with permission, from [3].

using CST thermal co-simulation, ignoring the thermal contact resistance (TCR) and variation in thermal conductivity with temperature. Figure 8.21 shows the simulated helix temperature in the absence of an electron beam as a function of dissipated RMS power per unit length at 90 GHz. The thermal conductivity values used in the simulation are 148 W K^{-1} m^{-1} for silicon, 1.4 W K^{-1} m^{-1} for silicon dioxide, and 401 W K^{-1} m^{-1} for copper. Based on the conductivity values used in figure 8.16, the highest temperature of the PH-SEC is captured. The highest-temperature points are the centers of the helix bridges, since these locations are farthest from the silicon substrate. It can be observed that the temperature increases significantly for helices with larger widths.

To form an idea of how these results may compare with those for circular helix TWTs, one may consider the results for average temperature vs. power dissipation of a tape helix reported in [11]; these results consider a copper-plated molybdenum tape helix held in compression inside a copper shell using three anisotropic pyrolytic boron nitride (APBN) or diamond support rods. The results for the helix temperature of the PH-SEC on thick silicon appear to be comparable to those for the circular helix using diamond support rods and significantly better than those for APBN support rods.

From the point of view of dielectric breakdown, the maximum value of E_x on the surface of silicon has also been examined. For 7 W_{rms} mm^{-1} dissipated power at 90 GHz, this value is 3.51×10^4, 3.62×10^4, and 4.06×10^4 V cm^{-1} for $2b = 190$, 230, and 310 μm, respectively. These values are at least an order of magnitude less than the breakdown fields for silicon and silicon dioxide, which are 3e5 and 1e6 V cm^{-1}, respectively.

8.3.6 Simulation results for hot-test parameters

We mentioned in section 8.3.1 that due to the presence of the thick silicon substrate beneath the PH-SEC structure, the variation in the interaction impedance perpendicular to the substrate is asymmetric and undesirable. Reducing the amount of silicon beneath the SWS results in better uniformity of the interaction impedance and an increase in its value at the center of the structure.

In view of the above considerations, the configuration used in simulations of the hot-test parameters is shown in figure 8.22; the PH-SEC is located on a thin silicon substrate, and a trench is present in the silicon beneath the structure. This configuration is similar to that in figure 8.13(b), with the difference that here the PH-SEC has a square cross section. All metal, including the shield, is copper with bulk conductivity. For this configuration, 3D particle-in-cell (PIC) simulation results are presented for the dimensions mentioned in the figure. A cylindrical e-beam, shown in figure 8.23, is assumed. The linear and nonlinear amplifications of the TWT are examined; the spectral purity and possibility of oscillations in the hot structure are also considered. The contents of this subsection closely follow [12].

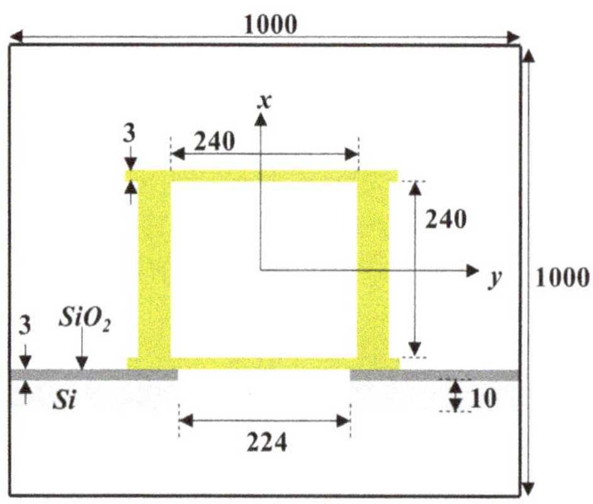

Figure 8.22. Dimensions of the square PH-SEC on a silicon substrate with a trench in silicon. All dimensions are in micrometers. © [2012] IEEE. Reprinted, with permission, from [12].

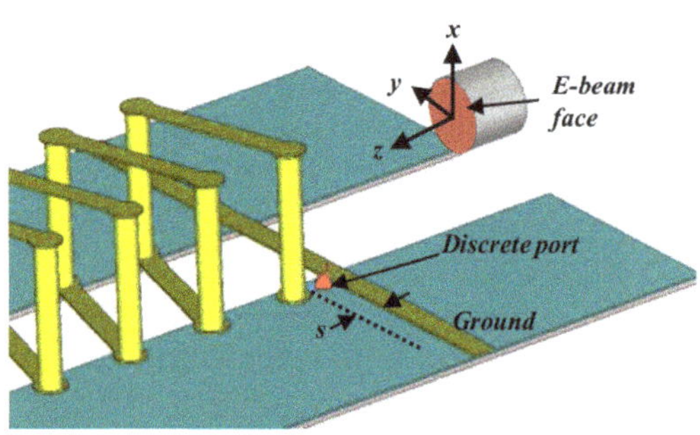

Figure 8.23. Perspective view of the PH-SEC with a cylindrical e-beam and incorporating a simplified CPW coupler. © [2012] IEEE. Reprinted, with permission, from [12].

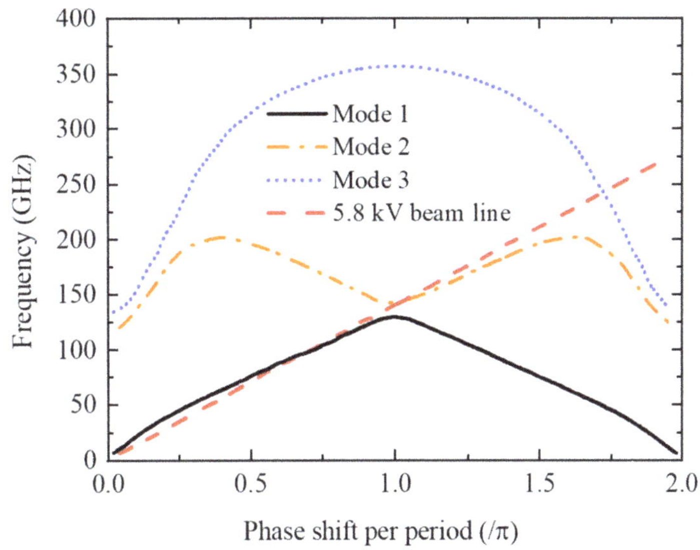

Figure 8.24. Dispersion curves of the fundamental mode and two higher-order modes. The 5.8 kV beam line is also included. © [2012] IEEE. Reprinted, with permission, from [12].

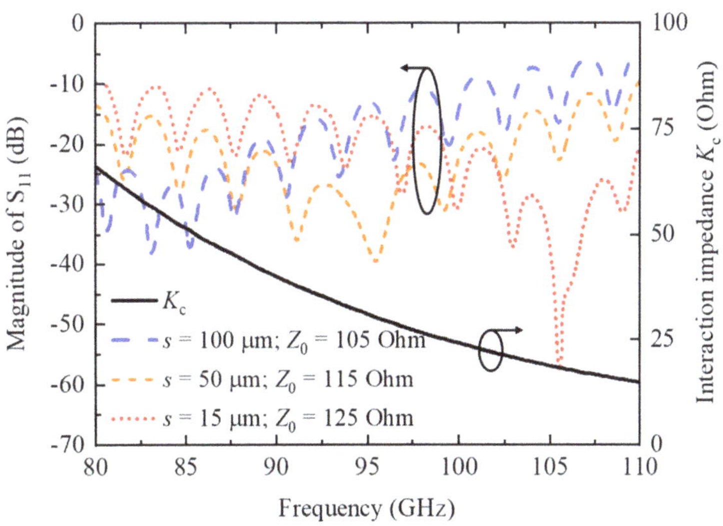

Figure 8.25. Return losses for different values of *s* (ground to ring-pad distance in figure 8.23). Interaction impedance is also included. © [2012] IEEE. Reprinted, with permission, from [12].

Figure 8.24 shows the dispersion curves for the fundamental mode 1 and two higher-order modes 2 and 3, obtained from the CST MWS Eigenmode Solver, together with a 5.8 kV beam line. The beam line intersects modes 1, 2, and 3 at 85, 139, and 235 GHz, respectively, showing possible beam–wave interactions at these frequencies. The interaction impedance of the forward fundamental mode obtained from the CST MWS Eigenmode Solver is shown in figure 8.25.

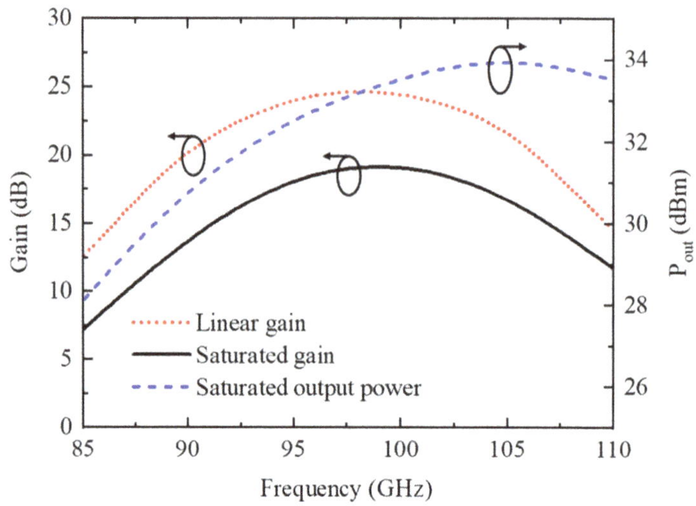

Figure 8.26. Linear gain, saturated gain, and saturated output power vs. frequency. © [2012] IEEE. Reprinted, with permission, from [12].

The structure is then designed with simplified input and output CPW couplers in CST MWS. As shown in figure 8.23, a ground strip is extended from the metal shield at a distance *s* from the edge of the ring pad. Discrete ports are used to define the impedance between the ground plane and the PH-SEC. As shown in figure 8.25, by changing *s* and the corresponding discrete port impedance value, a frequency range of low return loss (<-20 dB) can be obtained over the entire band. A similar discrete port is also utilized as an idealized sever. The sever is placed 58.5 periods from the input port and 137.5 periods from the output port.

The designed PH-SEC with the sever was simulated in the PIC Solver. The dielectric and conductor losses (with bulk resistivity) are considered in the simulation. As depicted in figure 8.23, a cylindrical e-beam with a beam voltage of 5.8 kV and a current of 8 mA is centered at the origin. The beam filling factor is 50%, and the current density is 70.7 A cm^{-2}. The beam propagates in the $+z$-direction under a homogeneous axial magnetic field of 0.5 T.

Figure 8.26 shows the simulated linear gain, saturated gain, and saturated output peak power vs. frequency curves. The maximum linear gain is 24.9 dB at 97.5 GHz. The saturated output power varies from 28 dBm (0.65 W) to 34 dBm (2.58 W) over the frequency range of 85–110 GHz. Figure 8.27 shows the output power and RF efficiency vs. input power curves at 97.5 GHz. The output power for a lossless circuit is also included. The maximum efficiency is about 7.3%.

The frequency spectra of the input, output, and reflected signals at $P_{\text{in}} = 17$ dBm (saturated) at 97.5 GHz are shown in figure 8.28. The peaks at 195 GHz in the output and reflected signals correspond to the 2nd harmonic; however, its power level is significantly lower than the power at the operating frequency of 97.5 GHz. Since the PH-SEC with the metal shield is a two-conductor structure, and the discrete port feed also supports DC, there is a peak near DC, but the output power

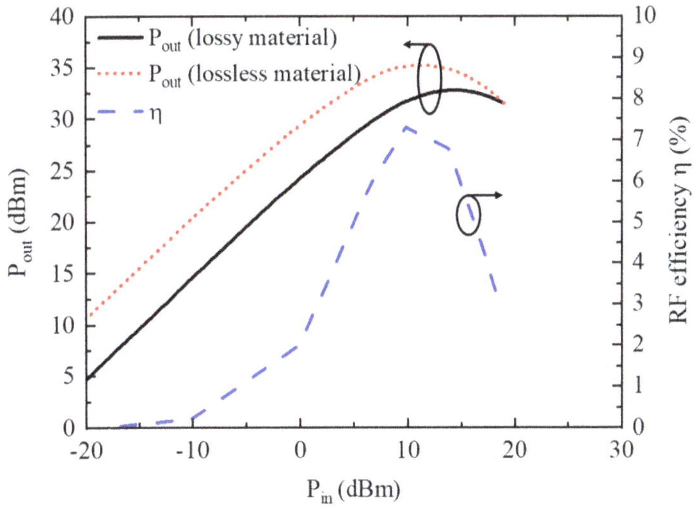

Figure 8.27. Output power and RF efficiency vs. input power at 97.5 GHz. © [2012] IEEE. Reprinted, with permission, from [12].

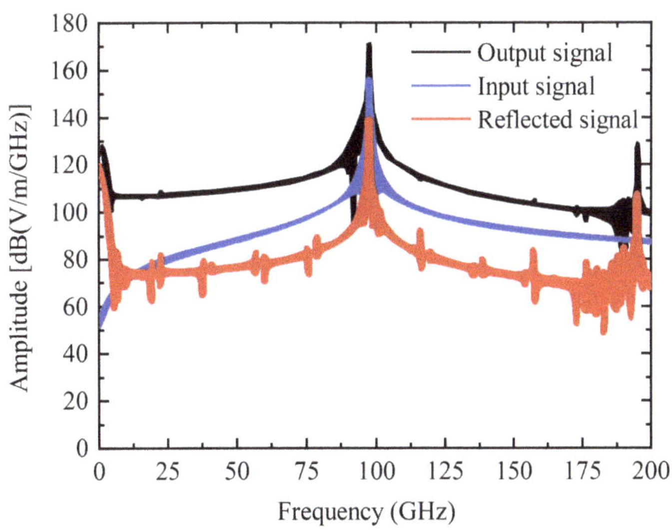

Figure 8.28. Frequency spectrum of input, output, and reflected signals at $P_{\text{in}} = 17$ dBm at 97.5 GHz. © [2012] IEEE. Reprinted, with permission, from [12].

level near DC is insignificant. Oscillations are unlikely to occur, since the reflected power is well below the input power level. There are insignificant backward-wave growths at several frequencies where the beam line crosses the higher-order backward modes.

Due to the narrow spacing between the CPW ground planes and the signal line, the CPW coupler has limited power handling capability. Hence, a rectangular waveguide coupler is also designed for high power handling. An idealized sever, containing two WR-10 waveguide ports, is used at the same location as the CPW coupler. The maximum saturated output power levels are similar for both types of couplers. The output power response is shifted toward lower frequencies for the rectangular waveguide couplers. The spectrum is also very clean compared to the case of the simplified CPW couplers. This indicates that the idealized sever using the rectangular waveguide couplers provides better isolation between the two sections separated by the sever. Such rectangular waveguide couplers are described for the rectangular ring–bar SWS with straight-edge connections (RRB-SEC SWS) in subsection 8.4.4.

8.4 Design and fabrication of a W-band RRB-SEC SWS

As mentioned in chapter 6, in high-power applications, the circular helix suffers from reduced interaction impedance of the fundamental forward-wave mode and increased interaction impedance of the backward-wave mode. These effects may result in oscillations [13]. A modification of the circular helix, the circular ring–bar structure, was proposed in [14] as a way to avoid these problems. The PH-SEC suffers from similar problems to those of the circular helix in high-power applications. To address these problems, the authors' research group has proposed a new structure: a rectangular ring–bar with straight-edge connections (RRB-SEC); the square shape is considered a special case of the rectangular shape.

The basic configuration of the RRB-SEC is shown again in figure 8.29. The RRB-SEC is a planar version of the circular ring–bar structure that is suitable for sheet beam applications. As described in chapter 6, analogous to the circular ring–bar, the new structure has the potential to enable high-power operation of TWTs operating at millimeter-wave and higher frequencies. This section describes the simulation results for phase velocity and interaction impedance obtained using the CST MWS Eigenmode Solver for two RRB-SEC configurations that can be microfabricated on a silicon wafer. We provide proof-of-concept microfabrication and measurement results in the W band for the simpler of the two RRB-SEC configurations. We also include a design for a CPW feed. Since the RRB-SEC structures are fabricated together with the PH-SEC structures described in section 8.3, the steps used for

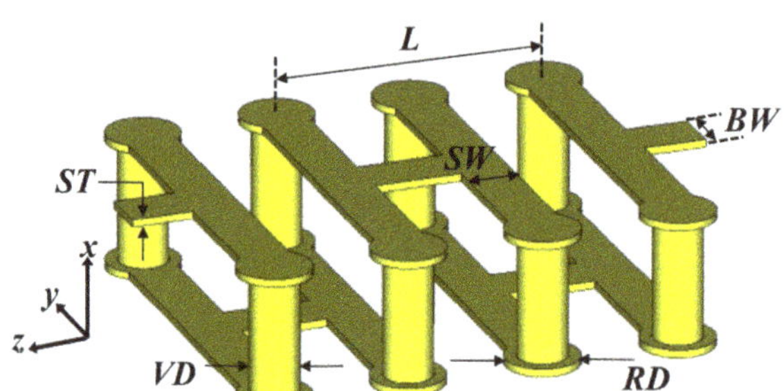

Figure 8.29. Perspective view of the basic configuration of RRB-SEC with cylindrical via. Reprinted with permission from [15].

microfabrication are identical to those mentioned in section 8.3.2. Cold-test S-parameters measured on-wafer, covering the 80–110 GHz frequency range, are also presented. The measured results are compared with the simulation results obtained using the CST MWS Transient solver. The contents of this section closely follow [15, 16].

8.4.1 Configuration and cold-test parameters

Cross-sectional views of the RRB-SEC with cylindrical vias, on a thick silicon substrate and on a thin silicon substrate with a trench, are shown in figures 8.30(a) and (b), respectively. Silicon has a dielectric constant of 11.9. The metal shield used in the simulations is not shown in this figure, since the shield is assumed to be placed far away from the structure. In addition, for simplicity, the effect of a thin layer of silicon dioxide (SiO_2), usually deposited on a silicon wafer to enhance electrical insulation, is ignored. Similar to the microfabrication of the PH-SEC [3, 5], both configurations shown in figure 8.30 can be microfabricated relatively easily, since, unlike many other structures, the fabrication process does not require bonding of the two halves of the structure. The configuration with thick silicon suffers from the effect of dielectric loading and an asymmetric field distribution. Such asymmetry may cause problems such as mode competition and gain loss in actual tube operation. The configuration with a trench in the silicon (figure 8.30(b)) reduces such problems; of course, this configuration requires additional silicon etching steps.

The phase velocity and interaction impedance for both RRB-SEC configurations mentioned above, with cylindrical vias and $b/a = 1$ and 23, are shown in figure 8.31. Both structures are assumed to have an identical P/L ratio (P is the perimeter of one turn); for the thick silicon case, h_1 is 350 μm, while for the case of the trench in silicon, h_1 and h_2 are 50 and 220 μm, respectively. The other dimensions of each structure are given in table 8.4.

As expected, due to dielectric loading, the bandwidth and phase velocity of the thick silicon configuration are lower than those of the trench in silicon. The interaction impedance values at $x = -a/2$, 0 and $a/2$, $y = 0$ (see figure 8.30) for $b/a = 1$ and 23 are plotted in figures 8.31(b) and (c), respectively. For the thick silicon configuration, the variation in the interaction impedance along the x-axis is asymmetric, with generally higher values for negative values of x for both aspect

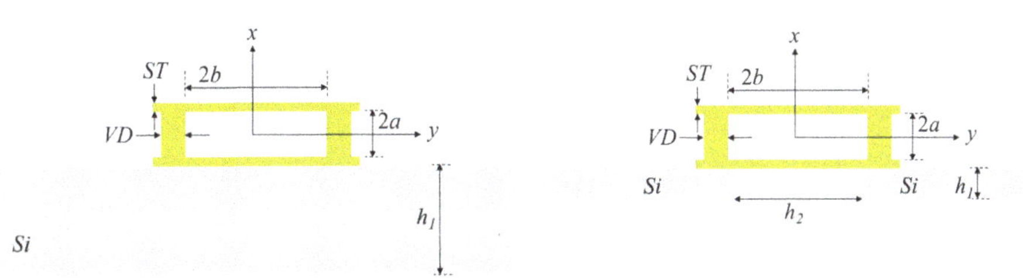

(a) (b)

Figure 8.30. Cross-sectional view of the RRB-SEC with cylindrical vias on (a) a thick silicon substrate and (b) a thin silicon substrate with a trench. Reprinted with permission from [15].

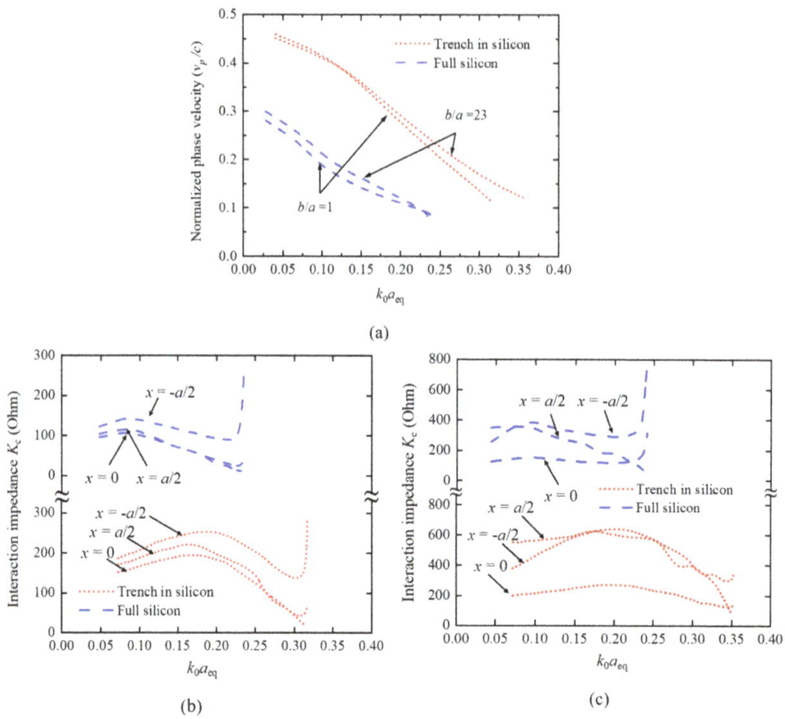

Figure 8.31. Simulated (a) normalized phase velocity of the RRB-SEC with cylindrical vias for *b*/*a* = 1 and *b*/*a* = 23. Simulated interaction impedance of the silicon configurations for (b) *b*/*a* = 1 and (c) *b*/*a* = 23. Reprinted with permission from [15].

Table 8.4. Summary of structural dimensions in figure 8.30 (μm) [15].

P	*L*	*ST*	*SW*	*RD*	*VD*	*BW*
944	150	3	40	56	40	40

ratios. On the other hand, for the trench-in-silicon configuration, the interaction strengths at ±*a*/2 are relatively more uniform. The uniformity is much better for the high-aspect-ratio structure, since the silicon is far away from the center of the structure; in addition, the interaction impedance at the center of the structure also increases compared to the thick silicon case.

8.4.2 Microfabrication process

For this proof-of-concept fabrication, we chose a configuration with a thick silicon substrate (figure 8.30(a)), with the standard thickness of 750 μm, due to its simpler

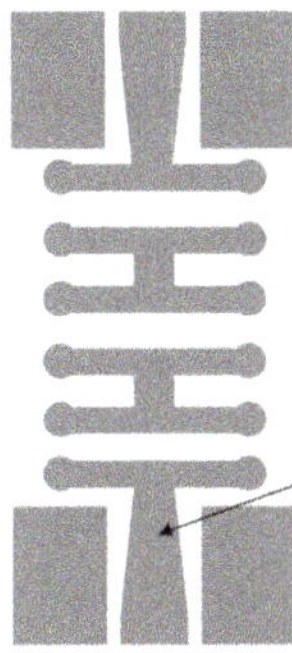

Figure 8.32. CPW feed for the RRB-SEC. Reprinted with permission from [15].

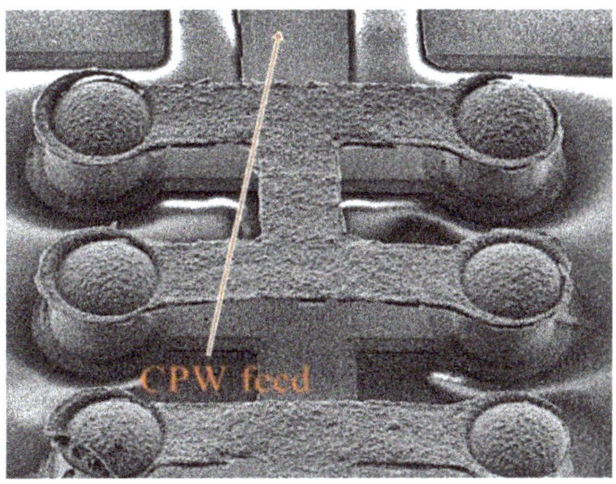

Figure 8.33. SEM micrographs of a fabricated RRB-SEC structure. Bird's-eye view showing the RRB-SEC with cylindrical vias and CPW feed. Reprinted with permission from [15].

microfabrication. The RRB-SEC with CPW feed is designed using CST MWS. A CPW section starts at the bottom of the first rectangular ring, as shown in figure 8.32, followed by a tapered section, with a taper length of 595 μm, which joins a 50 Ω section. To ensure that the symmetric mode is excited, the signal line of the CPW, with a width of 40 μm, is centered at $y = 0$ (see figure 8.30). The edges of the CPW ground planes are 31 μm away from the edges of the vias in the z-direction. The length of the tapered CPW section and the impedance of the initial CPW section are important parameters for achieving good impedance matching.

The fabrication process steps are the same as those used for the PH-SEC described in section 8.3.2. The capability of building high-density or high-aspect-ratio vias is also verified in this experiment. RRB-SEC structures with 22 periods, $b/a = 3.18$, $P = 368$ μm, $L = 192.85$ μm, $ST = 3$ μm, $SW = 40$ μm, $RD = 50$ μm, $VD = 40$ μm, and $BW = 40$ μm are fabricated on an 8-inch high-resistivity silicon wafer. SEM micrographs of the successfully fabricated RRB-SEC are shown in figure 8.33. The dimensions of the successfully fabricated structures are measured using a profiler and an SEM. The surface roughness of the fabricated structures is measured using an AFM. The thicknesses of the metal layers and the surface roughness values are identical to those mentioned in section 8.3.3.

8.4.3 Measurement results for S-parameters

Figure 8.34 shows a comparison of the measured and simulated S-parameters for the fabricated RRB-SEC structures. In simulations, the effective conductivities of the three metal layers are modified to consider the surface roughness, following the models presented in section 8.3.3 [3]. The effective conductivities of the bottom horizontal metal layer, the via layer, and the top horizontal metal layer are 5.75×10^7, 3×10^4, and 4×10^6 S m^{-1}, respectively. The results demonstrate that the RRB-SEC structure with a CPW feed has a wide -20 dB S_{11} bandwidth, covering the frequency range from 80 to 110 GHz. The simulated S_{21} results agree well with the measured results, since the effects of surface roughness and contact resistance are considered in the simulation. The deviation from the measured results is mainly attributed to dimensional errors arising during fabrication. The measured S_{21} for a 22-turn PH-SEC with an identical cross-sectional perimeter but a different period L (152.85 μm) is also included. It can be seen that the RRB-SEC has an insertion loss similar to that of the PH-SEC. However, as demonstrated in the next subsection, the RRB-SEC should be more suitable for high-power operation.

8.4.4 Simulation results for hot-test parameters

A square-shaped RRB-SEC is chosen for our examination of the hot-test parameters [16]. Figure 8.35 shows the cross section of the SWS and its dimensions. The shape and dimensions are chosen to be similar to those of the PH-SEC in section 8.3.6.

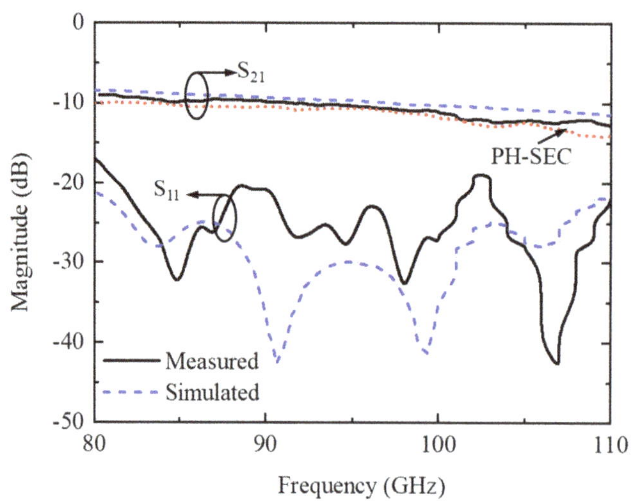

Figure 8.34. Measured and simulated S-parameters of the fabricated RRB-SEC with cylindrical vias. The surface roughness and contact resistance have been included in the effective conductivity of each metallized layer during simulation. Measured S_{21} of a PH-SEC with an identical cross-sectional perimeter is included for comparison. Reprinted with permission from [15].

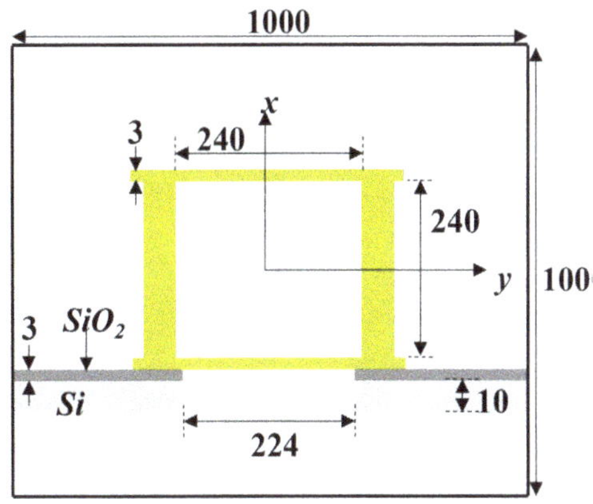

Figure 8.35. Cross-sectional view of the RRB-SEC SWS. © [2013] IEEE. Reprinted, with permission, from [16].

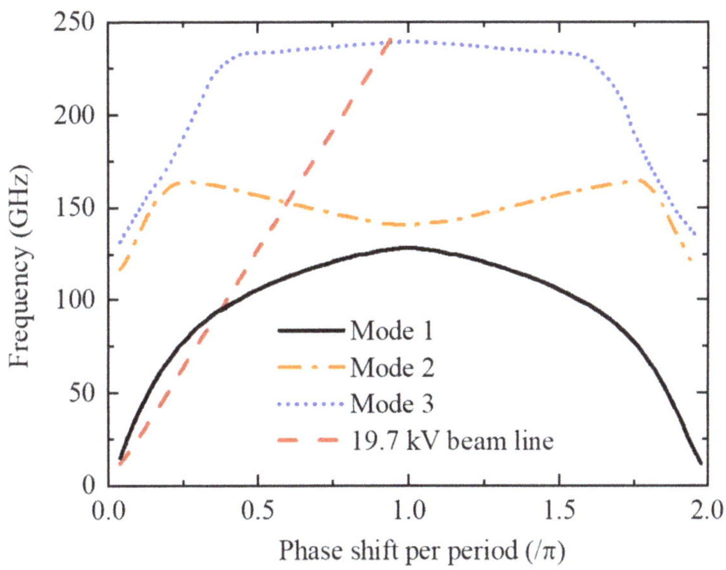

Figure 8.36. Dispersion curves of the fundamental mode and two higher-order modes. The 19.7 kV beam line is also included. © [2013] IEEE. Reprinted, with permission, from [16].

Figure 8.36 shows the dispersion curves for the fundamental mode 1 and two higher-order modes 2 and 3. To operate the RRB-SEC at W-band frequencies, a beam voltage of 19.7 kV is chosen for interaction with mode 1. The beam line intersects with modes 1, 2, and 3 at 90, 150, and 235 GHz, respectively, showing possible beam–wave interaction at these frequencies.

The RRB-SEC is designed with input/output couplers and a sever using WR-10 rectangular waveguides. Figure 8.37 shows the rectangular waveguide coupler together with the dimensions of the electric field probe inside the WR-10 waveguide (1270×2540 μm^2). Figure 8.38 shows the simulation model used for the TWT

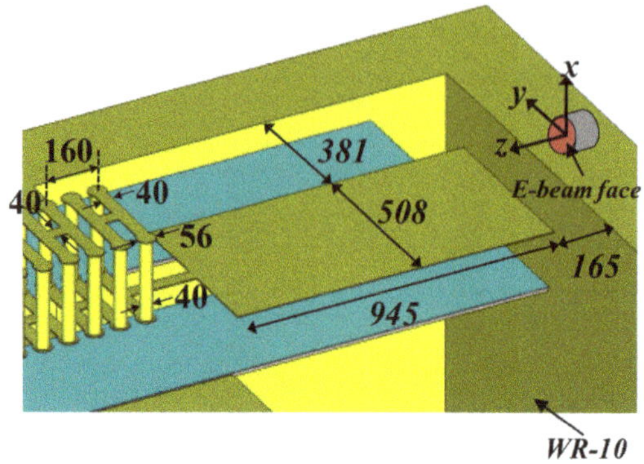

Figure 8.37. Electric field probe inside the WR-10 waveguide. © [2013] IEEE. Reprinted, with permission, from [16].

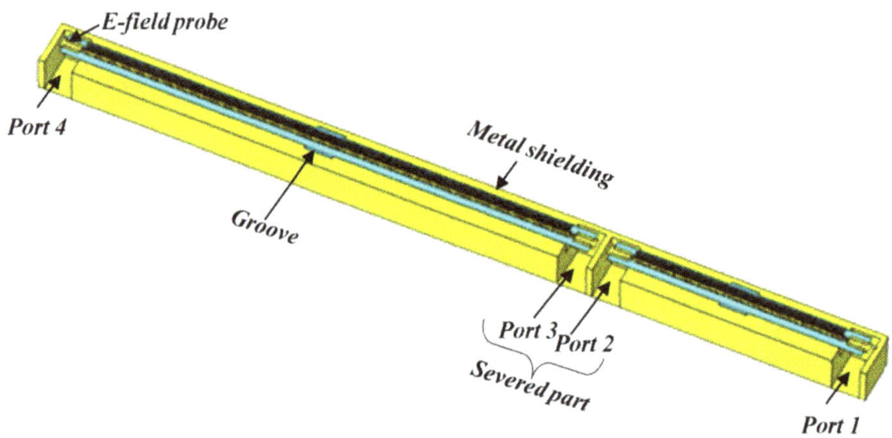

Figure 8.38. Simulation model for the travelling-wave tube amplifier (TWTA) incorporating the RRB-SEC and WR-10 couplers. © [2013] IEEE. Reprinted, with permission, from [16].

amplifier based on the RRB-SEC with rectangular waveguide couplers. The top cover of the waveguide and the metal shield are omitted from the figure. The sever, containing two WR-10 waveguide ports (Port 2 and 3), is placed 58.5 periods away from the input port (Port 1) and 137.5 periods away from the output port (Port 4).

The return loss at the input port and the interaction impedance of the forward fundamental mode are shown in figure 8.39. A return loss of −20 dB is achieved over the 88–105 GHz frequency range. The interaction impedance values are considerably higher than those shown in figure 8.25 for the corresponding PH-SEC.

As depicted in figure 8.37, for PIC simulations, a cylindrical electron beam centered at the origin is used to interact with the square RRB-SEC. A beam current of 20 mA with a current density of 177 A cm^{-2} is selected. The beam filling factor is 50%, and the beam is assumed to be under a homogeneous axial magnetic field of 0.5 T. Figure 8.40 shows the simulated saturated output power, gain, and RF efficiency vs. frequency curves.

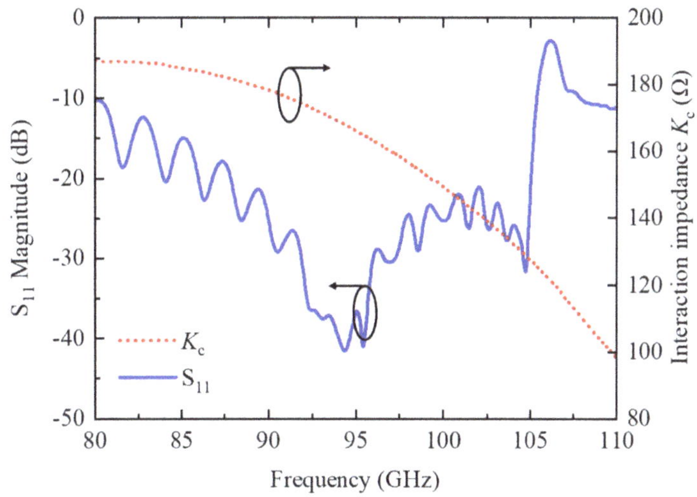

Figure 8.39. Return loss at port 1 in figure 8.38. Interaction impedance is also included. © [2013] IEEE. Reprinted, with permission, from [16].

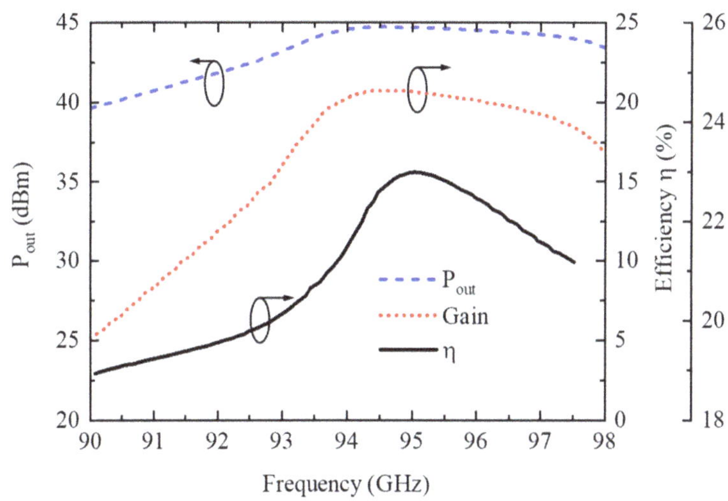

Figure 8.40. Saturated output power, saturated gain, and RF efficiency vs. frequency. © [2013] IEEE. Reprinted, with permission, from [16].

The saturated output peak power varies from 39 dBm (8 W) to 44.6 dBm (28.6 W) over the frequency range of 90–98 GHz. The saturated peak power for the RRB-SEC is higher than that for the PH-SEC, since the beam current is higher. On the other hand, the 6% 3 dB bandwidth of the saturated gain is less than that of the PH-SEC. The maximum RF efficiency is 15.7% at 95 GHz. This is significantly higher than the RF efficiency of 7.3% observed for the PH-SEC.

The frequency spectra of the input, output, and reflected signals at $P_{\text{in}} = 20$ dBm (saturated input power) at 93.75 GHz are shown in figure 8.41. There are small

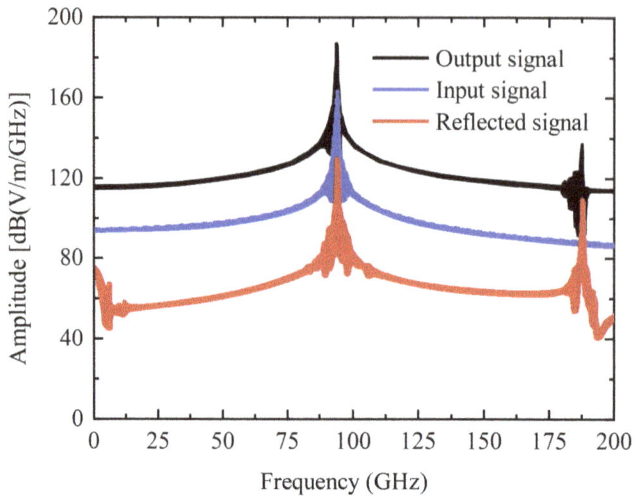

Figure 8.41. Frequency spectra of the input, output, and reflected signals for $P_{out} = 44.3$ dBm at 93.75 GHz. © [2013] IEEE. Reprinted, with permission, from [16].

peaks at the 2nd harmonic, 187.5 GHz, in the output and reflected signals. The reflected power at the 2nd harmonic is well below the fundamental input power level; hence, it is unlikely to cause oscillations. The spectrum is very clean, since rectangular waveguide couplers are used for the sever in this design.

8.5 Design and fabrication of a wideband Ka-band PH-SEC SWS

As mentioned in chapter 5, a flat SWS dispersion curve is important for the wideband operation of a TWT, and negative dispersion can help to reduce the in-band harmonic content of a wideband TWT. A number of dispersion-shaping techniques described in chapter 5 were applied to the design of a wideband PH-SEC SWS. The design of the SWS, reported by the authors' research group in [2, 17], was described in the same chapter, together with its cold-test and hot-test parameters. In addition to achieving wideband performance, the design specifically considers the feasibility of microfabrication.

For easy reference, the configuration of the designed wideband PH-SEC SWS is shown again in figures 8.42(a) and (b). This design achieves an S_{11} value better than −20 dB from 23.9 to 36.9 GHz, corresponding to a 42.8% cold-test bandwidth with discrete input and output ports. An electron beam of elliptical cross section at a beam voltage of 3.72 kV and a current of 50 mA yields a 3 dB small-signal gain bandwidth of 48.5% and a maximum gain of 42 dB. A saturated power of 26.5 W is achieved by using pitch tapering and 143 periods.

The following subsections present the thermal properties and electric field values at some critical locations in the SWS to examine the possibility of dielectric breakdown, the microfabrication process, and the results of microfabrication. This work was supported by grants to the authors' research group at Nanyang

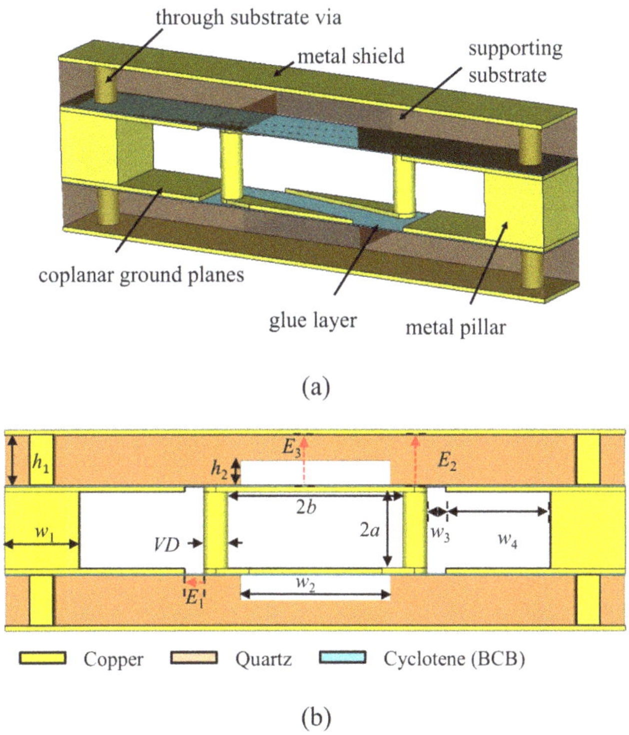

Figure 8.42. (a) Perspective view of the wideband PH-SEC SWS. (b) Cross section of the SWS. © [2016] IEEE. Reprinted, with permission, from [2].

Technological University (NTU), Singapore, by the Economic Development Board, Singapore, and the Office of Space Technology and Industry, Singapore.

Parts of this section have been reprinted, with permission, from [2]. © [2016] IEEE.

8.5.1 Thermal properties and power handling

The thermal characteristics of an SWS are very important for the reliable operation of a TWT that incorporates this component. Heat in a TWT may originate from two sources: (i) ohmic loss in the structure and (ii) interception of a part of the electron beam. Here, we study the temperature rise due to these effects using the CST Thermal Steady State Solver.

A tapered-pitch SWS with a length of 42 500 μm is used for the thermal analysis. Table 8.5 lists the electrical and thermal conductivities of different SWS materials. The background temperature is set to 300 K. The TCR between different materials is not considered here, since TCR values for quartz do not appear to be available. Our previous studies show that the inclusion of the TCR may increase the temperature by ~25% [18]. The dissipated powers resulting from ohmic loss and beam interception can be approximated from the insertion loss and beam energy, respectively.

Figure 8.43 shows the RF power vs. longitudinal position at 30 GHz for an SWS with 143 periods and pitch tapering. In order to calculate the circuit loss as the RF

Table 8.5. Material properties used in the simulation. © [2016] IEEE. Reprinted, with permission, from [18].

Material	Conductivity (S m^{-1})	Thermal conductivity (W K^{-1} m^{-1})
Copper	2.9×10^7	401
Quartz	1×10^{-14}	1.46
Bisbenzocyclobutene (BCB)	1×10^{-17}	0.29

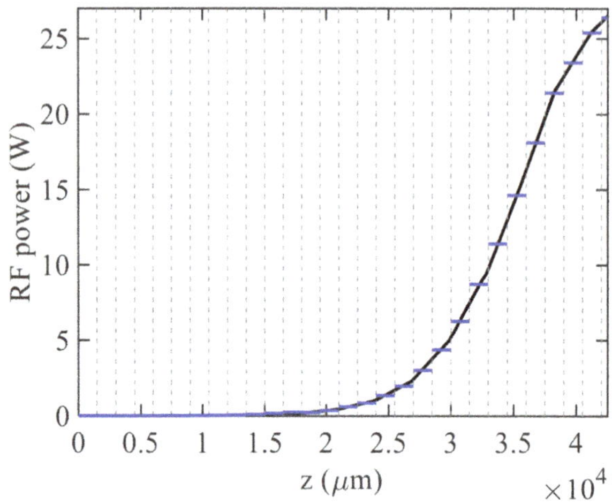

Figure 8.43. RF power vs. longitudinal position at 30 GHz for an SWS with 143 periods and pitch tapering. © [2016] IEEE. Reprinted, with permission, from [2].

power increases, we divide the tapered-pitch SWS into 29 small sections. The first 28 small sections have equal lengths of 1500 μm each. The last section has a length of 500 μm. To estimate the power dissipation due to RF circuit loss, we assume that the power flow in each small section is locally constant and equal to the average value of power flow in that section. Thus, the short straight blue lines in figure 8.43 are used to approximate the power flow in each section. Next, since the insertion loss at 30 GHz is −4 dB for the 130-period structure that is 39 000 μm long, we estimate the loss per mm to be 0.1025 dB. Thus, the insertion loss is 0.154 dB for each of the first 28 sections, and it is 0.051 dB for the last section.

With the assumption of constant power flow in each section ($P_{\text{transmitted}}$), the power loss in each section (P_{loss}) can be calculated from S_{21} as follows:

$$S_{21\text{section}} = \frac{P_{\text{transmitted}}}{P_{\text{transmitted}} + P_{\text{loss}}}. \tag{8.3}$$

From (8.3), the power loss in each section can be calculated. The calculated power loss is presented in table 8.6 for each section. The total loss for all sections is 5.32 W. To verify this method of loss calculation, we calculated the difference between the total RF power drawn from the beam, as calculated directly by CST (32.3 W), and

Table 8.6. Power loss for each section. © [2016] IEEE. Reprinted, with permission, from [2].

No.	1	2	3	4	5	6
P_{loss}(W)	0.0002	0.0001	0.0001	0.0001	0.0002	0.0005
No.	7	8	9	10	11	12
P_{loss}(W)	0.009	0.0014	0.0021	0.0035	0.006	0.0086
No.	13	14	15	16	17	18
P_{loss}(W)	0.009	0.0129	0.0216	0.0302	0.0474	0.0691
No.	19	20	21	22	23	24
P_{loss}(W)	0.1060	0.1533	0.2204	0.3063	0.4000	0.5123
No.	25	26	27	28	29	
P_{loss}(W)	0.6351	0.7509	0.8211	0.8913	0.308	

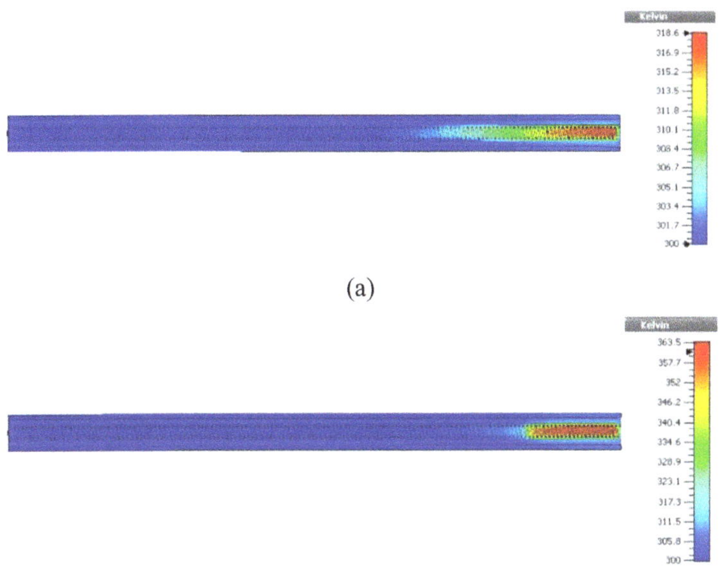

(a)

(b)

Figure 8.44. Temperature distribution in the SWS: (a) without beam interception; (b) with beam interception. © [2016] IEEE. Reprinted, with permission, from [2].

the RF output power for the lossy structure (26.5 W). This difference is 5.8 W, which is quite close to the total loss estimated in table 8.6.

The calculated power loss is applied as thermal power dissipation in the thermal simulation for each section. In addition, a beam interception loss of 5% of the beam energy (9.3 W), distributed in the last five sections of the tube (0.7 W in the last section and 2.15 W each in the remaining four sections), is also considered. Thermal simulations were carried out with and without beam interception, and the temperature distribution in the structure is presented in figure 8.44. We can see that the temperature is high only in the last few sections, where the power loss is significant.

The maximum temperature in the structure is 318.6 K without beam interception and 363.5 K with beam interception. For each section, the temperature is highest at the centers of the helix metal strips, since these parts of the strips are not in contact with the quartz. It is believed that the temperature rise is modest due to the significant contact area between the planar helix and quartz, as well as the excellent contact between the quartz, the outer metal shield, and the CPW ground planes. Thus, the proposed SWS may not suffer from an excessive temperature rise, even when subjected to RF input powers of tens of watts.

8.5.2 Consideration of the breakdown electric field

As in other millimeter-wave SWSs, many features of the considered structure are only hundreds of micrometers in size. The distance d between the planar helix metal strips and the metal shield on the outer surfaces of the quartz substrates is 205 μm, and the gap w_3 between the helix and the CPW ground planes is 100 μm. This can result in high values of RF electric field strength at these locations and may lead to breakdown, adversely affecting the power handling capability. This problem may become more severe as the operating frequency increases [19]. Therefore, the values of the RF electric field at such locations need to be examined. In general, for a given power level, if the electric field is too high, it can be reduced by increasing the distance or gap between the different parts of the SWS. In this subsection, values of RF electric field strength are examined with respect to the distance d and the gap w_3.

The RF electric field values at three different locations where the field values are rather high have been investigated using CST MWS. These locations, as shown in figure 8.42(b), are: (i) the gap w_3 between the helix and the coplanar ground planes (E_1); (ii) the distance d between the top (or bottom) right (or left) edges of the planar helix metal strips and the metal shield (E_2); (iii) the distance d between the top (or bottom) midpoints of the planar helix metal strips and the metal shield (E_3). The symbols E_1, E_2, and E_3 respectively stand for the average amplitude of the RF electric field at these three locations for an input power of 26.5 W. Figure 8.45 shows the variation in E_1, E_2, and E_3 with the gap w_3 when $d = 205$ μm. We can see that E_1 reduces as w_3 increases, and it reduces by 70% when w_3 increases from 100 to 400 μm. E_2 and E_3 remain relatively unaffected by this change. Figure 8.46 shows the variation in electric field values with d when $w_3 = 300$ μm. It can be seen that E_2 and E_3 reduce as d increases, while E_1 remains almost the same.

Figures 8.45 and 8.46 show that E_1 can be controlled by adjusting w_3, and E_2 or E_3 can be controlled by adjusting d. In addition, it is quite interesting to note that the E_1 curve intersects the other two curves at the position where the gap between the helix and the coplanar ground planes (w_3) and the distance between the helix and the metal shield (d) are nearly the same.

The breakdown E-field values for quartz at low frequencies are quite high (~5e7 V m^{-1}), but these values do not appear to be available in the literature for the Ka-band frequencies. On the other hand, breakdown field values at millimeter-wave frequencies

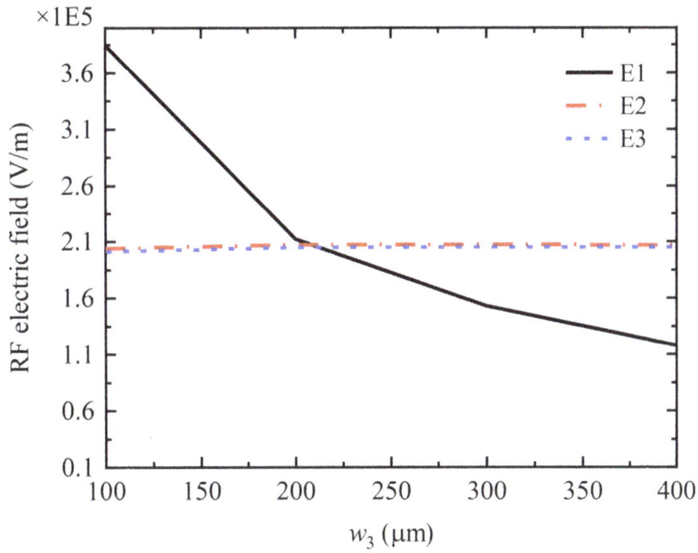

Figure 8.45. Average RF electric field strengths at three different positions versus the gap width w_3 for $d = 205$ μm. © [2016] IEEE. Reprinted, with permission, from [2].

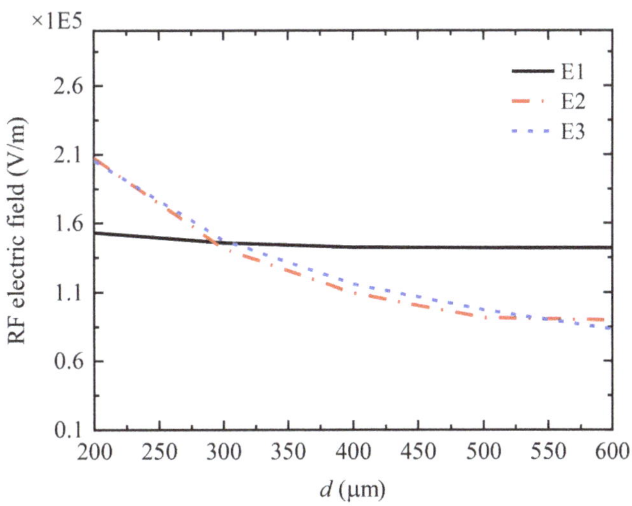

Figure 8.46. Average RF electric field strengths at three different positions versus the distance between the metal strip and the shield for $w_3 = 300$ μm. © [2016] IEEE. Reprinted, with permission, from [2].

may be much smaller than those at low frequencies [19]. Furthermore, whether the *E*-field values observed here cause breakdown in an actual SWS also depends on other factors such as material purity, surface finish, the presence of sharp metal edges, etc.

There is a trade-off between preventing breakdown and controlling dispersion. As we increase the separation between the helix and the metal shield or the CPW ground planes, the dispersion increases. If improvement in power handling is the main concern, one possible approach would be to prioritize separation values based on breakdown considerations and only then consider the dispersion issue.

8.5.3 Microfabrication process

The microfabrication of a W-band PH-SEC on a thick Si substrate without a metal shield has been described in section 8.3. Since the wideband Ka-band PH-SEC structure consists of two symmetric substrates and a metal shield, a much more complex fabrication process is required.

The fabrication steps used to produce the proposed structure are shown in figure 8.47. The PH-SEC layer and the quartz substrate layers are fabricated separately. As shown in figure 8.47(a), two Si substrates are used. First, beginning with a sacrificial Si wafer, the vertical straight-edge connections of the PH-SEC and the side walls are fabricated using the through-silicon-via (TSV) process. Next, we use photolithography and copper deposition to fabricate the horizontal parts of the PH-SEC and the coplanar ground planes on one side (figures 8.47(b) and (c)). Then, the device and the sacrificial Si wafers are separated. The device wafer is turned around, after which the horizontal parts of the PH-SEC on the other side are

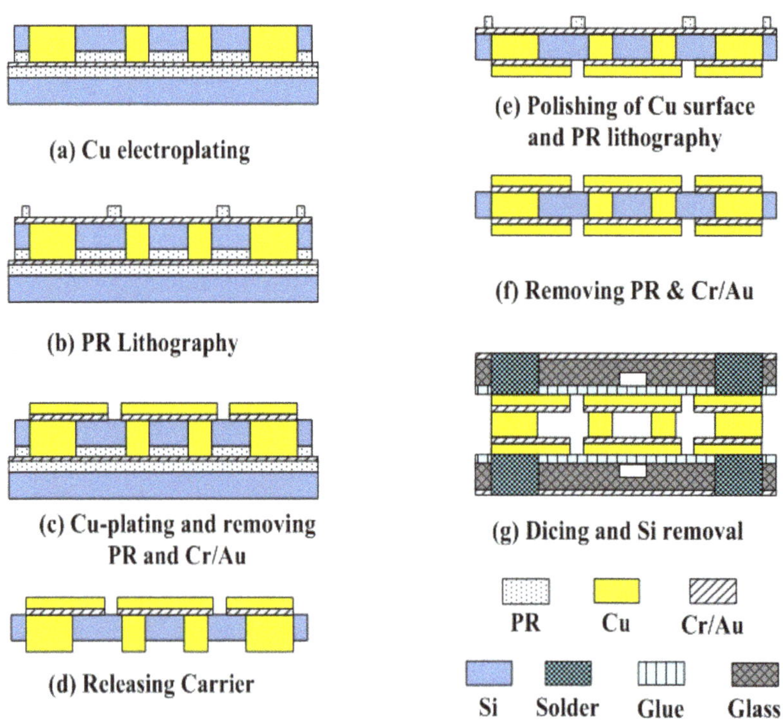

Figure 8.47. Fabrication process steps for the K_a-band PH-SEC SWS.

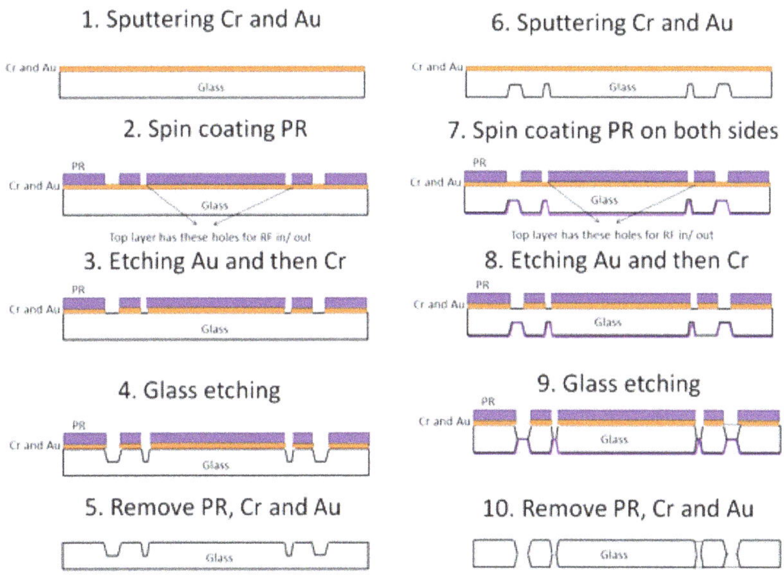

Figure 8.48. Initial process flow used to produce the through-wafer holes in the quartz substrates.

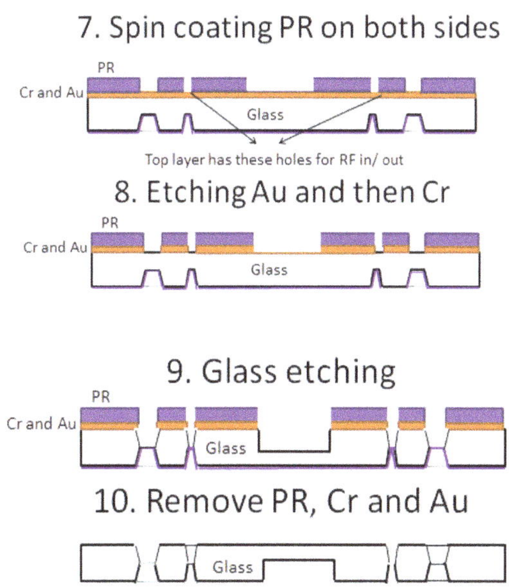

Figure 8.49. Modified process flow steps 7–10 used to produce a trench in the quartz substrates.

fabricated using the same copper deposition and photolithography process (figures 8.47(e) and (f)).

The through-wafer holes in the quartz substrates are fabricated using glass etching techniques; the process flow is shown in figure 8.48. To realize the trench in these wafers, steps 7–10 of figure 8.48 are modified as shown in figure 8.49. After

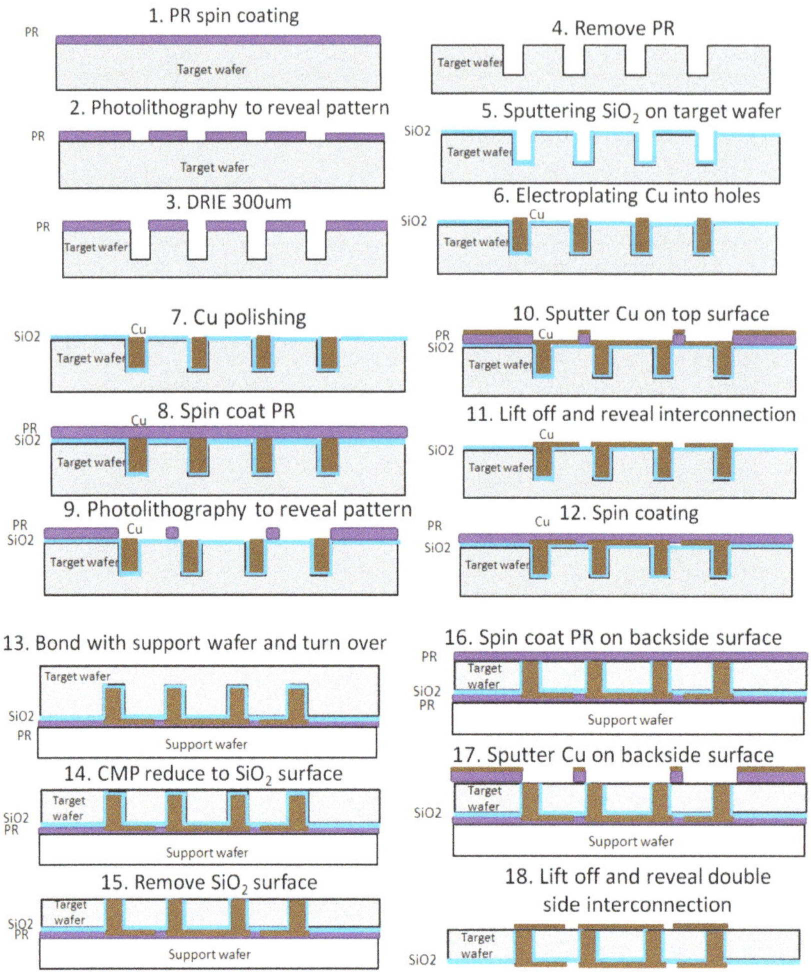

Figure 8.50. Process flow used for the fabrication of PH-SEC structures on Si wafer by SINANO.

completing the steps in figures 8.48 and 8.49, the metal shield is sputter deposited on the outer surfaces of the quartz substrates. The PH-SEC and the two quartz substrates are bonded together using a low-loss nonconductive bonding material, Cyclotene 3022 bisbenzocyclobutene (BCB) [20], which has a low dielectric constant of 2.65. The thickness of the bonding layer is 5 μm. Finally, the Si wafer is etched away (figure 8.47(g)).

Based on the microfabrication experience at NTU, the research group collaborated with SINANO, China, to fabricate the PH-SEC SWS on a silicon wafer (TSV electroplating with double-sided horizontal interconnection) using their advanced facilities. The design dimensions and fabrication recipes shown in figure 8.50 were provided by NTU.

8.5.4 Results of microfabrication

Due to the complex fabrication process, the complete SWS shown in figure 8.47(g), i.e. a PH-SEC with top and bottom quartz wafers, could not be realized. Presented below are some parts of the SWS that were fabricated successfully.

Images of the double-sided processed quartz wafer (200 μm thick, with a four-inch diameter) before and after Au/Cr etching, as well as after metallic layer sputtering, are presented in figures 8.51 and 8.52. These images depict a good-quality wafer surface with uniform trench and hole dimensions.

Another issue with the quartz wafer is the electrical connection between the outer shield and inner ground in the PH-SEC-SWS. To realize this electrical connection, solder balls 0.1 mm in diameter were heated to melt in the etched-through holes on

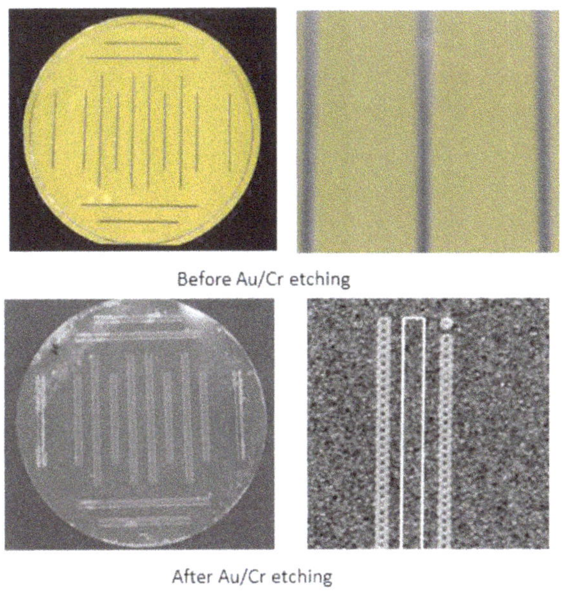

Figure 8.51. Photos of the bottom quartz wafer before and after Au/Cr etching.

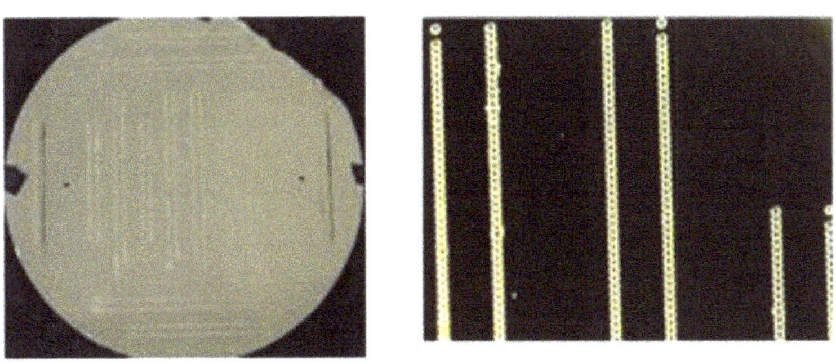

Figure 8.52. Photos of the metallic layer sputtering on the bottom quartz wafer.

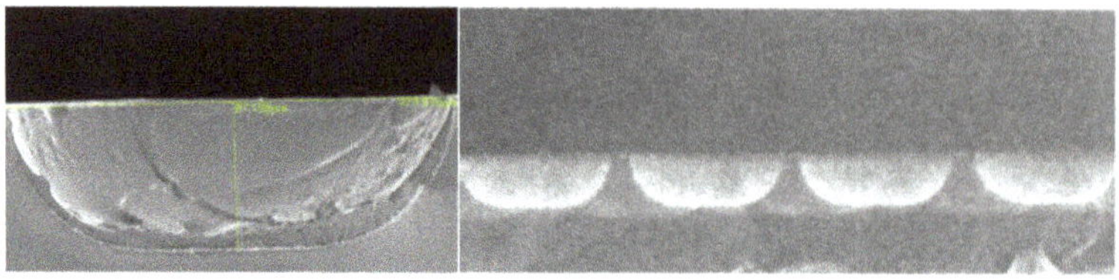

Figure 8.53. Results of solder ball melting in quartz holes; (left) without solder ball; (right) with molten solder in holes.

Figure 8.54. The horizontal interconnections and the CPW ground planes on Si wafer fabricated at NTU.

the quartz shield wafer. Figure 8.53 shows SEM images of the quartz holes before and after they were filled with solder balls.

The TSVs (straight-edge connections in the PH-SEC) in the silicon wafer are connected on both sides of the wafer by horizontal interconnections. The high-resistivity (1000 Ω-cm) silicon wafer is 300 μm thick and four inches in diameter and has a crystal orientation of <100>. Figure 8.54 shows the horizontal interconnections without TSVs and the CPW ground planes fabricated at NTU.

Images of the PH-SEC SWS fabricated by SINANO with good TSV Cu electroplating and double-sided horizontal interconnections on the Si wafer are shown in figure 8.55. However, the measured resistance of the PH-SEC was high. The Si successfully etched away using wet KOH etching and the complete SWS structure are shown in figure 8.56. However, the horizontal interconnections did not form a strong joint with the TSVs.

8.6 Summary

This chapter describes the design and fabrication of PH-SEC structures for operation in the C/X (7–10 GHz) and Ka bands (26–40 GHz) as well as PH-SEC and RRB-SEC structures for operation in the W band. A vacuum-compatible printed circuit fabrication technique is used for the C/X-band structure. For the

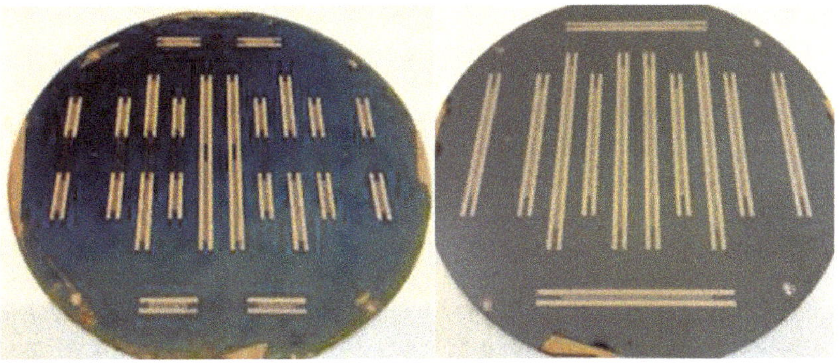

Figure 8.55. The horizontal TSV Cu electroplating and double-sided horizontal interconnections on the Si wafer (SINANO).

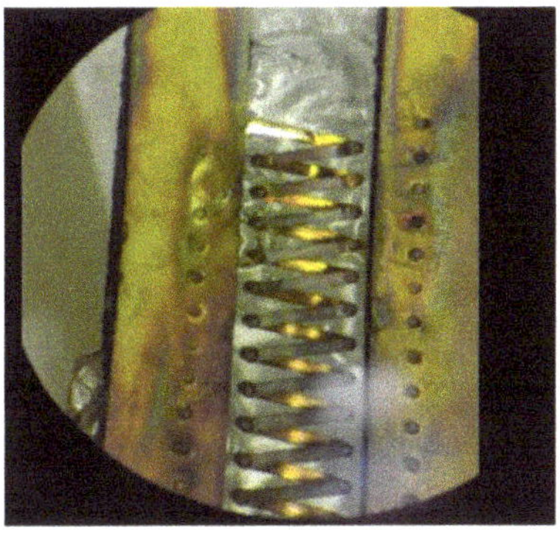

Figure 8.56. Etched-away Si and the complete SWS structure (SINANO).

Ka-band structure, which achieves wideband performance, the design specifically considers the feasibility of microfabrication. For the W-band structures, thousands of which are microfabricated on a silicon substrate, a detailed assessment of the effects of the dimensions and surface roughness of the fabricated structures is carried out. For the Ka-band and W-band structures, thermal and dielectric breakdown issues are also described. A highlight of the work on W-band structures is the 'on-wafer' measurement of the cold-test S-parameters using a CPW probe station. This avoids the need to dice individual SWSs, enabling speedy evaluation of a large number of microfabricated SWSs.

References

[1] Chua C S, Zhao C and Aditya S 2017 Design and fabrication of a planar helix slow-wave structure for C/X-band TWT *IEEE Trans. Compon. Packag. Manuf. Technol.* **7** 1663–9

[2] Zhao C, Aditya S, Wang S, Miao J and Xia X 2016 A wideband microfabricated Ka-band planar helix slow-wave structure *IEEE Trans. Electron Devices* **63** 2900–6

[3] Chua C *et al* 2011 Microfabrication and characterization of W-band planar helix slow-wave structure with straight-edge connections *IEEE Trans. Electron Devices* **58** 4098–105

[4] Chua C S 2012 *Studies on Planar Helical Slow-Wave Structures for Travelling-Wave Tube Applications* (Singapore: SingaporeNanyang Technological University)
[5] Chua C S, Tsai M L J, Tang M, Aditya S and Shen Z X 2011 Microfabrication of a planar helix with straight-edge connections slow-wave structure *Adv. Mater. Res.* **254** 17–20
[6] Hammerstad E O and Bekkadal F 1975 *Microstrip Handbook* (Trondheim, Norway: University of Trondheim)
[7] Zhang P, Lau Y Y and Gilgenbach R M 2009 Analysis of radio-frequency absorption and electric and magnetic field enhancements due to surface roughness *J. Appl. Phys.* **105** 114908
[8] Fowler A M 1970 Radio frequency performance of electroplated finishes *Proc. IREE* 148–64
[9] Datta S K, Kumar L and Basu B N 2009 A simple and accurate analysis of conductivity loss in millimeter-wave helical slow-wave structures *J. Infrared, Millimeter, Terahertz Waves* **30** 381–92
[10] Sengele S, Jiang H, Booske J H, Kory C L, van der Weide D W and Ives R L 2009 Microfabrication and characterization of a selectively metallized W-band meander-line TWT circuit *IEEE Trans. Electron Devices* **56** 730–7
[11] Yao L, Yang Z, Li B, Liao L, Zeng B and Zhu X 2006 Thermal analysis of novel helix TWTs *IEEE Int. Vacuum Electronics Conf.* 139–40
[12] Chua C, Aditya S, Tsai J and Shen Z 2012 PIC simulation for W-band planar helix with straight-edge connections *IVEC 2012* 459–60
[13] Chodorow M and Chu E L 1955 Cross-wound twin helices for traveling-wave tubes *J. Appl. Phys.* **26** 33–43
[14] Birdsall C K and Everhart T E 1956 Modified contra-wound helix circuits for high-power traveling-wave tubes *IRE Trans. Electron Devices* **3** 190–204
[15] Chua C, Aditya S, Tsai J M, Tang M and Shen Z 2011 Microfabricated planar helical slow-wave structures based on straight-edge connections for THz vacuum electron devices *Terahertz Sci. Technol.* **4** 208–29
[16] Chua C, Aditya S and Lau Y Y 2013 W-band rectangular ring-bar structure with straight-edge connections *2013 IEEE 14th Int. Vacuum Electronics Conf. (IVEC)* 1–2
[17] Zhao C 2016 *Planar Helix-Based Slow-Wave Structures For Millimeter Wave Traveling-Wave Tubes* (Singapore: Nanyang Technological University)
[18] Wang S, Zhao C, Aditya S and Chua C 2015 Thermal characteristics of a Ka-band planar helix slow-wave structure *2015 IEEE Int. Vacuum Electronics Conf. (IVEC)* pp 1–2
[19] Gladkov S O 1992 Theory of the breakdown of porous dielectrics with a high frequency electric field *Phys. Lett.* A **161** 559–61
[20] *Cyclotene Advanced Electronics Resins* http://dow.com/cyclotene/

Chapter 9

Design and fabrication of meander-line slow-wave structures for TWTs

9.1 Introduction

Chapters 2, 5, and 6 describe microstrip meander-line (MML) slow-wave structures (SWSs) and their variations, covering the methods used for their analysis and simulation and their application in traveling-wave tubes (TWTs). Among the various SWSs available for application in TWTs, MML SWSs stand out due to their simple fabrication and can thus easily be realized, even at millimeter-wave frequencies. MML SWSs can readily accommodate a sheet electron beam. Moreover, an MML is typically in very good contact with its substrate, and thus thermal dissipation can be very effective. Therefore, MML SWSs are prime candidates for low-cost, moderate-gain TWTs.

This chapter describes examples of the fabrication of MML SWSs as reported in the literature by various research groups. The examples comprise structures operating in the Ka band (26–40 GHz), the V band (50–75 GHz), and the W band (75–110 GHz). The two Ka-band structures described in section 9.2 were designed and microfabricated by the authors' research group. For one of these Ka-band structures (section 9.2.1), measurement of the cold-test S-parameters was carried out 'on-wafer,' using CPW probe stations. This avoided the need to dice individual SWSs and enabled fast evaluation of a large number of microfabricated SWSs. Similar to the previous chapter, this aspect is also a highlight of this chapter. For the second Ka-band structure (section 9.2.2), an assembly for hot tests, including an electron gun, focusing magnetic circuit, and collector, was also completed.

doi:10.1088/978-0-7503-5764-7ch9

9.2 Design and fabrication of Ka-band meander-line SWSs

Chapter 6 describes a number of variations of the basic MML SWS. These variations improve one or more properties of the SWS, which enhances the performance of TWTs based on such an SWS. This section describes two of these variations, namely, the V-shaped MML and the double-V or symmetric-V MML, both designed and fabricated by the authors' research group at Nanyang Technological University (NTU). The work of a few other research groups on MML SWSs is described in the next section.

9.2.1 On-wafer V-shaped meander-line SWSs

The V-shaped MML was first proposed in [1], which showed that it offers a wider cold bandwidth and generally higher interaction impedance compared to the U-shaped MML. The goal of this work is to design a V-shaped MML SWS that operates in the Ka band and is suitable for microfabrication. An important design requirement is the ability to carry out on-wafer cold-test measurements, since this permits speedy assessment of multiple SWSs. The contents of this subsection closely follow [2].

9.2.1.1 Configuration and cold-test parameters

A period of the V-shaped MML chosen for this work is shown in figure 9.1. As shown in figure 9.1(a), the sharp inner bends are replaced by a circular arc, which is easier to produce accurately using photolithography. Figures 9.1(b) and (c) show the side view and the 3D model of the SWS, respectively. As shown in figure 9.1(c), to avoid dicing to extract individual SWSs and using a metal enclosure for testing, a patterned ground plane on the back side and an inverted U-shaped metal cover on top are included to provide electrical shielding. The input and output of the ML SWS transition to 50 Ω coplanar waveguide (CPW) pads for on-wafer measurement by a probe station. With such a configuration, all the SWSs fabricated on a wafer can be cold tested in a single setup. The substrate material chosen is quartz, since it has a relatively low dielectric constant of 4.43 and a very low loss tangent of 3×10^{-5} at 30 GHz. Although the thermal conductivity of quartz is modest (1.4 W $°K^{-1}$ m^{-1}), its heat dissipation can still be very effective, since the SWS is in intimate contact with the quartz substrate. In particular, the cost of quartz is much lower than that of boron nitride, and microfabrication based on quartz wafers is well developed.

The characteristics of the proposed SWS were studied using the Eigenmode Solver of CST Microwave Studio (MWS). In this work, the dimensions are dictated by the fabrication accuracy achievable by our research group and the requirement to make the corresponding TWT operate at a beam voltage of less than 4 kV.

Figure 9.2(a) shows the normalized phase velocity of the proposed configuration for different widths W_sws of the meander line, with the other parameters fixed as follows: W_trench = 2000 μm, H_trench = 300 μm, W_strip = 40 μm, W_sub = 6000 μm, H_sub = 200 μm, and H_str = 2 μm. It can be seen that the

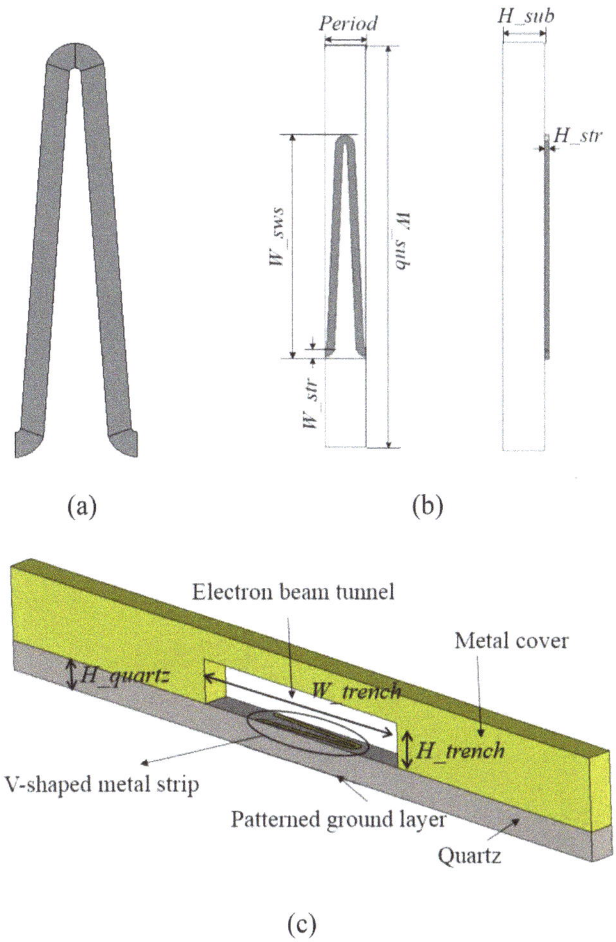

Figure 9.1. (a) A period of the V-shaped MML; (b) top and side view and (c) 3D model with dimensional parameters. © [2018] IEEE. Reprinted, with permission, from [2].

phase velocity depends on the width of the meander line. When *W_sws* is 1100 μm, the normalized phase velocity is 0.1 at 30 GHz, and the corresponding operating voltage is ~3.5 kV.

Figure 9.2(b) shows the interaction impedance at the center of the electron-beam tunnel, which is at a height of 150 μm above the meander line. In addition to the phase velocity, the width of the SWS also has an effect on the electric field strength; both of these effects impact the interaction impedance. The interaction impedance is about 14 Ω at 30 GHz for *W_sws* = 1100 μm.

9.2.1.2 CPW feed and electrical shielding

To enable on-wafer measurements using CPW probes, tapered CPW sections are added to achieve an impedance match between the input/output of the SWS and 50 Ω CPW pads. CST MWS is used to check the cold-test S-parameters of the

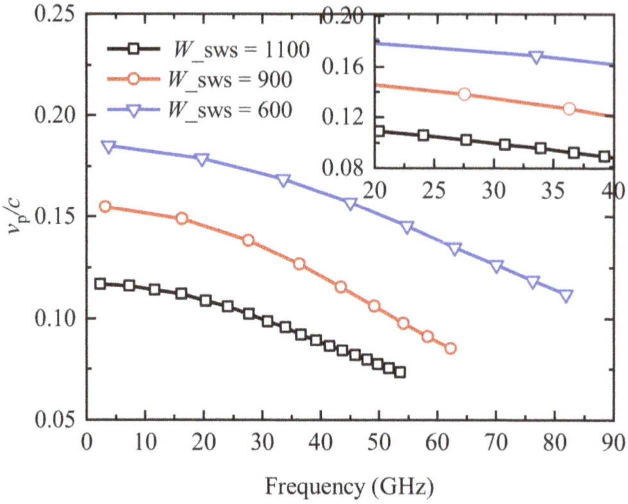

(a)

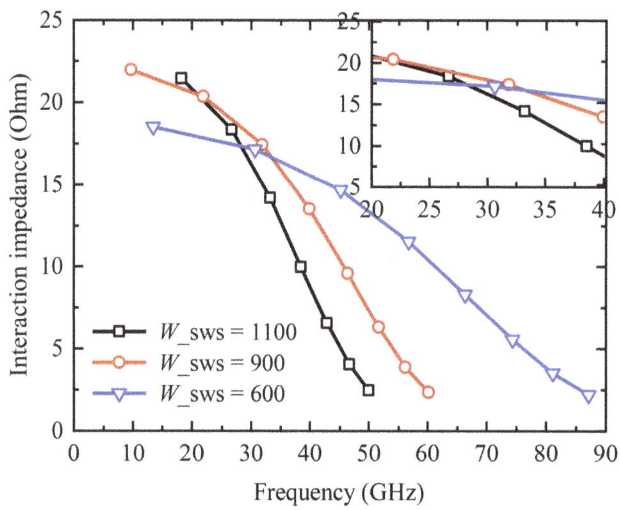

(b)

Figure 9.2. (a) Dispersion and (b) interaction impedance characteristics of the Ka-band V-shaped ML SWS for different widths *W_sws* (in μm). The inset figures show the results for the 20–40 GHz frequency range. © [2018] IEEE. Reprinted, with permission, from [2].

V-shaped meander-line SWS together with the microstrip-to-CPW transitions. Figure 9.3(a) shows the front side of the SWS pattern. As shown, A and A′ are CPW pads with a port impedance of 50 Ω, B and B′ are tapered CPW sections, C and C′ are straight microstrip sections, which have the same strip width as the microstrip meander line, D and D′ are tapered periods, and E is the main V-shaped meander-line

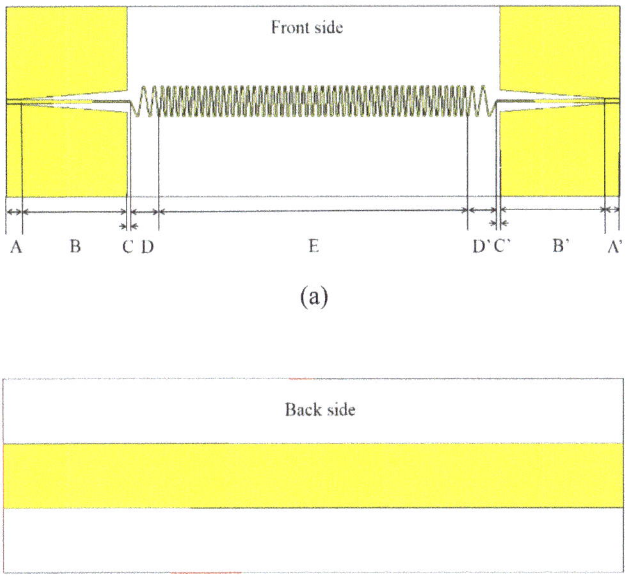

Figure 9.3. V-shaped meander-line SWS together with CPW-to-microstrip transitions: (a) front side and (b) back side. © [2018] IEEE. Reprinted, with permission, from [2].

SWS. The optimized lengths of the segments are as follows: $LA = LA' = 500$ μm, $LB = LB' = 3400$ μm, $LC = LC' = 100$ μm, and $LD = LD' = 940$ μm.

Since the maximum separation between input/output probes is restricted on the available wafer probe station, the number of periods is limited to 50. The overall length of the SWS, including the CPW transitions, is about 19.9 mm. Figure 9.3(b) shows the metal pattern on the back side of the quartz substrate. The metal ground plane is truncated to cut off the transverse leakage of the signal through the substrate mode. Effective electrical shielding can be realized by adjusting the width of the truncated ground plane. In our design, this width is 3400 μm.

Figure 9.4 shows the S-parameters of the typical CPW symmetric electric field mode. The background is set to vacuum, and the boundary is set to open (add space). In the 22–36 GHz frequency range, S_{11} is less than −15 dB, and S_{21} is about −7 dB at 30 GHz.

9.2.1.3 Microfabrication and assembly

The fabrication model of the SWS is shown in figure 9.5. The chosen thickness of the metal strip *H_str* is 2 μm, since this can be easily realized using copper sputtering. Otherwise, electroplating may be required to produce a thicker metal strip, which makes the fabrication complex. We also use a 20 nm thick titanium layer below the Cu and a 20 nm thick gold layer over the Cu. The meander-line pattern with the CPW transitions on the front surface of the quartz wafer and the truncated ground plane on the back surface are realized by metal sputtering, followed by

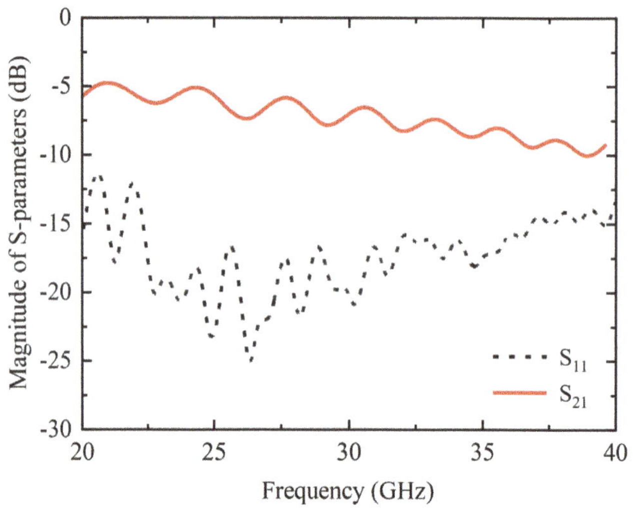

Figure 9.4. S-parameters of the V-shaped meander-line SWS. © [2018] IEEE. Reprinted, with permission, from [2].

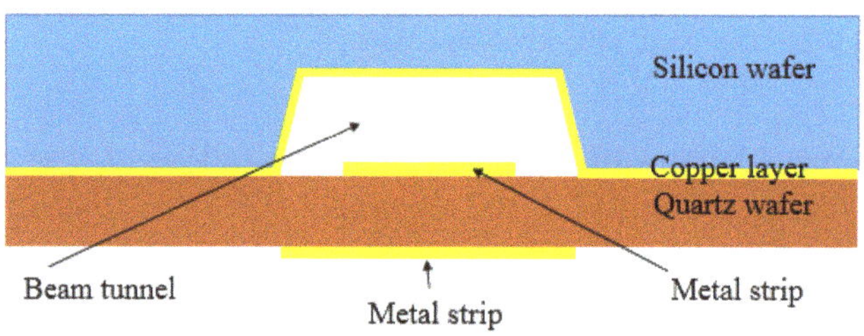

Figure 9.5. Fabrication model of the V-shaped meander-line SWS. © [2018] IEEE. Reprinted, with permission, from [2].

photolithography and etching of multiple metal layers. The top metal cover is realized using a silicon wafer in which a trench is made using wet etching, and a copper layer is sputtered on the trench-side surface. The two wafers are bonded using a low-loss conductive glue. Alignment marks on both wafers guarantee the accuracy of alignment.

Figure 9.6 shows the fabricated MML SWSs on a quartz wafer. SWSs were fabricated with two different lengths, corresponding to 25 and 50 periods. The separation between input/output CPW pads is kept the same for all SWSs by including microstrip sections of appropriate length. There are 14 short SWSs and 15 long SWSs on a 4″ quartz wafer. The measurements show that the fabricated strip width is 36–38 μm (design value: 40 μm), the CPW signal line width is 136–138 μm (design value: 140 μm), while the CPW gap width is 19–21 μm (design value: 18 μm). Figure 9.7 shows the assembled quartz wafer and silicon cover. The meander-line

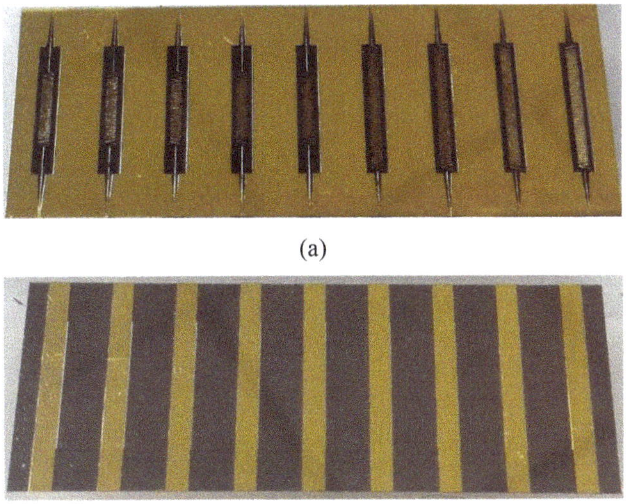

(a)

(b)

Figure 9.6. Fabricated V-shaped ML SWSs on quartz: (a) front side and (b) back side. © [2018] IEEE. Reprinted, with permission, from [2].

Figure 9.7. Assembly of the quartz wafer and silicon cover. © [2018] IEEE. Reprinted, with permission, from [2].

SWSs are covered by the metallized silicon cover, while the CPW pads are exposed and accessible for on-wafer measurement.

9.2.1.4 Measurement results for S-parameters

The S-parameters of the fabricated V-shaped meander-line SWSs are obtained by on-wafer measurements and compared with simulation results. In figures 9.8, a1 and a2 are results for the 25-period SWSs, while b1 and b2 are results for the 50-period SWSs. For example, 'simulated a' and 'simulated b' are the simulated results for the 25-period and 50-period structures, respectively. Figure 9.8(a) shows that all the measured structures have very similar S_{11} values. S_{11} is better than -10 dB in

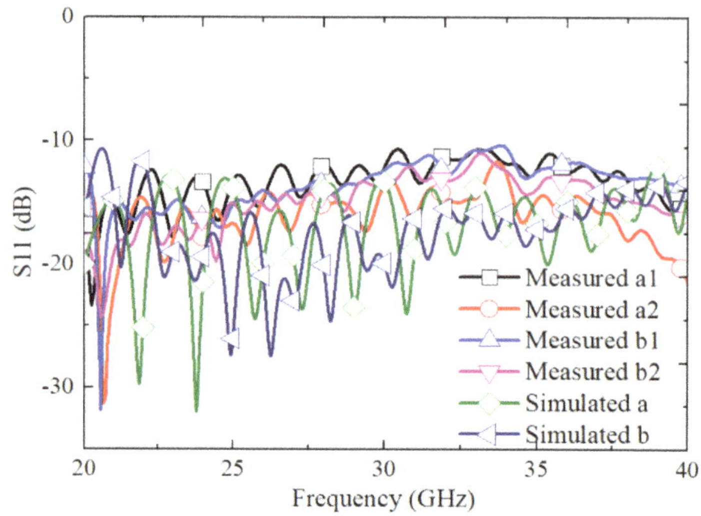

(a)

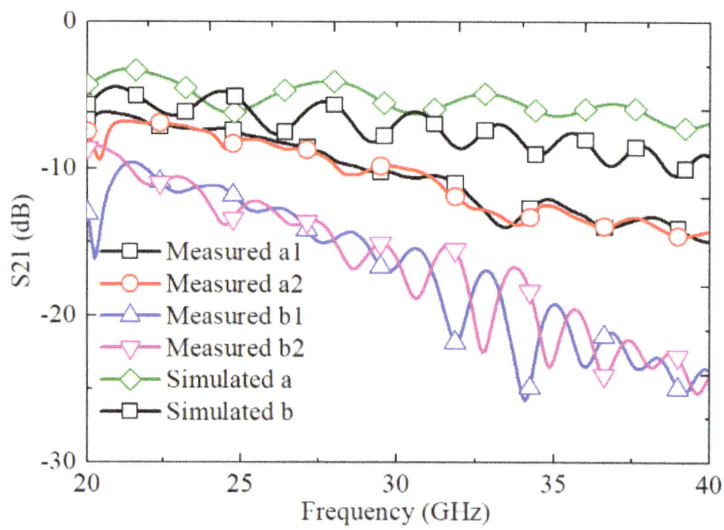

(b)

Figure 9.8. Comparison between experimental and simulated S-parameters. (a) S_{11}, and (b) S_{21}. © [2018] IEEE. Reprinted, with permission, from[2].

the frequency range of 20–40 GHz and better than −14 dB and −15 dB, respectively, in the frequency range of 25–36 GHz.

Figure 9.8(b) shows the measured and simulated S_{21} values. In general, the curves show that S_{21} gets worse as the frequency increases. For structures with the same number of periods, the measured S_{21} values are quite similar. The average difference between the simulated S_{21} values for the short and long structures is about 2.0 dB,

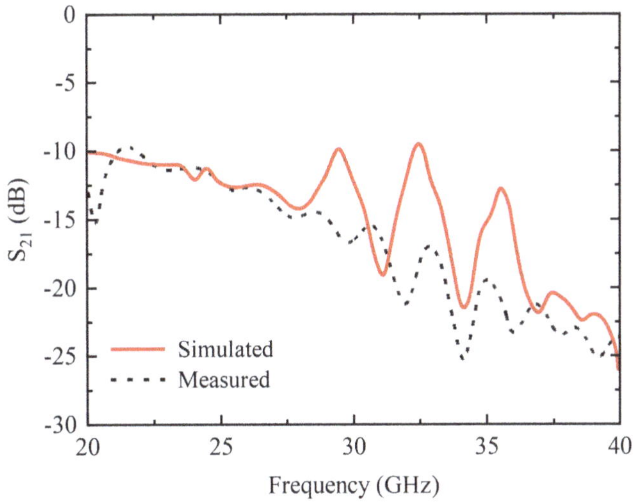

Figure 9.9. Comparison between simulated (for a conducting layer with conductivity 2.38×10^6 S m^{-1}) and measured S_{21} values of the 50-period microfabricated V-shaped ML SWS. © [2018] IEEE. Reprinted, with permission, from [2].

which should represent the theoretical insertion loss for the 25-period V-shaped meander line. On the other hand, the average difference between the measured S_{21} values is about 5.5 dB. It is clear that the measured insertion loss is much higher than the simulated loss.

The reason for the higher loss is attributed to the conductor loss of the meander-line strip. As the tanδ of quartz is very small (3×10^{-5}), the dielectric loss is negligible compared to the conductor loss. The conductor loss is determined by the dimensions and conductivity of the microstrip line. The fabricated meander line consists of three layers: 20 nm of titanium, 2 μm of copper, and 200 nm of gold. The overall conductivity of the fabricated microstrip line can be calculated using the formulas listed in [3], and the result is about 2.9×10^6 S m^{-1}. This value is quite close to that of titanium, which is 2.38×10^6 S m^{-1}. On the other hand, the simulation results in figure 9.8 assume just a copper layer with a conductivity of 2.2×10^7 S m^{-1}. Thus, it is likely that the extra loss is caused by the titanium layer.

Figure 9.9 shows the simulated S_{21} values for a conducting layer with a conductivity of 2.38×10^6 S m^{-1} together with the measurement results for the 50-period SWS. The results match very well up to 28 GHz; beyond that, the average values of the two results differ by only 1–2 dB.

9.2.1.5 Simulation results for hot-test parameters

The hot-test parameters of the proposed V-shaped meander-line SWS for TWT application are estimated using the PIC Solver of CST Particle Studio for the structure with discrete ports. The S-parameters of a 50-period SWS are first determined using CST MWS. The results show that S_{11} is less than −20 dB in the frequency range of 15–40 GHz when the port impedance is 106 Ω; S_{21} is about

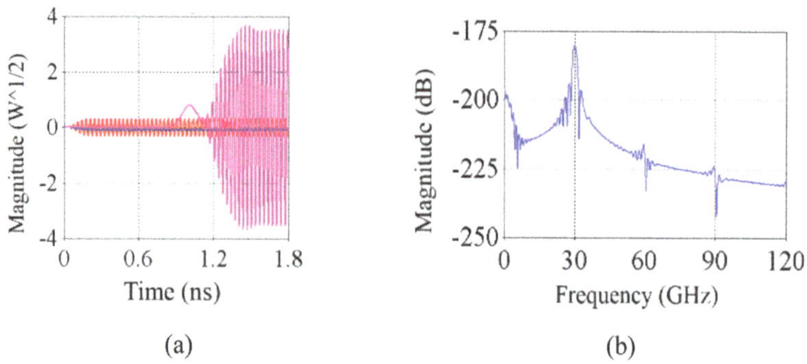

Figure 9.10. PIC simulation results at 30 GHz. (a) Input and output signals; (b) Fourier transform of the output signal. © [2018] IEEE. Reprinted, with permission, from [2].

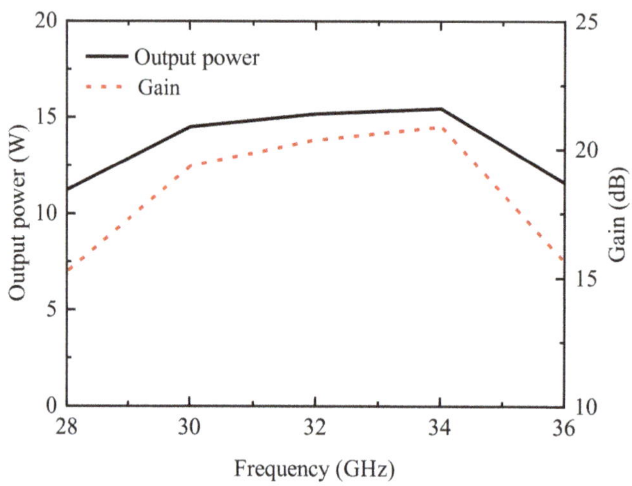

Figure 9.11. Output power and gain of the microstrip V-shaped ML SWS at different frequencies. © [2018] IEEE. Reprinted, with permission, from [2].

−5 dB at 30 GHz when the metal strip is set to 2 μm copper with a conductivity of 2×10^7 S m^{-1}. This is the insertion loss value that can be expected if single-layer aluminum or copper metallization is used. The number of periods used in the PIC simulation is 170. An electron beam (e-beam) with a beam voltage of 3.6 kV and a current of 50 mA is centrally applied in the electron-beam tunnel. The rectangular cross section of the e-beam has an optimized width of 700 μm and a thickness of 150 μm. A uniform focusing magnetic field of 0.4 T is used.

Figure 9.10(a) shows input and output signals with magnitudes of 0.32 $\sqrt{W}$ and 3.54 $\sqrt{W}$, respectively. The corresponding peak output power is 12.53 W for an input peak power of 0.1 W, indicating a gain of 21 dB. Figure 9.10(b) presents the spectrum of the output signal, showing a clean signal at 30 GHz.

The output power vs. frequency curve is shown in figure 9.11. The maximum output power is 14.5 W at 34 GHz, with a gain of 21.6 dB. In the frequency range of

28–36 GHz, the output power is higher than 7.1 W, showing a 3 dB bandwidth of about 25% centered at 32 GHz.

9.2.2 Symmetric double-V-shaped meander-line SWS

The symmetric double-V-shaped meander-line SWS (ML SWS) is another variation of the meander-line SWS that has been designed and fabricated by the authors' research group at NTU. It was first proposed in [4]; as mentioned in chapter 6, it offers higher interaction impedance and larger output power compared to the corresponding single-V structure. A study exploring the use of this structure in TWTs designed for Ka-band operation was reported in [5]. In addition to design, microfabrication, and cold-test measurements, some other important aspects covered are input–output couplers and an assembly suitable for hot tests. The contents of this subsection closely follow the report in [5].

9.2.2.1 Configuration and cold-test parameters

Figure 9.12 shows the simulation model and the main dimensions of the symmetric V-shaped ML SWS; *W_str* and *H_str* are, respectively, the strip width and thickness of the microstrip, *W_sws* and *W_sub* are the widths of the SWS and quartz substrate, respectively, and *H_sub* is the thickness of the substrate. The dimensions listed in table 9.1, including the

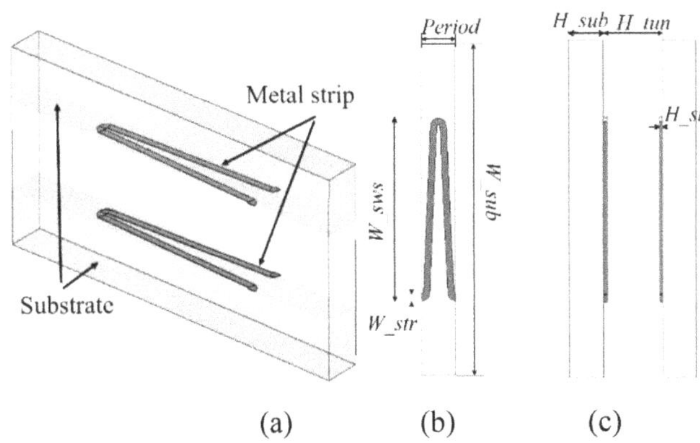

Figure 9.12. (a) Three-dimensional model of one period, (b) top view, and (c) side view with dimensional parameters of the symmetric V-shaped ML SWS. © [2019] IEEE. Reprinted, with permission, from [5].

Table 9.1. Dimensions of the symmetric V-shaped ML SWS. © [2019] IEEE. Reprinted, with permission, from [5].

Parameter	Value (μm)	Parameter	Value (μm)
W_SWS	900	*H_tun*	300
W_str	40	*H_str*	2
W_sub	2000	*H_sub*	200
Period (l)	200		

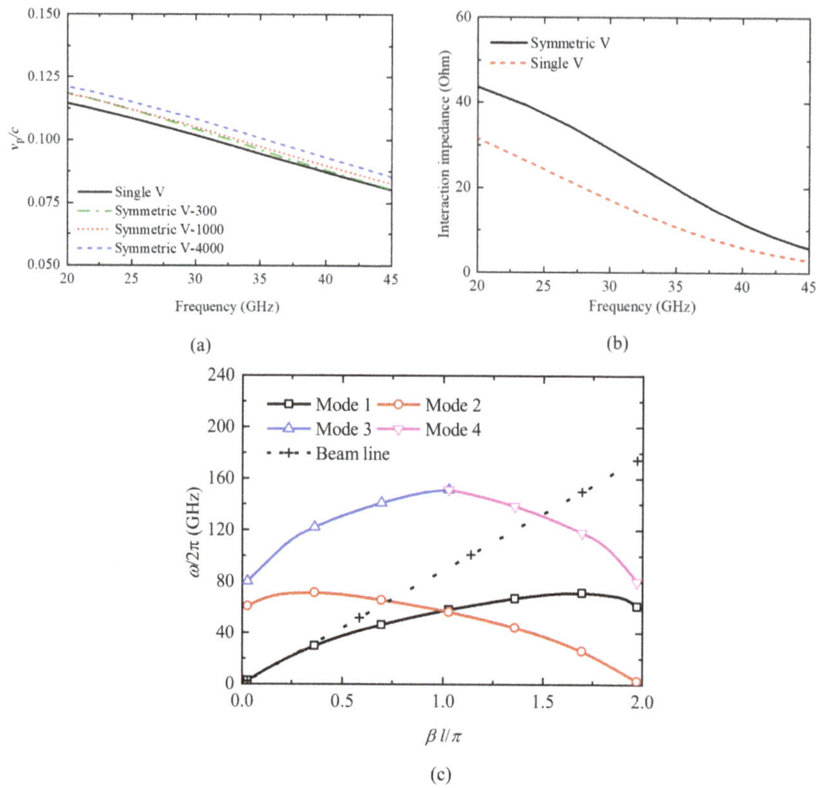

Figure 9.13. (a) Dispersion characteristics, (b) interaction impedance, and (c) ω-β characteristics of the Ka-band single-V and symmetric V-shaped ML SWSs. © [2019] IEEE. Reprinted, with permission, from [5].

height of the beam tunnel, are selected to yield the same dispersion characteristics for both the symmetric and single V-shaped ML SWSs. The properties of the materials in the SWS are set as follows: the dielectric constant and loss tangent of quartz are 4.43 and 3.0×10^{-5}, respectively; the conductivity of copper for our fabrication is estimated to be 2×10^7 S m^{-1}.

The dispersion and interaction impedance characteristics are obtained using CST MWS. Figures 9.13(a) and (b) show the dispersion and interaction impedance vs. frequency, respectively, for the single and symmetric V-shaped ML SWSs. Figure 9.13(a) includes results for different separations of the ML SWSs in the symmetric-V configuration. It is clear from this figure that the dispersion curve of the symmetric structure is very close to that of the single V-shaped ML SWS. As the separation between SWSs (i.e. the height of the beam tunnel) increases, the phase velocity increases, but this increase is noticeable only when the height increases to thousands of micrometers. As shown in figure 9.13(b), the on-axis interaction impedance of the symmetric structure is about 30 Ω at 30 GHz; this is due to the stronger electric field in the beam-tunnel region of the symmetric structure. Figure 9.13(c) shows the ω-β diagram of the symmetric V-shaped ML SWS. The diagram includes a beam line corresponding to 3.6 kV, which indicates wideband beam–wave interaction around 30 GHz.

9.2.2.2 Waveguide input–output couplers

The standard WR-28 waveguide is used as the port for the input–output couplers. The design involves two steps: (A) transition from the symmetric V-shaped ML SWS to a symmetric stripline and (B) transition from the symmetric stripline to the waveguide. For step A, shown in figure 9.14, the stripline input–output feeds consist of approximately three tapered periods of the symmetric ML SWS (section 2) and sections of symmetric stripline at both ends (section 1) to achieve a good match. The symmetric stripline feeds are in the same layer as the strips of the V-shaped ML; thus, both features can be fabricated in the same process step. It should be noted that there is no ground plane above or below the stripline.

For step B, shown in figures 9.15(a) and (b), the symmetric stripline feed designed in step A is extended and inserted into the WR-28 rectangular waveguide to act as a probe to excite the TE_{10} mode in the waveguide. The probe is inserted through the middle of the long wall of the rectangular waveguide. The probe has the same substrate as the symmetric V-shaped ML SWS shown in figure 9.12. The main dimensional parameters of the stripline probe are marked in figure 9.15(b), while the parameter values are listed in table 9.2.

When the waveguide port supports the TE_{10} mode, the first mode of the dual stripline port is an even mode and the second one is an odd mode. The odd mode is

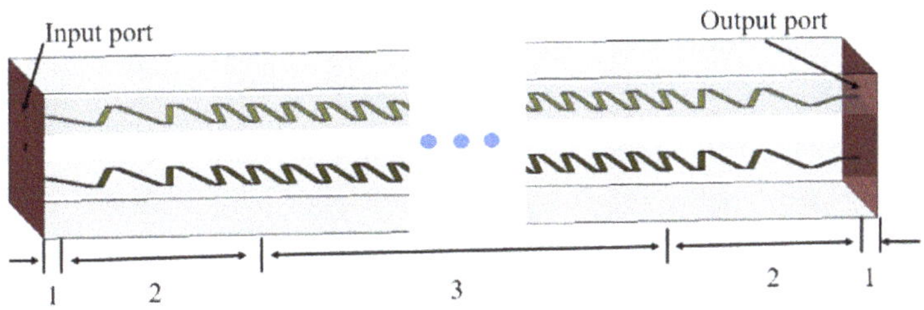

Figure 9.14. The symmetric V-shaped ML SWS with 50 periods and input and output stripline feeds. The metal enclosure is not shown. © [2019] IEEE. Reprinted, with permission, from [5].

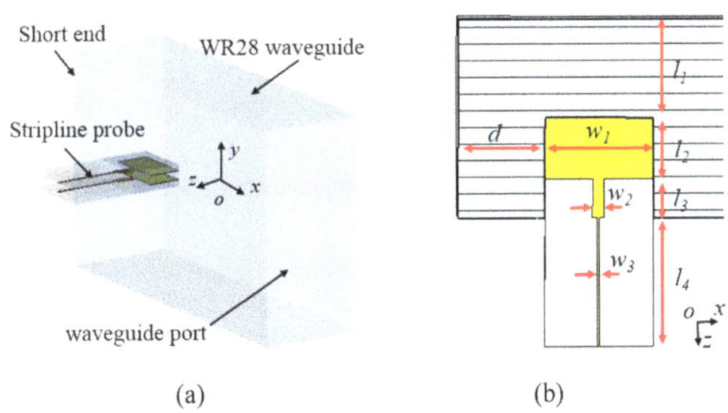

Figure 9.15. Ka-band waveguide-to-stripline transition. (a) 3D view, (b) dimensional parameters. © [2019] IEEE. Reprinted, with permission, from [5].

Table 9.2. Main structural dimensions of the waveguide-to-symmetric-stripline transition. © [2019] IEEE. Reprinted, with permission, from [5].

Parameter	Value (μm)	Parameter	Value (μm)
w_1	2000	l_1	1756
w_2	220	l_2	1100
w_3	40	l_3	700
d	1800	l_4	2100

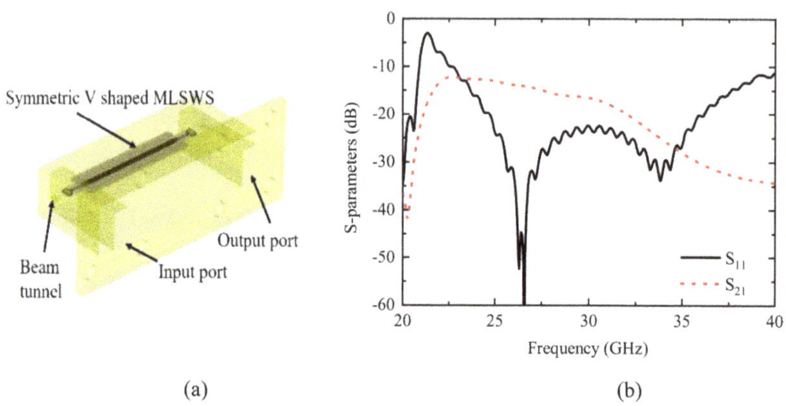

Figure 9.16. (a) Model of the Ka-band symmetric V-shaped ML SWS (150 periods) with WR-28 waveguide ports and (b) the simulation S-parameters. © [2019] IEEE. Reprinted, with permission, from [5].

not desired, since it does not contribute to beam–wave interaction. Simulations indicate that misalignment in the x-direction increases the strength of the odd mode, while misalignment in the y-direction causes a shift in the central frequency. The simulation results show that even for a misalignment in the x-direction as large as 3 mm, the excited odd mode is still about 20 dB weaker than the even mode.

The transitions are assembled with the symmetric V-shaped ML SWS. Figure 9.16(a) shows the simulation model built in CST MWS to estimate the overall transmission characteristics. The S-parameters of the entire assembly with 150-period-long SWS are shown in figure 9.16(b); S_{11} is better than -15 dB in the frequency range of 25–37 GHz, while S_{21} is about -15 dB at 30 GHz.

9.2.2.3 Microfabrication and assembly

The symmetric V-shaped ML SWS is realized by first fabricating a number of V-shaped ML SWSs on a four-inch quartz wafer using the standard photolithographic process. As shown in figure 9.17(a), there are ten ML SWSs and four straight lines on one wafer; six ML SWSs have 210 periods, and four ML SWSs have 150 periods. Individual SWSs and straight lines are subsequently laser diced. A diced V-shaped ML SWS is shown in figure 9.17(b). The substrate width of the diced

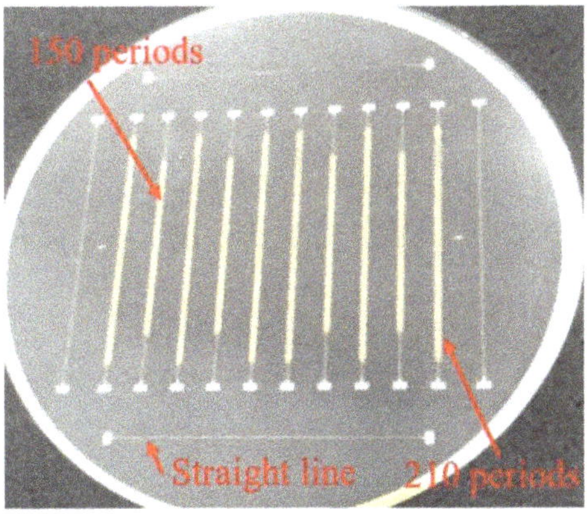

(a)

(b)

Figure 9.17. Microfabricated V-shaped ML SWS with tapered periods and probes at both ends. (a) Quartz wafer with a number of ML SWSs and straight microstrip lines; (b) One ML SWS after laser dicing. © [2019] IEEE. Reprinted, with permission, from [5].

pieces is extended by 2 mm on either side of the ML SWS to enable easy placement and a good fit inside the metal enclosure, which is described later on.

The chosen thickness of the metal strip *H_str* is 2 μm, since this can easily be realized using copper sputtering. There is also a 20 nm thick titanium layer below the Cu and a 20 nm thick gold layer over the Cu. This metallization scheme is identical to that used in [2] for the single V-shaped ML SWS, except that there is no metallization on the back surface of the quartz wafer. The metal enclosure acts as the ground plane for the microstrip meander line.

The metal enclosure is fabricated in two symmetric halves; each half has three 'trenches': one for the SWS and two for the input–output waveguide couplers. There are also holes and pins for alignment and for bolting together the two halves. The outline of the diced ML SWSs ensures that these fit snugly inside the metal enclosure. A pair of spacers on either side of the SWSs fixes them inside the metal enclosure and ensures a gap between the SWSs equal to the height of the beam tunnel. Once the SWSs and the spacers are properly placed and the two halves of the enclosure are aligned and bolted together, the various pieces do not shift laterally or vertically.

The metal enclosure and the spacers are fabricated out of stainless steel by a computer numerical control (CNC) machine with a tolerance of 0.01 mm. Figure 9.18(a) shows the fabricated metal enclosure with spacers and a pair of ML SWSs placed inside. Figure 9.18(b) shows the metal enclosure after assembly. The S-parameters of the assembled symmetric V-shaped ML SWS with 150 periods

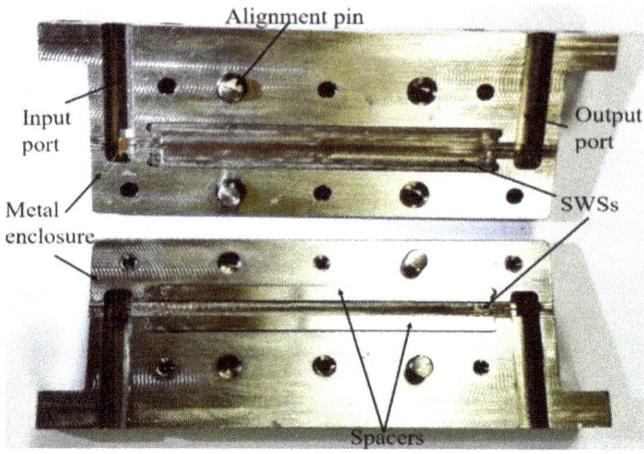

(a)

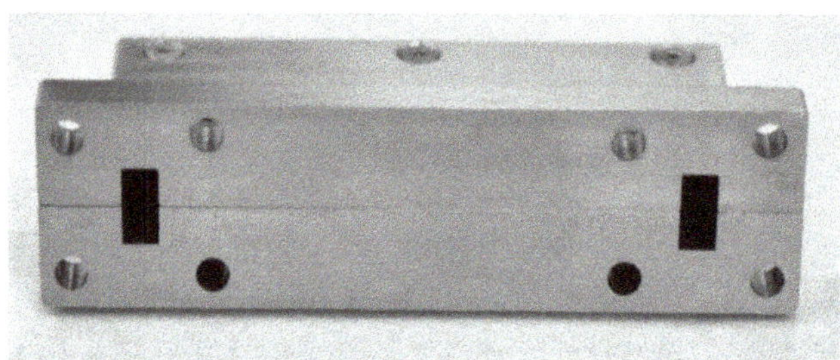

(b)

Figure 9.18. Fabricated components of the symmetric V-shaped ML SWS: (a) before and (b) after assembly.

are obtained by measurements made using a performance network analyzer (PNA). To connect the assembled SWS and PNA, a pair of adapters from a waveguide-to-K connector is used. However, the measurement results show an insertion loss 30 dB higher than the simulation results.

As mentioned in the case of the single-V ML SWS, the large discrepancy between the simulated insertion loss and the measured insertion loss is attributed to the conductor loss of the fabricated meander-line strip, which could be significantly improved by avoiding the use of a low-conductivity thin metal layer (e.g. Ti) at the bottom, since it has strong surface currents due to the skin effect. It should be noted that the simulation results use a single layer of copper with a conductivity of 2×10^7 S m^{-1}. Increasing the strip thickness and strip width, if permitted by the fabrication process and interaction impedance considerations, respectively, could also reduce the strip resistance.

To verify these arguments, a symmetric V-shaped ML SWS is designed and fabricated, avoiding the use of high-resistivity thin film and a thin copper layer. For speedy fabrication, the SWS is fabricated on low-cost copper-clad FR4 substrates with a dielectric constant of 4.4, a loss tangent of 0.02, and a thickness of 0.19 mm. The thickness and width of the copper strip are increased to 17.5 and 80 μm,

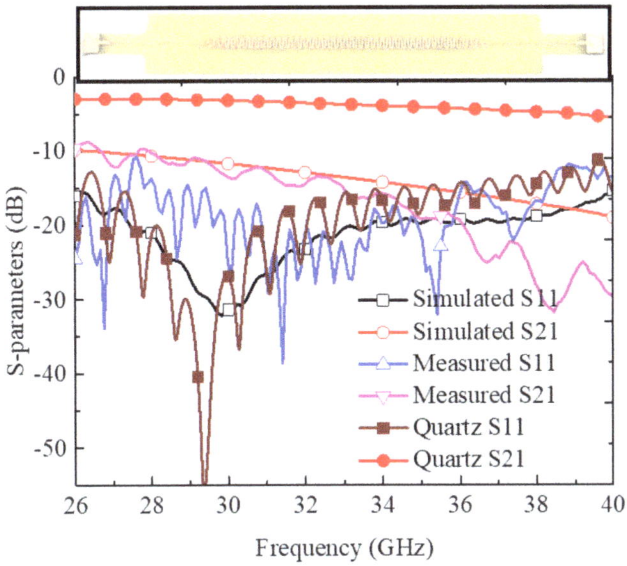

Figure 9.19. Simulated and measured S-parameters of the 50-period symmetric V-shaped ML SWS fabricated on FR4 substrates. Also included are simulated S-parameters for quartz substrates. © [2019] IEEE. Reprinted, with permission, from [5].

respectively. The copper conductivity is again considered to be 2×10^7 S m^{-1} in simulations. The number of periods is reduced to 50, and the metal enclosure length is reduced from 71.5 to 53.5 mm. The substrates are separated by 0.3 mm.

Figure 9.19 shows the simulated and measured S-parameters of the FR4 ML SWS. The SWS is also shown in the inset. The simulated and measured S_{11} are better than −15 dB over the frequency range of 28–36 GHz and better than −10 dB over the frequency range of 26–40 GHz. The measured S_{21} is a good match for the simulated values over the frequency range of 26–34 GHz. The simulated and measured S_{21} are −12.39 dB and −13.43 dB at 30 GHz, respectively. Further study shows that the loss is mainly caused by the relatively high loss tangent of FR4 and can be improved by selecting materials with a lower loss tangent. For example, as shown in figure 9.19, if we were to replace FR4 with quartz, the insertion loss would be only approximately 3.2 dB at 30 GHz. Thus, it is clear that the insertion loss can be significantly improved by avoiding the use of high-resistivity thin film, using a thick copper layer, and using a low-loss substrate material.

9.2.2.4 Assembly for hot-tests

Figure 9.20 shows the assembly of the various components of a TWT using the symmetric V-shaped ML SWS. This assembly can be used for hot tests.

The metal enclosure shown in figure 9.18, containing the symmetric V-shaped ML SWS, is placed at the center of the magnetic circuit. The magnetic circuit is designed and tested to produce a solenoidal magnetic field of ~0.4 T using a modification of the configuration presented in [6]. The magnetic circuit in [6] produced a solenoidal

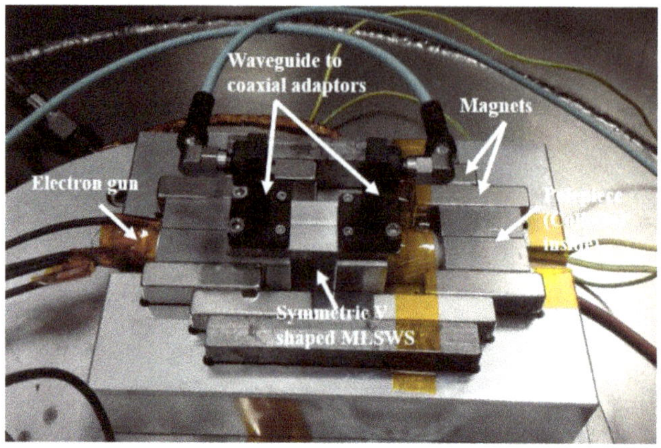

Figure 9.20. Assembled symmetric V-shaped meander-line TWT. © [2019] IEEE. Reprinted, with permission, from [5].

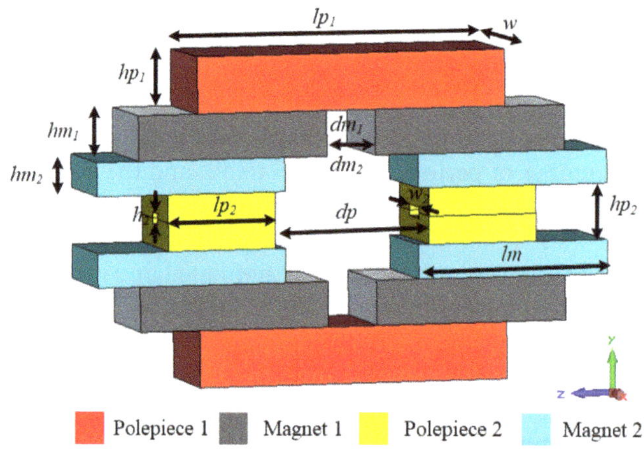

Figure 9.21. Magnetic circuit used to provide a uniform solenoidal focusing magnetic field [6].

magnetic field of ~0.22 T. This design, shown in figure 9.21, has been optimized using CST EM Studio. Its various dimensions are listed in table 9.3. Neodymium magnets and Consumet® Electronic Iron were selected as the magnet and pole piece materials in this design, respectively. To aid in the assembly of the magnetic circuit, an aluminum fixture is designed and fabricated to hold the magnets and pole pieces together.

The electron gun and collector are placed on the left and right sides of the SWS, respectively, passing through the pole pieces of the magnetic circuit. The electron gun, shown in figure 9.22, consists of a thermionic cathode, a beam-forming electrode, and an anode to produce a rectangular sheet beam of the required dimensions.

Table 9.3. Dimensions of the magnetic circuit [6].

Parameter	Value (mm)	Parameter	Value (mm)
lp_1	80	dm_2	50
lp_2	26	dp	60
hp_1	10	hm_1	9
hp_2	20	hm_2	9
w	25	w_2	8
lm	37	h_2	8
dm_1	17		

Figure 9.22. Thermionic electron gun.

A pair of waveguide-to-coaxial adapters is used for the input and output signals. For hot tests, it was planned to place the entire assembly inside a vacuum chamber that had suitable feedthroughs for the RF signals and various DC voltages for the e-gun. Unfortunately, the hot tests could not be completed due to the following problems: (i) one of the high-voltage feedthroughs could not handle the applied high voltage, (ii) one of the power supplies did not have enough isolation to withstand the high voltage it was subjected to, and (iii) there was an inadequate vacuum level in the vacuum chamber.

9.2.2.5 Simulation results for hot-test parameters

The hot-test parameters of the proposed symmetric V-shaped ML SWS for TWT application are estimated using the PIC Solver of CST Particle Studio for the structure with waveguide ports (figure 9.16(a)). An e-beam with a beam voltage of 3600 V and a current of 50 mA is applied centrally in the electron-beam tunnel. The rectangular cross section of the e-beam is assumed to have a width of 700 μm and a thickness of 150 μm. A uniform focusing magnetic field of 0.4 T is used.

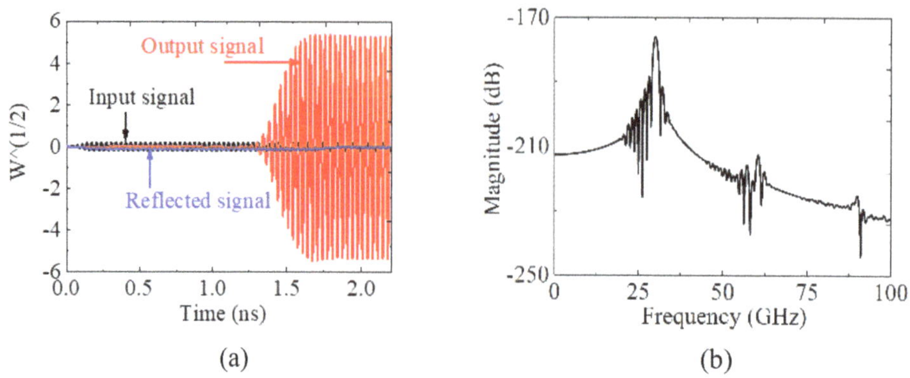

Figure 9.23. PIC simulation results at 30 GHz for the symmetric V-shaped ML SWS. (a) Input and output signals; (b) Fourier transform of the output signal. © [2019] IEEE. Reprinted, with permission, from [5].

The PIC simulation results show that with the same electron beam, number of periods, and input power of 0.1 W as used for the single-V ML SWS, the symmetric-V ML SWS achieves a peak output power of 18 W at 30 GHz. This output is about 26.3% higher than the saturated power output of the single-V ML SWS. However, 0.1 W of input power oversaturates the symmetric structure.

An input power of 0.04 W avoids saturation of the symmetric V-shaped ML SWS and produces the maximum output power. Figure 9.23(a) shows the input, reflected, and output signals, which have magnitudes of 0.2 $\sqrt{W}$, 0.02 $\sqrt{W}$, and 5.3 $\sqrt{W}$, respectively. The corresponding output power is 28 W, indicating a gain of 28.5 dB. Figure 9.23(b) presents the spectrum of the output signal, showing a relatively clean signal at 30 GHz. The presence of signals at 60 and 90 GHz can be understood by referring to figure 9.13(c), which indicates additional beam–wave interactions at 60 and 120 GHz.

The output power vs. frequency curve is shown in figure 9.24. The maximum output power is 28.13 W at 34 GHz, with a gain of 28 dB. In the frequency range of 28–36 GHz, the output power shows a 3 dB bandwidth of about 25% centered at 32 GHz. When compared to the single V-shaped ML SWS, the symmetric structure gives a 96.6% higher output power in the frequency range of 28–36 GHz and requires a 60% lower saturation input power.

Due to the small dimensions of the SWS, there could be some concern regarding dielectric breakdown of the SWS at the power levels predicted by simulations. However, for a transmission power of 28 W, the maximum axial electric field on the surface of the SWS is observed to be $\sim 2 \times 10^5$ V m^{-1}; this field value is much smaller than the breakdown electric field strength of quartz ($\sim 5 \times 10^7$ V m^{-1}).

9.3 Design and fabrication of W- and V-band meander-line SWSs

In addition to the fabrication of the single V-shaped and symmetric V-shaped ML SWSs described in the previous two sections, the fabrication of many other ML SWSs has been reported in the literature. These ML SWSs cover both W-band and

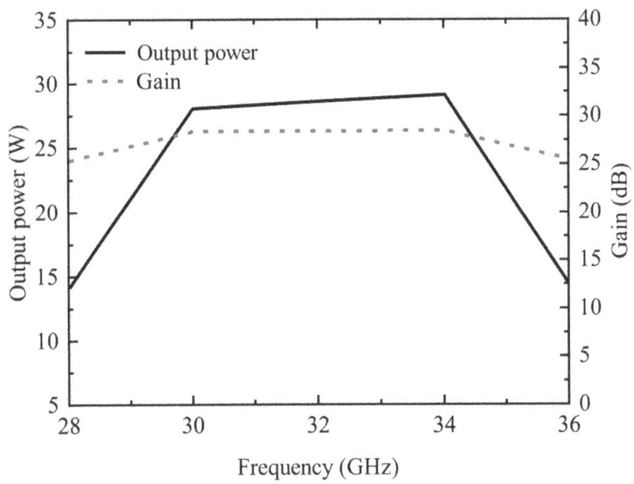

Figure 9.24. Output power and gain vs. frequency for the symmetric V-shaped ML SWS. © [2019] IEEE. Reprinted, with permission, from [5].

V-band frequencies. Some of the early attempts that have reported novel technologies are described in the following subsections.

Parts of this section have been reprinted, with permission, from [7]. © [2009] IEEE.

9.3.1 W-band meander-line SWSs

Three W-band ML SWSs are described in this subsection. The first is a meander-line SWS fabricated on top of a serpentine silicon dielectric ridge in order to increase interaction impedance and reduce dielectric losses [7]. This was an early attempt at the microfabrication of a W-band ML SWS; due to the complex fabrication process, it was only partially successful. The second ML SWS included here was printed on a boron nitride substrate and yielded good cold-test results [8]. The third ML SWS is a more recent fabrication [9]; although the configuration follows the main idea of [7], the authors call it a conformal microstrip meander line (CMML). The CMML has a ridge between the quartz substrate and the metal meander line. The ridge exactly follows the shape of the meander-line pattern and has a much lower height than that used in [7]. The structure exhibits low cold-test attenuation.

9.3.1.1 Selectively metallized raised meander-line SWS on silicon substrate

The MML SWS has a simple configuration, but it exhibits lower efficiency of beam–wave interaction due to low interaction impedance and large attenuation. To overcome these shortcomings, a raised meander-line configuration was proposed in [7]. This configuration, shown in figure 9.25, minimizes the amount of dielectric beneath the meander-line pattern. The reduced amount of dielectric contributes to increased electric flux density above the meander line where the beam propagates and thus increases interaction impedance. Second, it also contributes to reduced

Figure 9.25. SEM image of the raised meander-line geometry. The structure was made from silicon using deep reactive ion etching (DRIE). To complete the fabrication, only the top of the ridge needed to be metallized. © [2009] IEEE. Reprinted, with permission, from [7].

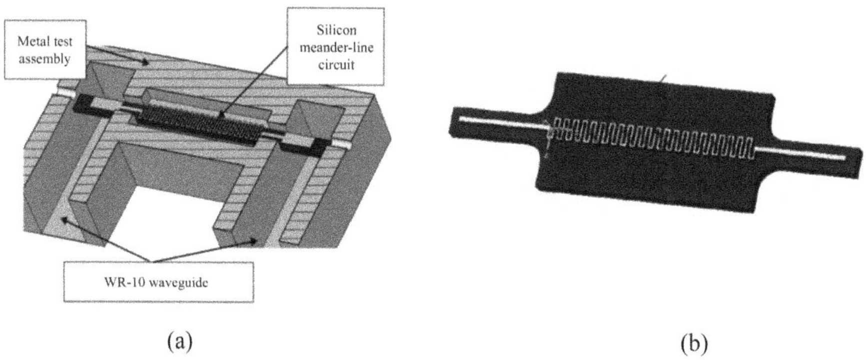

Figure 9.26. (a) Illustration of the raised meander-line circuit positioned in the bottom half of a metal test assembly. When covered with the top half of the metal assembly (not shown), the circuit is positioned approximately at the center of a conducting waveguide. (b) Actual shape of the fabricated meander-line circuit. The extra silicon 'wings' above and below the meander-line metallization are used to mechanically capture and position the circuit in the metal test assembly. © [2009] IEEE. Reprinted, with permission, from [7].

attenuation due to dielectric loss. For typical material properties and beam configuration, the attenuation of the raised meander-line circuit is approximately halved, and the interaction impedance is about three times that of the simple configuration.

Design

The design of the raised meander-line SWS targeted a 10 W continuous-wave (CW) TWT operating at 83.5 GHz with approximately 4% bandwidth. Figure 9.26(a) shows an illustration of the circuit situated within a conducting waveguide housing; this figure includes E-plane probe antennas for coupling into a WR-10 waveguide. Figure 9.26(b) shows a fabricated circuit. The material chosen for the raised serpentine dielectric ridge was Si due to the mature microfabrication technology available for this material. The serpentine Si ridge was approximately 30 μm wide and 120 μm tall.

For a 9 kV, 28 mA, round beam 200 μm in diameter with its lower edge 25 μm above the circuit, simulations predicted a saturated output power greater than 10 W. A solenoidal magnetic field with a peak value of 0.213 T was used for beam focusing.

Fabrication

The raised meander-line SWSs were fabricated on four-inch, 350 μm thick Si wafers with 14 SWSs per wafer. To reduce cost and time, off-the-shelf Si wafers were used instead of high-resistivity intrinsic Si. The first step in the fabrication was deep reactive ion etching (DRIE) processing to make the raised serpentine ridge. The second step consisted of selective metallization of the top of the ridge to make the raised conductive meander line; in this step, a thin seed layer was sputter coated and the metal thickness was built up to 2 μm using gold electroplating. In the third and final step, individual SWSs were extracted using through-wafer DRIE.

Cold-test assembly

A precision-machined cold-test assembly (CTA) was built for measuring the S-parameters of the fabricated SWSs. The CTA, as shown in figure 9.27, consisted of two copper blocks that had grooves to accommodate the SWS. The grooves facilitated the alignment of the SWS within the block, acted as a waveguide surrounding the SWS, and allowed coupling of energy into and out of the CTA.

The dimensions of the so-formed waveguide surrounding the SWS were those of a typical W-band waveguide. The waveguide groove also contained choke points located between the antenna probes on either side and the ML circuit. The choke points ensured that W-band signals propagated in the CTA only in the presence of the SWS. It was experimentally verified that no power traveled through the CTA in the absence of an installed SWS.

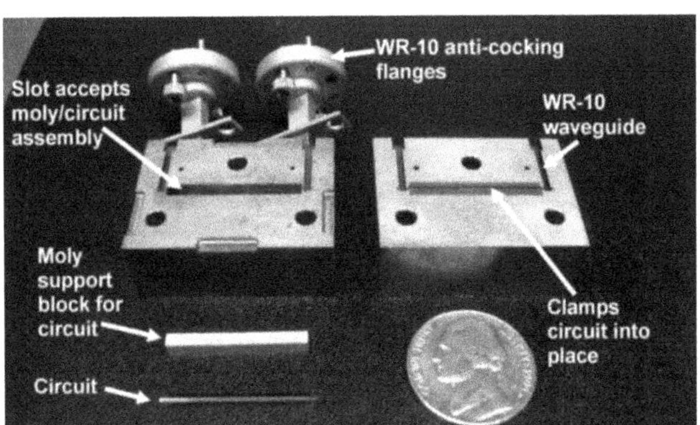

Figure 9.27. Exploded view of the CTA. A molybdenum block sits beneath the circuit to properly position the SWS inside the structure. The two halves clamp together, and WR-10 anticocking flanges are attached for coupling. © [2009] IEEE. Reprinted, with permission, from [7].

Cold-test results

Figure 9.28 shows the results of the CTA with a SWS inside. An S_{11} of approximately < -20 dB and an S_{12} of > -35 dB were observed. The frequency of maximum power throughput, 91 GHz, was approximately 7.5 GHz above the design frequency. Also, the simulated S_{12} was only -16 dB. Dimensional and electrical factors accounted for these discrepancies.

Measurement of the fabricated dimensions revealed a systematic relative error of up to 15% when compared to the design dimensions. In particular, the width of the metallization on top of the serpentine ridge was 7 μm larger than the design value. When the increased width of the metallization was taken into account in the simulations, the simulation and the experimental results matched better. The use of low-resistivity Si wafers instead of intrinsic high-resistivity Si (loss tangent of about 0.0005) also caused a significant deviation from the design. Simulations showed that a loss tangent value of 0.01 allowed for a good match with the measured S_{21} as well as the frequency of maximum power transmission.

9.3.1.2 Meander-line SWS on boron nitride substrate

M. Sumathy *et al* reported a W-band ML SWS fabricated on a 0.5 mm thick boron nitride substrate [8]. The authors first presented an analysis of the SWS that considered the dielectric substrate and the metallic shield. A waveguide coupler was designed to efficiently couple the ML SWS to the WR-10 waveguide. A schematic of a typical ML SWS with the relevant dimensional parameters is shown in figure 9.29(a). A 3D solid model of the ML SWS, along with the waveguide couplers, is shown in figure 9.29(b).

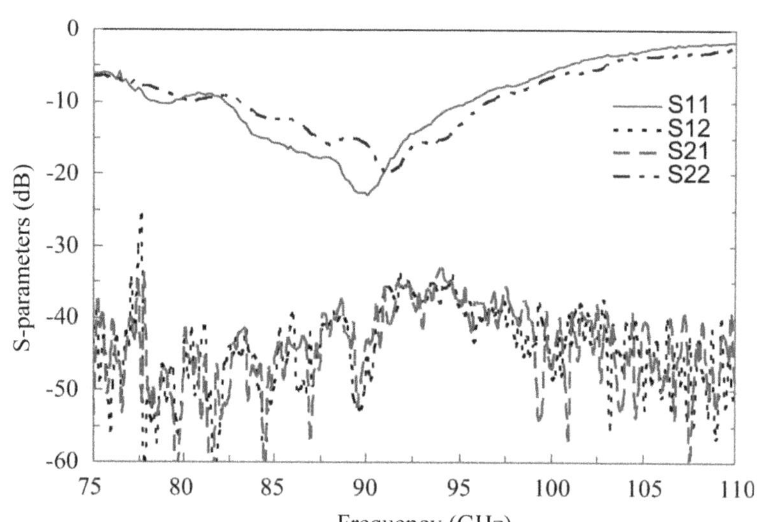

Figure 9.28. Measured S-parameters of the CTA with an ML SWS. © [2009] IEEE. Reprinted, with permission, from [7].

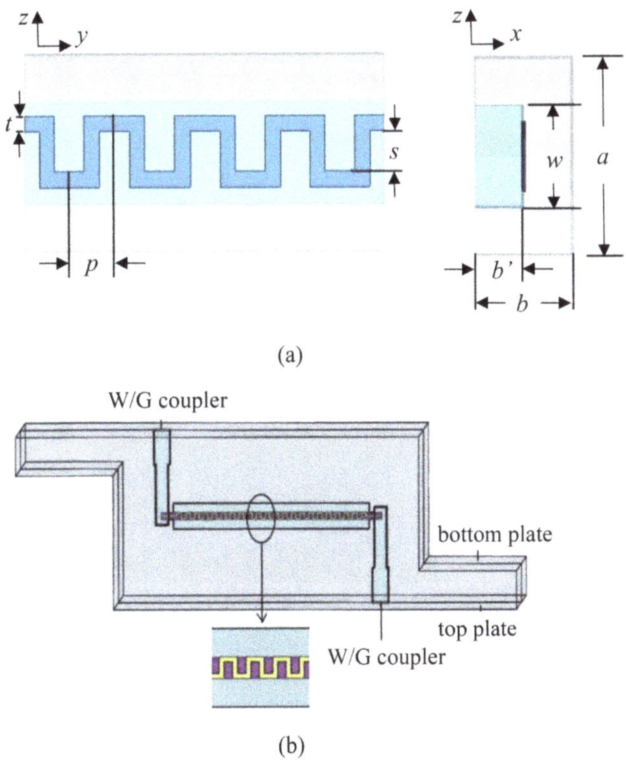

Figure 9.29. (a) Two-dimensional schematic of a typical ML SWS, showing the top and side views with the relevant dimensional parameters. (b) Three-dimensional solid model of the ML SWS with input–output waveguide couplers. © [2013] IEEE. Reprinted, with permission, from [8].

Figure 9.30. Printed ML-pattern on the boron nitride substrate. © [2013] IEEE. Reprinted, with permission, from [8].

For the fabrication of the meander-line circuit, the meander-line pattern was printed on a boron nitride substrate (figure 9.30) using a mask made out of a thin copper sheet. The printed circuit was placed inside a precision-fabricated metallic shield which included input and output waveguide couplers.

The measured S-parameters, presented in figure 9.31(a), showed an S_{11} value better than -10 dB and an S_{21} value better than -8 dB over the frequency range of 92–97 GHz. Figures 9.31(b) and (c) respectively compare the measured voltage standing wave ratio (VSWR) and measured dispersion characteristics with those obtained through analysis and simulation. Interaction impedance characteristics calculated from the analytical results were compared against numerical simulation only; the simulations were carried out using CST MWS. The interaction impedance was computed at the beam axis, which was assumed to be placed at a height of

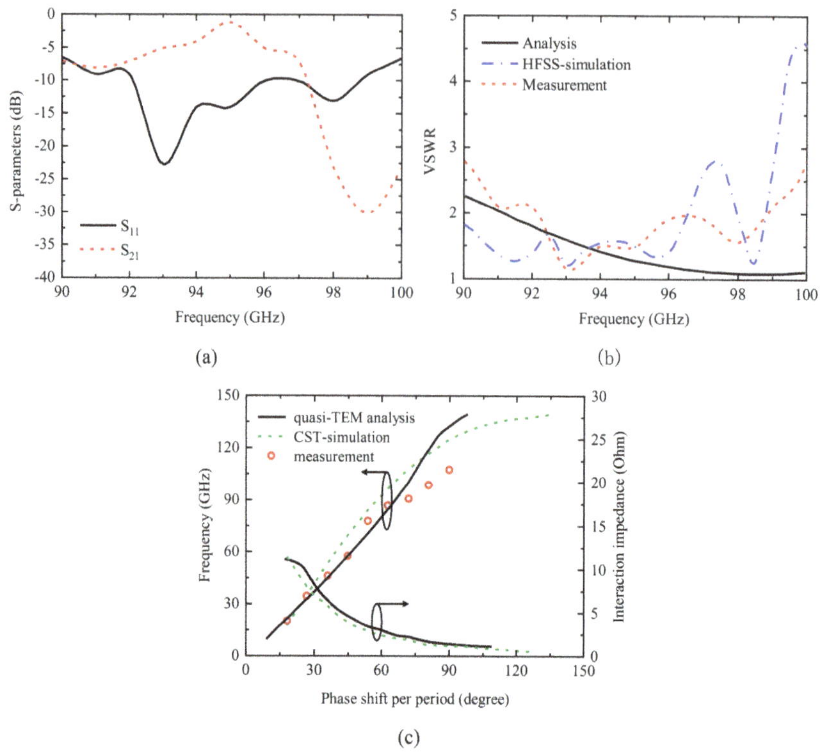

Figure 9.31. (a) Measured S-parameters. (b) Comparison of voltage standing wave ratio (VSWR) characteristics. (c) Comparison of dispersion and interaction impedance characteristics of W-band ML SWS for $a/p = 5.0$, $b/p = 2.5$, $b/p = 1.25$, $s/p = 0.75$, $t/p = 0.30$, $w/p = 1.35$, and $\varepsilon_r = 3.5$. © [2013] IEEE. Reprinted, with permission, from [8].

0.25 mm above the surface of the ML. The measured results show reasonable agreement with simulation and analysis for dispersion characteristics around the lower cutoff frequency. There is higher circuit loss around the upper cutoff frequency.

9.3.1.3 Meander-line SWS with conformal dielectric layer on a quartz substrate

Parts of this section have been reprinted, with permission, from [9]. © [2022] IEEE.

A U-shaped microstrip meander line is investigated in [9] for a W-band TWT. The structure includes an additional dielectric layer or ridge with exactly the same shape as the metal microstrip between the metal microstrip and the main dielectric substrate. Hence, the structure is called the CMML. Similar to [7], the motivation for the additional conformal dielectric layer is to increase interaction impedance and reduce attenuation due to dielectric loss. In addition, this layer can, to some extent, increase the separation between the electron beam and the substrate, thereby reducing the chance of electrons landing on the dielectric substrate.

Design

Figure 9.32 shows the schematic of the CMML SWS together with the dimensional parameters. In particular, the thickness of the conformal dielectric layer is t_1. The metal enclosure housing the planar structure is not shown in this figure. Quartz is selected as the substrate due to its relatively low dielectric constant (3.78), relatively low cost, and modest thermal conductivity. Also, a 3 μm thick conformal dielectric layer can be realized on the quartz substrate. The other optimized dimensions of the structure are listed in table 9.4.

For the dimensions in table 9.4, the U-shaped CMML exhibits an interaction impedance of about 5 Ω at 96 GHz, measured at a height of 7 μm above the microstrip. This value is 29% higher than the impedance observed without the conformal dielectric layer. If t_1 increases, the interaction impedance also increases further. The quartz substrate is kept approximately 2.5 mm wide, allowing for easy assembly, but this can cause unwanted modes in a normal metal waveguide enclosure connected to standard WR-10 waveguides. Therefore, the top half of the two-part enclosure has reduced width and height at the locations close to the WR-10 waveguides. This helps to avoid unwanted modes.

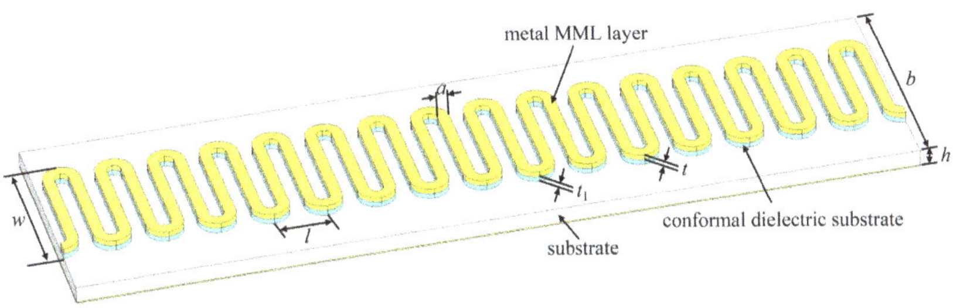

Figure 9.32. 3D view of the CMML SWS with the conformal dielectric layer; also shown are dimensional parameters. © [2022] IEEE. Reprinted, with permission, from [9].

Table 9.4. Optimized dimensions of the U-shaped CMML structure. © [2022] IEEE. Reprinted, with permission, from [9].

Parameter	Value (mm)
a	0.03
b	2.49
w	0.41
t	0.003
t_1	0.003
l	0.132
h	0.19

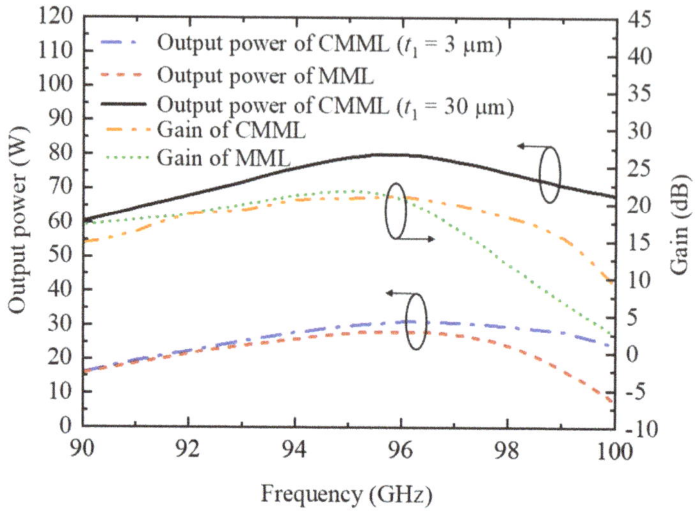

Figure 9.33. Saturated output power and gain versus frequency for the proposed U-shaped CMML and U-shaped MML. © [2022] IEEE. Reprinted, with permission, from [9].

The PIC Solver of CST Particle Studio was used to assess the beam–wave interaction of a TWT based on the proposed U-shaped CMML with waveguide ports. The simulations used a 6550 V, 100 mA electron beam with a 410 × 50 μm^2 cross section. Figure 9.33 shows the saturated output power and gain versus frequency for the CMML as well as the corresponding conventional MML (i.e. without the conformal dielectric layer). Within the 3 dB bandwidth of 92–98 GHz, the gain and output power exceed 20.1 dB and 22.3 W, respectively. While the maximum output power is 31.4 W for the 3 μm thick conformal quartz layer, for a 30 μm thick conformal quartz layer, the output power can exceed 80 W. On the other hand, the 7 μm height of the beam above the metal microstrip surface (mentioned for the interaction impedance) may not be practically feasible.

Fabrication

The designed SWSs were fabricated on a 300 μm thick, two-inch quartz wafer. As shown in figure 9.34(a), a 60 nm thick titanium tungsten (TiW) film was first sputter coated, followed by a 100-nm-thick gold coating. Figure 9.34(b) shows that a 3 μm thick gold MML was realized next. Figures 9.34(c) and (d) show the conformal dielectric layer which was formed by DRIE; this restricted the thickness of the conformal quartz layer to ~3 μm. Next, chemical mechanical polishing (CMP) was used to reduce the thickness of the quartz wafer to 190 μm, and individual SWS samples were obtained by laser dicing. The fabricated SWS is shown in figure 9.35.

The metal waveguide housing was precision fabricated in two asymmetric halves with a tolerance of 0.005 mm. The two halves are shown in figure 9.36; the two parts can be bolted together through the four holes on the waveguide housing.

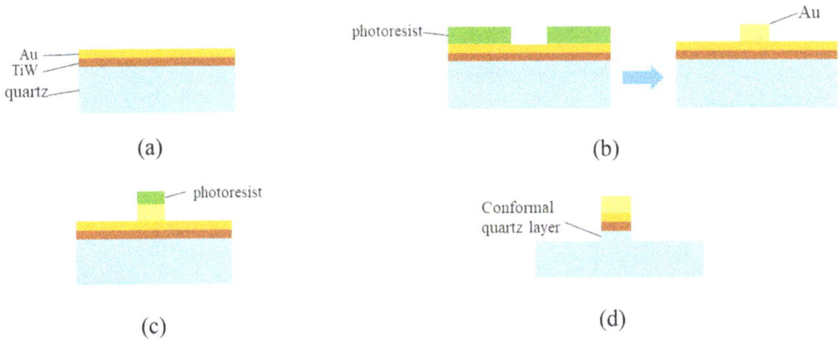

Figure 9.34. Main fabrication process steps used to produce the CMML. (a) Sputtered metal layers. (b) Photolithography and electroplating. (c) Covered by photoresist. (d) DRIE. © [2022] IEEE. Reprinted, with permission, from [9].

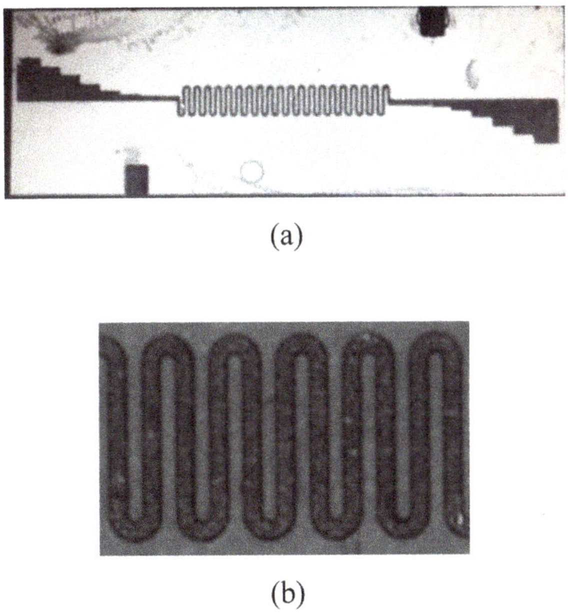

Figure 9.35. (a) Microscope image of a single SWS after laser dicing, and (b) an enlarged view of a section of SWS. © [2022] IEEE. Reprinted, with permission, from [9].

Cold-test results

The S-parameters for a 20-period CMML SWS are shown in figure 9.37. In the frequency range of 88–102 GHz, the simulated and measured S_{11} values are below -14.3 and -12.8 dB, respectively, while the simulated and measured S_{21} values are better than -4.75 and -5.3 dB, respectively. The simulations assume an equivalent conductivity of 1.6×10^7 S m^{-1} for the CMML and the waveguide. The measured return loss and transmission loss are good matches for the simulated values. The measured attenuation of the CMML SWS, shown in figure 9.38, is about 5.9–7.7 dB cm^{-1} and is better than the values reported in [7, 8].

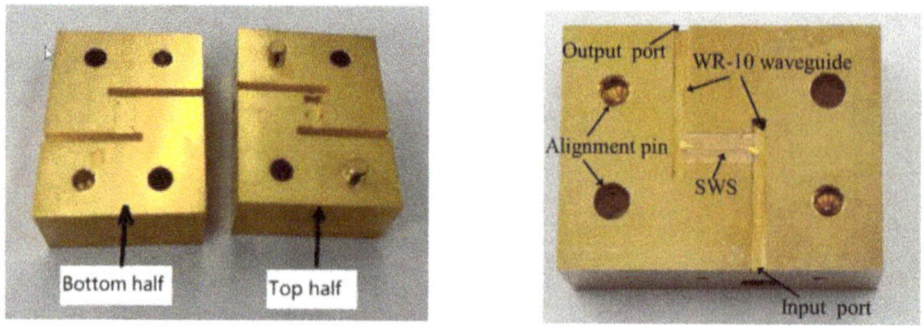

(a) (b)

Figure 9.36. (a) Fabricated metal housing of the U-shaped CMML before assembly. (b) Enlarged photograph of the bottom half. © [2022] IEEE. Reprinted, with permission, from [9].

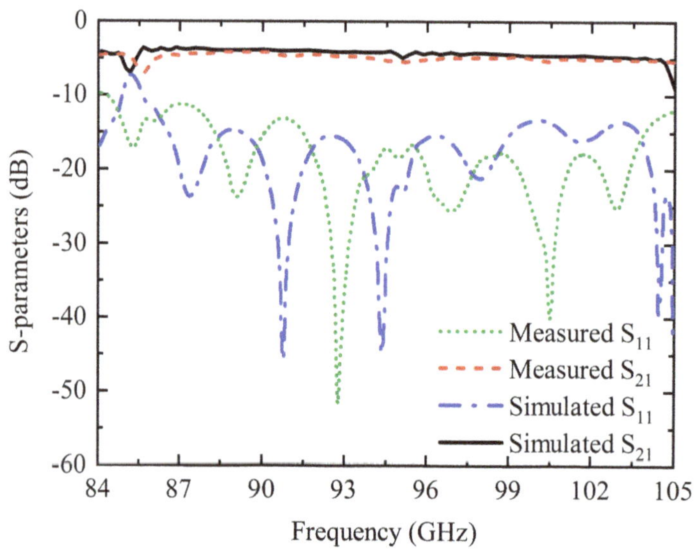

Figure 9.37. Simulated and measured S-parameters. © [2022] IEEE. Reprinted, with permission, from [9].

9.3.2 Meander-line SWS for V band using laser ablation

While photolithography has been used for the fabrication of microstrip planar SWSs (see, for example, [8]), it suffers from high cost, low speed, and poor flexibility. A new method for the fabrication of microstrip SWSs utilizing magnetron sputtering and laser ablation for V-band meander-line SWSs was reported in [10]. The following subsections briefly describe the design, fabrication, and cold-test results of this work.

Design

Figure 9.39 shows the schematic of the SWS. The design assumes a 1 μm thick copper microstrip layer on a quartz substrate ($\varepsilon_r = 3.75$), which is placed into a WR-15 rectangular metallic waveguide. All dimensions of the SWS are listed in table 9.5.

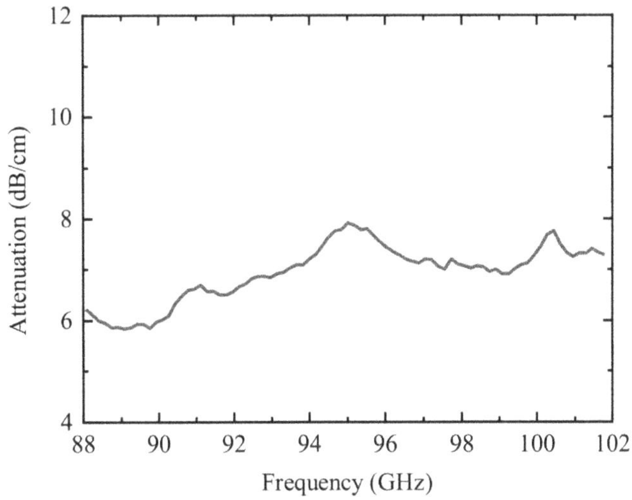

Figure 9.38. Measured attenuation versus frequency. © [2022] IEEE. Reprinted, with permission, from [9].

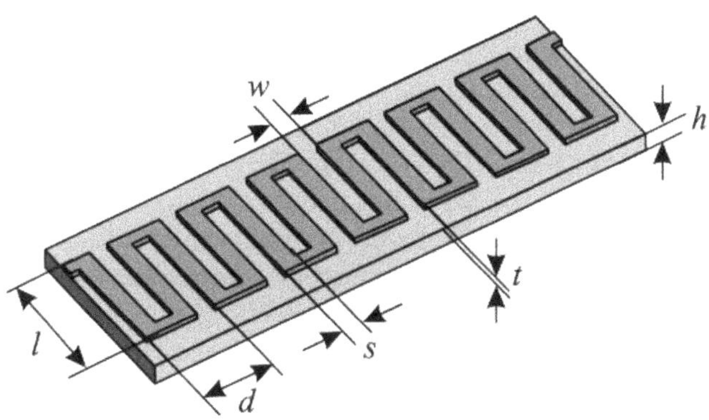

Figure 9.39. Schematic diagram of the V-band meander-line SWS. © [2018] IEEE. Reprinted, with permission, from [10].

Table 9.5. Optimized dimensions of the SWS. © [2018] IEEE. Reprinted, with permission, from [10].

Parameter	Value (μm)
SWS period, *d*	200
SWS width, *l*	650
Strip width, *s*	50
Slit width, *w*	50
Metal thickness, *t*	1
Substrate thickness, *H*	500

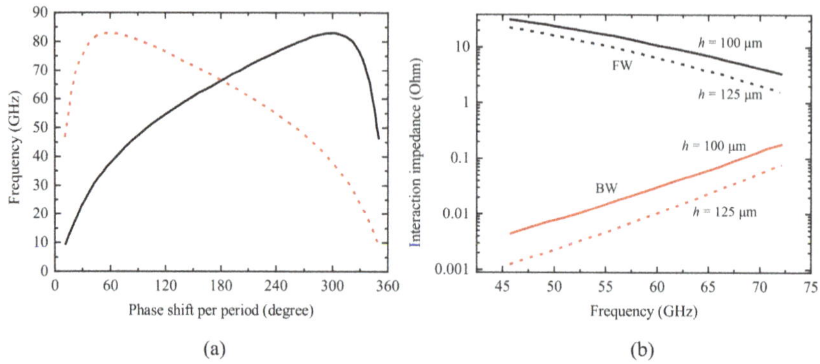

Figure 9.40. (a) Dispersion diagram and (b) interaction impedance for two different distances from the beam to the SWS surface. © [2018] IEEE. Reprinted, with permission, from [10].

Figure 9.40 shows the simulated dispersion and interaction impedance characteristics of the SWS. The interaction impedance K_c of the forward-wave (FW) mode exceeds that of the backward-wave mode (BW) by several orders of magnitude. In the simulations, a sheet electron beam with a cross section of 500 × 50 μm^2 is assumed to propagate along the structure with the beam axis at a height *h* above the microstrip. The interaction impedance, presented in figure 9.40(b), is averaged over the beam cross section. Due to the rapid decay of electromagnetic fields with height above the SWS, K_c also rapidly decreases with *h*. However, values of K_c around 1–10 Ω are still obtained over the 50–70 GHz frequency range at $h = 100$–125 μm. The beam voltage V_s ranges from 2.5 to 5.5 kV for beam–wave synchronism in this frequency range.

The small-signal gain of a TWT based on the meander-line SWS was calculated using 1D parametric code, which was verified by comparison with 3D particle-in-cell (PIC) code. The simulations were based on an SWS 25 mm long and a 0.1 A, 500 × 50 μm^2 sheet electron beam with its axis 100 μm above the SWS. Based on the S-parameter measurements, a 0.1 dB mm^{-1} distributed loss was included in the simulations. The simulation results, presented in figure 9.41, show a peak gain exceeding 35 dB. While the bandwidth of the amplifier is only 2–3 GHz, the central frequency is tunable from 50 to 70 GHz by varying the beam voltage from 3.5 to 5.5 kV.

Fabrication

The fabrication of SWSs consisted of three stages. First, magnetron sputtering was used to deposit a copper coating onto the substrate. In the second stage, laser ablation was used to realize meander-line patterns. Finally, the substrate was diced into individual SWS samples fitting inside the width of the waveguide. Optical and scanning electron microscopy confirmed that the coating had a uniform thickness without cracks and defects and without detachment from the substrate. Figure 9.42 shows optical microscopy images of the SWS. The SWS consists of 50 periods with input/output couplers at both ends. The strip thickness was restricted to 1 μm in this work due to the limitations imposed by the magnetron sputtering; the same research group has continued to report further improvements, e.g. in [11].

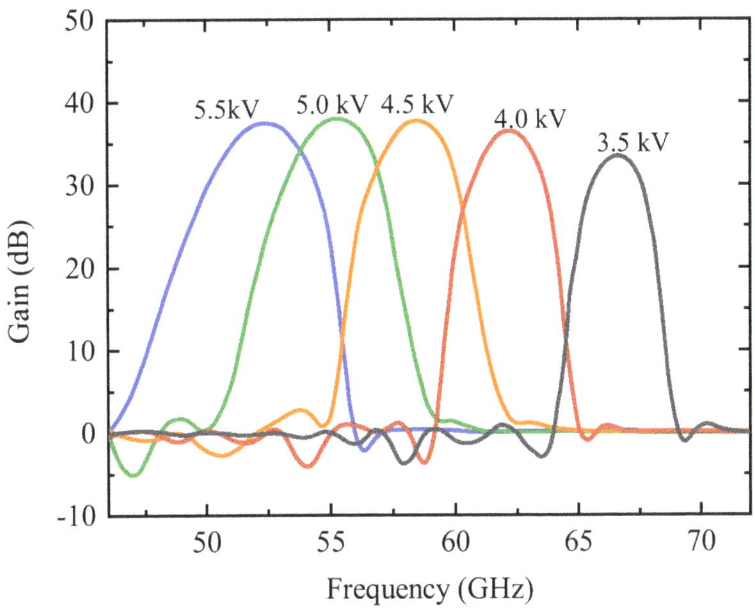

Figure 9.41. Gain versus frequency for the TWT with the meander-line SWS for different values of the DC beam voltage. © [2018] IEEE. Reprinted, with permission, from [10].

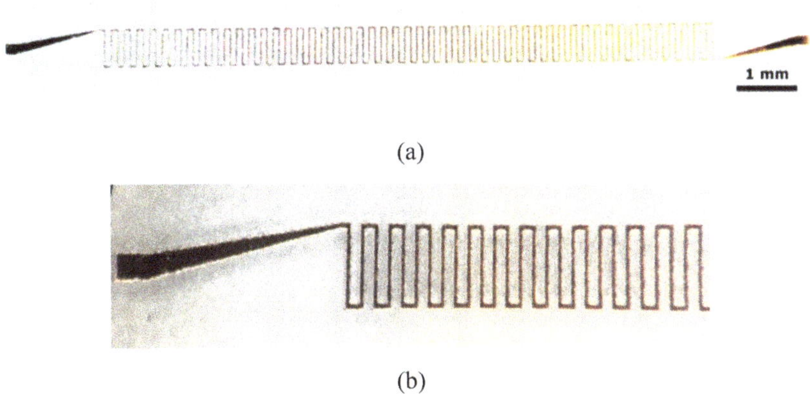

Figure 9.42. (a) Optical microscopy image of the fabricated SWS, (b) and an enlarged segment. © [2018] IEEE. Reprinted, with permission, from [10].

Cold-test results

Figure 9.43 shows the measured as well as calculated return (S_{11}) and transmission (S_{21}) losses; the measured transmission loss of the SWS placed inside the waveguide section was better than 2.5 dB. The simulations assumed a metal conductivity value of $\sigma = 3.0 \times 10^7$ S m^{-1}.

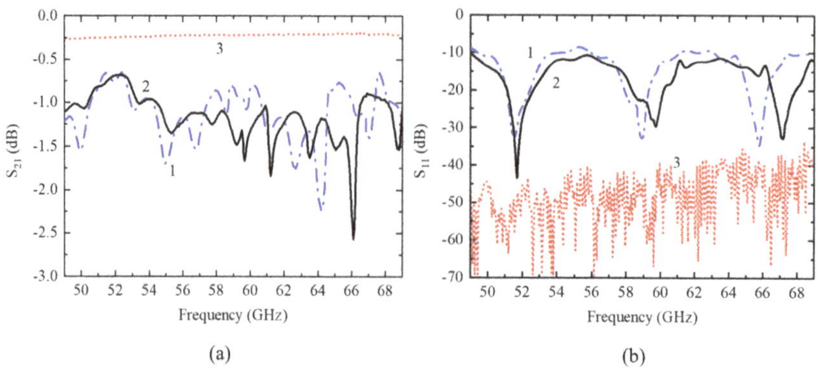

Figure 9.43. Transmission loss (a) and reflection loss (b) of the ML SWS. (1) Calculated results, (2) measured results, (3) empty waveguide (measured). © [2018] IEEE. Reprinted, with permission, from [10].

9.4 Summary

MML SWSs have advantages of easy fabrication, compatibility with a sheet electron beam, and, in general, good thermal contact with the substrate. This chapter has described several examples of the fabrication of MML SWSs that operate in the Ka band (26–40 GHz), V band (50–75 GHz), and W band (75–110 GHz). For one of the Ka-band structures, measurement of the cold-test S-parameters was carried out 'on-wafer,' using CPW probe stations. For the other Ka-band structure described here, an assembly suitable for hot tests was completed. The V-band structure described here was the first to utilize laser ablation for fabrication. The three W-band structures described here demonstrated innovative configuration and/or fabrication techniques: the first was an ML SWS fabricated on top of a serpentine silicon dielectric ridge, which helped to increase interaction impedance and reduce dielectric losses. The second ML SWS was printed on a boron nitride substrate and yielded good cold-test results. The third W-band ML SWS has a ridge between the quartz substrate and the metal meander line; the structure showed low cold-test attenuation.

References

[1] Shen F *et al* 2012 A novel V-shaped microstrip meander-line slow-wave structure for W-band MMPM *IEEE Trans. Plasma Sci.* **40** 463–9

[2] Wang S, Aditya S, Xia X, Ali Z and Miao J 2018 On-wafer microstrip meander-line slow-wave structure at Ka-band *IEEE Trans. Electron Devices* **65** 2142–8

[3] Bahl. I and Bhartia P 2003 *Microwave Solid State Circuit Design* 2nd edn (New York: Wiley)

[4] Shen F *et al* 2012 Symmetric double V-shaped microstrip meander-line slow-wave structure for W-band traveling-wave tube *IEEE Trans. Electron Devices* **59** 1551–7

[5] Wang S, Aditya S, Xia X, Ali Z, Miao J and Zheng Y 2019 Ka-band symmetric V-shaped meander-line slow wave structure *IEEE Trans. Plasma Sci.* **47** 4650–7

[6] Wang S and Aditya S 2016 Magnetic circuit for a sheet electron beam Ka-band microfabricated traveling wave tube *2016 IEEE Region 10 Conf. (TENCON)* 1794–7

[7] Sengele S, Jiang H, Booske J H, Kory C L, van der Weide D W and Ives R L 2009 Microfabrication and characterization of a selectively metallized W-band meander-line TWT circuit *IEEE Trans. Electron Devices* **56** 730–7

[8] Sumathy M, Augustin D, Datta S K, Christie L and Kumar L 2013 Design and RF characterization of W-band meander-line and folded-waveguide slow-wave structures for TWTs *IEEE Trans. Electron Devices* **60** 1769–75

[9] Yue L *et al* 2022 A high interaction impedance microstrip meander-line with conformal dielectric substrate layer for a W-band traveling-wave tube *IEEE Trans. Electron Devices* **69** 5826–31

[10] Ryskin N M *et al* 2018 Planar microstrip slow-wave structure for low-voltage V-band traveling-wave tube with a sheet electron beam *IEEE Electron Device Lett.* **39** 757–60

[11] Ryskin N M *et al* 2021 Development of microfabricated planar slow-wave structures on dielectric substrates for miniaturized millimeter-band traveling-wave tubes *J. Vac. Sci. Technol.* B **39** 013204

www.ingramcontent.com/pod-product-compliance
Ingram Content Group UK Ltd.
Pitfield, Milton Keynes, MK11 3LW, UK
UKHW051903100926
8556IPUK00005B/743

9 780750 357623